*Convex Functions*

# Pure and Applied Mathematics

A Series of Monographs and Textbooks

Editors **Paul A. Smith and Samuel Eilenberg**

Columbia University, New York

RECENT TITLES

44: ERNST-AUGUST BEHRENS. Ring Theory. 1972
45: MORRIS NEWMAN. Integral Matrices. 1972
46: GLEN E. BREDON. Introduction to Compact Transformation Groups. 1972
47: WERNER GREUB, STEPHEN HALPERIN, AND RAY VANSTONE. Connections, Curvature, and Cohomology: Volume I, De Rham Cohomology of Manifolds and Vector Bundles, 1972. Volume II, Lie Groups, Principal Bundles, and Characteristic Classes, 1973
48: XIA DAO-XING. Measure and Integration Theory of Infinite-Dimensional Spaces: Abstract Harmonic Analysis. 1972
49: RONALD G. DOUGLAS. Banach Algebra Techniques in Operator Theory. 1972
50: WILLARD MILLER, JR. Symmetry Groups and Their Applications. 1972
51: ARTHUR A. SAGLE AND RALPH E. WALDE. Introduction to Lie Groups and Lie Algebras. 1973
52: T. BENNY RUSHING. Topological Embeddings. 1973
53: JAMES W. VICK. Homology Theory: An Introduction to Algebraic Topology. 1973
54: E. R. KOLCHIN. Differential Algebra and Algebraic Groups. 1973
55: GERALD J. JANUSZ. Algebraic Number Fields. 1973
56: A. S. B. HOLLAND. Introduction to the Theory of Entire Functions. 1973
57: WAYNE ROBERTS AND DALE VARBERG. Convex Functions. 1973

*In preparation*

SAMUEL EILENBERG. Automata, Languages, and Machines: Volume A
H. M. EDWARDS. Riemann's Zeta Function
WILHELM MAGNUS. Non-Euclidean Tesselations and Their Groups
A. M. OSTROWSKI. Solution of Equations and Systems of Equations. Third Edition

# CONVEX FUNCTIONS

## A. Wayne Roberts
Macalester College
St. Paul, Minnesota

## Dale E. Varberg
Hamline University
St. Paul, Minnesota

ACADEMIC PRESS   New York and London   1973
*A Subsidiary of Harcourt Brace Jovanovich, Publishers*

COPYRIGHT © 1973, BY ACADEMIC PRESS, INC.
ALL RIGHTS RESERVED.
NO PART OF THIS PUBLICATION MAY BE REPRODUCED OR
TRANSMITTED IN ANY FORM OR BY ANY MEANS, ELECTRONIC
OR MECHANICAL, INCLUDING PHOTOCOPY, RECORDING, OR ANY
INFORMATION STORAGE AND RETRIEVAL SYSTEM, WITHOUT
PERMISSION IN WRITING FROM THE PUBLISHER.

ACADEMIC PRESS, INC.
111 Fifth Avenue, New York, New York 10003

*United Kingdom Edition published by*
ACADEMIC PRESS, INC. (LONDON) LTD.
24/28 Oval Road, London NW1

Library of Congress Cataloging in Publication Data

Roberts, Arthur Wayne, DATE
    Convex functions.

    (Pure and applied mathematics; a series of monographs and textbooks, v. 57)
    Bibliography: p.
    1.  Convex functions.   I.  Varberg, Dale E.,
joint author.   II.  Title.   III.  Series.
QA3.P8 vol. 57   [QA331.5]    515'.88    72-12186
ISBN 0–12–589740–5

AMS (MOS) 1970 Subject Classifications: 26A51, 26A24, 26A86, 90C05

PRINTED IN THE UNITED STATES OF AMERICA

**Willamette University Library**
**Salem,    Oregon**

*To R. C. Buck and R. H. Cameron,
our teachers
and to Dolores and Idella,
our wives*

It seems to me that the notion of convex function is just as fundamental as positive function or increasing function. If I am not mistaken in this, the notion ought to find its place in elementary expositions of the theory of real functions.

J. L. W. V. JENSEN

# Contents

*Preface*, xi
*Acknowledgments*, xiii
*Introduction*, xv
*List of Notation*, xix

## I  Convex Functions on the Real Line

        10. Introduction, 2
        11. Continuity and Differentiability, 3
        12. Characterizations, 9
        13. Closure under Functional Operations, 15
        14. Differences of Convex Functions, 22
        15. Conjugate Convex Functions, 28

## II  Normed Linear Spaces

        20. Introduction, 38
        21. Normed Linear Spaces, 38
        22. Functions on Normed Linear Spaces, 54
        23. Derivatives in a Normed Linear Space, 62

## III  Convex Sets

30. Introduction, 73
31. Convex Sets and Affine Sets, 73
32. Hyperplanes and Extreme Points, 81

## IV  Convex Functions on a Normed Linear Space

40. Introduction, 89
41. Continuity of Convex Functions, 91
42. Differentiable Convex Functions, 97
43. The Support of Convex Functions, 104
44. Differentiability of Convex Functions, 113

## V  Optimization

50. Introduction, 122
51. Maxima and Minima, 122
52. Minimax Theorems and the Theory of Games, 128
53. Linear Programming, 138
54. The Simplex Method, 154
55. Convex Programming, 170
56. Approximation, 179

## VI  Inequalities

60. Introduction, 189
61. The Classical Inequalities, 189
62. The Generalized Geometric Mean–Arithmetic Mean Inequality and Norms, 194
63. Matrix Inequalities, 200

## VII  Midconvex Functions

70. Introduction, 211
71. Midconvex Functions on a Normed Linear Space, 211
72. Midconvex Functions on **R**, 218

## VIII  Related Classes of Functions

80. Introduction, 227
81. Quasiconvex Functions, 228
82. Completely Convex Functions, 233
83. Convex Functions of Higher Order, 237
84. Generalized Convex Functions, 240
85. More about Generalized Convex Functions, 246
86. Other Related Topics, 253

## Appendix, 263

Introduction, 264
Independent Study Projects, 264
Unsolved Problems, 271

## Bibliography, 273

*Author Index,* 289

*Subject Index,* 295

# *Preface*

The idea of writing this book grew out of an Undergraduate Research Program cooperatively directed by the authors. It has been well recognized, especially by Russian mathematicians, that the topic of convex sets affords an excellent means of introducing high school students to the beauty and fascination of advanced mathematics. (See for example Yaglom and Boltyanskiĭ, "Convex Figures," 1961.) We came to feel that the study of convex functions was similarly suited to undergraduate college students of mathematics. There are several reasons for this: (a) it has high geometric and intuitive content, (b) it uses and reinforces in a significant way ideas the student has learned in his linear algebra and calculus courses, (c) it permits easy generalization to an abstract setting, (d) it naturally illuminates a number of mathematical gems, developing a taste for beauty and elegance, and (e) it is easy to pose challenging and even unsolved problems. With this in mind, we set out to collect in one place all the basic ideas and present them at a level appropriate to a college senior, to put together an extensive list of problems ranging from easy to very difficult, and to indicate areas that can be investigated as individual or group projects.

Selected elementary properties of convex functions are often developed as needed in texts by authors whose main interest is another topic. Other properties appear as applications of theorems in real analysis or in the geometry of convex sets. As a result, known facts about convex functions are scattered throughout the literature. Where efforts have been made to draw them together, it has been done in a highly sophisticated way directed primarily to the specialist. As we began to draw things together for our intended undergraduate audience, we decided that we might render a service to social scientists, engineers, people in operations research, and even mathematicians by providing a handy reference book containing all the central facts about convex functions. We aimed, therefore, to write so that someone with even a modest training in college mathematics could open our book to the area of his particular interest with the expectation of at least understanding the statement of the theorems.

Given these intentions, we have used as little special terminology and notation as possible. We have also tried to minimize demands on the reader's mathematical preparation, assuming only that material associated with undergraduate courses in linear algebra, advanced calculus, and introductory measure theory (the latter necessary only for isolated theorems). Wherever our experience as undergraduate teachers suggested it to be advisable, we have provided references in these three subjects to one of three standard texts as explained in the Introduction.

The book was conceived as a source book for a senior seminar on convex functions. Actually it could serve as a text for a course dealing with the role of convexity in mathematical analysis, or a course in optimization. We also have tried to provide a good introduction to normed linear spaces with an emphasis on Fréchet differentiation.

# Acknowledgments

The Undergraduate Research Program, already mentioned in the Preface as having given impetus to the writing of this book, was sponsored by the National Science Foundation. For support of this initial program and some of the subsequent research necessary to this project, we are in debt to the National Science Foundation. We also acknowledge support for the actual writing made available through the sabbatical leave programs of Macalester College and Hamline University.

We first discussed the idea of this monograph with Professor Victor Klee, and it is a pleasure to thank him for his encouragement, a number of helpful conversations and letters, for his careful reading of the manuscript, and for the privilege of using, in his absence, the materials he has assembled relating to all aspects of convexity. The value of this latter resource can only be appreciated by someone who has had opportunity to use it. Preliminary versions of at least some part of our manuscript were read by Professors R. P. Boas, H. J. Greenberg, V. L. Klee, R. M. Mathsen, R. T. Rockafellar, C. A. Rogers, J. H. M. Whitfield, and D. E. Wulbert. The staff of Academic Press provided valued editorial assistance, compiled the Author Index, and were particularly helpful in preparing the Bibliography. While responsibility for the final version of the manuscript rests entirely with the authors, we wish to acknowledge the helpful comments of all these people. Finally, we express our deep appreciation for the continued interest and many conversations with Professors Rockafellar and Grünbaum during that delightful year at the University of Washington in Seattle where much of this was written.

# Acknowledgments

The list below gives the source of each quotation used in this book. The number in square brackets indicates the page in our book where the quote appears. In each case, permission to quote has been requested from the publishers. We gratefully acknowledge their generous cooperation.

N. BOURBAKI
    The architecture of mathematics, *American Mathematical Monthly*, Vol. 57 (1950), p. 231 [37].

H. J. BREMERMANN
    Complex convexity, Reprinted with permission of The American Mathematical Society, from *Transactions of The American Mathematical Society*, Copyright © 1956, Vol. 82, p. 38 [255].

C. H. CHAPMAN
    The theory of transformation groups, Reprinted with permission of The American Mathematical Society, from *Bulletin of The American Mathematical Society*, Copyright © 1892, Vol. 2, p. 61 [226].

RICHARD COURANT
    Mathematics in the modern world, *Scientific American*, Copyright © 1964 by Scientific American, Inc., All rights reserved, Vol. 211, p. 43 [88].

G. B. DANTZIG
    "Linear Programming and Extensions," Reprinted by permission of Princeton Univ. Press, Princeton, New Jersey, Copyright © 1963 by the Rand Corporation, p. 1 [153].

    In "Applications of Mathematical Techniques," NATO Scientific Affairs Committee (E. M. Beale, ed.), Academic Press, New York, 1968, p. 15 [147].

    Linear programming and its progeny, *In* "Applications of Mathematical Programming Techniques" (E. M. Beale, ed.), English Univ. Press, London, 1970, p. 12 [121].

HERMANN HANKEL
    Quoted in "A History of Mathematics," by C. B. Boyer, Wiley, New York, 1968, p. 598 [210].

HEINRICH HERTZ
    Quoted in "Men of Mathematics," by E. T. Bell, Dover, New York, 1937, p. 16 [188].

J. L. W. V. JENSEN
    Sur les fonctions convexes et les inégalités entre les valeurs moyennes, *Acta Mathematica*, Vol. 30 (1906), p. 191 [vi].

V. L. KLEE
    What is a convex set? *American Mathematical Monthly*, Vol. 57 (1950), p. 231 [72].

MORRIS KLINE
    Geometry, *Scientific American*, Copyright © 1964 by Scientific American, Inc., All rights reserved, Vol. 211, p. 69 [72].

D. S. MITRINOVIĆ
    "Analytic Inequalities," Springer-Verlag, Berlin and New York, 1970, p. 23 [188].

C. S. PIERCE
    The essence of mathematics, In "Collected Papers of Charles Sanders Pierce," (C. Hartshorne and P. Weiss, ed.), Harvard Univ. Press, Cambridge, Massachusetts, Vol. 4, 1933, p. 196 [226].

HEINRICH TIETZE
    "Famous Problems of Mathematics," Graylock Press, Baltimore, Maryland, 1965, p. xv [263].

F. VALENTINE
    The dual cone and Helly type theorems, Reprinted with permission of The American Mathematical Society, from "Proceedings of Symposia in Pure Mathematics," Copyright © 1963, Vol. 7, p. 492 [29].

A. WEIL
    The future of mathematics, *American Mathematical Monthly*, Vol. 57 (1950), p. 297 [263].

    The future of mathematics, *American Mathematical Monthly*, Vol. 57 (1950), p. 304 [121].

HERMANN WEYL
    Emmy Noether, *Scripta Mathematica*, Vol. 3 (1935), p. 214 [1].

L. C. YOUNG
    "Lectures on the Calculus of Variations and Optimal Control Theory," Saunders, Philadelphia, 1969, p. 94 [37].

*Introduction*

Deciding where to begin is a major step. One procedure is to lay out all necessary preliminary material, introduce the major ideas in their most general setting, prove the theorems, and then specialize to obtain classical results and various applications. Both of us strongly resisted this approach. We believe that mathematics is best learned by examining in depth the major ideas in as familiar a setting as possible, and then seeing how they can be generalized. Thus, without fanfare, Chapter I takes up the very concrete (and historic) case of convex functions defined on an interval of the real line. This material appears as a logical extension of what the student has learned in his calculus course, has immediate applications (for example, most of Chapter VI on Inequalities), motivates most of the generalizations to come later, raises most of the important questions that will be studied in the rest of the book

When we do come to generalizing the material in Chapter I, we face the question of what kind of spaces to allow in the discussion. One appropriate decision would be to restrict attention to finite dimensional spaces. This has the advantage of eliminating many difficulties that occur in infinite-dimensional spaces, but it also eliminates consideration of many examples of great importance in mathematical analysis. This choice therefore seemed inconsistent with our desire to provide a reference useful in a wide variety of applications, and to use convex functions as a vehicle for introducing the student to advanced topics in mathematics. An alternate decision would be to use the setting of linear topological spaces, in a certain sense the natural domain for the study of convexity. This route requires, however, the introduction of technical terminology that runs counter to the intent of making theorems in the middle of the book immediately accessible to our intended audience.

We have chosen an intermediate path; our spaces will be normed linear spaces. All students of mathematics at least know the language of normed linear spaces; thus the number of new terms to be assimilated is minimized. Moreover, with regard to differentiability questions, we believe the point of view natural to normed linear spaces (namely the Fréchet derivative) is really the most illuminating way to approach the topic for functions defined in $n$-dimensional Euclidean space.

A function is convex if and only if the set of points above its graph is convex. Thus, all questions about convex functions can be phrased in terms of convex sets. Since most mathematicians interested in convexity come at it from a geometric point of view, and since the subject of convex geometry is well developed, this geometric approach to convex functions is the one many prefer. Our subject, however, is in the mainstream of mathematical analysis, and we address ourselves to an audience presumably more familiar with analysis than with geometry. We wished therefore to avoid reliance on the results of convex geometry, and chose to state theorems and use methods cast in the language of functions.

Finally we mention another choice we have made. Our convex functions will be finite valued unless specifically stated to the contrary. While some elegance and generality can be gained by allowing infinite valued functions, we are not willing to pay the price. Exceptional cases eliminated by this device usually reappear, perhaps in disguised form. Moreover, it requires a careful reinterpretation of a great number of mathematical words and concepts that may not be familiar (or at any rate second nature) to our intended audience.

Some comments on the format of the book are in order. Sections are numbered consecutively within chapters; thus Section 32 is the second section of Chapter III. A reference in Section 32 to Theorem A refers to the first theorem of that section. Otherwise the reference will identify the section and the theorem, as in Theorem 21A. References to numbered expressions are either to (1) in the same section, or to (21.1), expression (1) in Section 21.

Each section in the first seven chapters ends with Remarks and Problems. Problems are for the most part stated as assertions that may be taken either as an exercise to be worked out, or as a fact related to the previous section, but not deemed important enough to be proved in the main body of the text. Some of the problems summarize related papers, in which case a reference is given. Most of these and some others as well are marked with a * indicating that they require more work to establish than would be expected in a normal exercise. Occasionally a question is posed without indication of whether or not the authors think they know the answer.

We have already mentioned the use of three books that serve as more or less standard references to which we turn for results needed in our exposition. We have selected Buck, "Advanced Calculus," Halmos, "Finite Dimensional Vector Spaces," and Natanson, "Theory of Functions of a Real Variable" for this purpose. The reader is not expected to have mastered all that is in these books; indeed it is the expectation of his need to be refreshed that motivates the inclusion of a reference. Citation of these three books and other papers and books about convex functions is always made by author and date. We write either of the original work of Jensen [1906] or of results available in the literature [Valentine, 1964, p. 129], suiting the form to the literary construction of the sentence. The complete reference is to be found in the Bibliography in the back of the book.

In addition to the complete Bibliography just mentioned, many sections close with an abbreviated bibliography. The purpose of this is to collect in one spot references that relate to a specific topic and to indicate by a chronological listing the historical development of that topic. All papers so cited are of course included in the complete Bibliography at the end of the book.

# List of Notation

| Symbol | Definition | See |
|---|---|---|
| $x \in U$ | $x$ is a member of set $U$ | |
| $\{x : x \text{ satisfies } P\}$ | The set of all elements that satisfy condition $P$ | |
| $A \subseteq B$ | Set $A$ is contained in (perhaps is equal to) set $B$ | |
| $A \setminus B$ | Difference of sets $= \{x : x \in A, x \notin B\}$ | Problem 21C |
| $U^0$ | The interior of set $U$ | Section 21 |
| $\bar{U}$ | The closure of set $U$ | Section 21 |
| $\text{seg}[\mathbf{x}, \mathbf{y}]$ | In a linear space, the set $\{\mathbf{z} = \mathbf{x} + t(\mathbf{y} - \mathbf{x}) : t \in [0, 1)\}$ | |
| $\mathbf{R}$ | The real numbers | |
| $\mathbf{R}^n$ | Euclidean $n$-space in which points $\mathbf{x} = (x_1, ..., x_n)$ have length $\|\mathbf{x}\| = (\sum_1^n x_i^2)^{1/2}$ | |
| $\mathbf{R}_+^n$ | The nonnegative orthant of $\mathbf{R}^n$; the points $(x_1, ..., x_n)$ with $x_i \geq 0, i = 1, ..., n$ | |
| $\mathbf{L}$ | A normed linear space | Section 21 |
| $\mathbf{L}^n$ | An $n$-dimensional normed linear space with unspecified norm | Theorem 21G |
| $\langle \mathbf{x}, \mathbf{y} \rangle$ | The inner product | Section 21 |
| $\mathscr{L}(\mathbf{L}, \mathbf{M})$ | The space of continuous linear transformations $T : \mathbf{L} \to \mathbf{M}$ | Theorem 22E |
| $\mathbf{L}^*, \mathbf{R}^{n*}$ | Dual spaces; $\mathscr{L}(\mathbf{L}, \mathbf{R}), \mathscr{L}(\mathbf{R}^n, \mathbf{R})$ | Section 22 |
| $l_n^p$ | $n$-tuples $\mathbf{x} = (x_1, ..., x_n)$ with $\|\mathbf{x}\| = (\sum_1^n |x_i|^p)^{1/p}, p \geq 1$ | Theorem 62C |
| $l_n^\infty$ | $n$-tuples with $\|\mathbf{x}\| = \max\{|x_1|, ..., |x_n|\}$ | |
| $l^p$ | Sequences $\mathbf{x} = \{x_i\}_1^\infty$ such that $\|\mathbf{x}\| = (\sum_1^\infty |x_i|^p)^{1/p} < \infty$ | Problem 62D |
| $l^\infty$ | Sequences such that $\|\mathbf{x}\| = \sup_i |x_i| < \infty$ | |
| $C[0, 1]$ | Space of functions continuous on $[0, 1]$ | Example 21D |

| Symbol | Definition | See |
|---|---|---|
| $C^\infty[0, 1]$ | Space of functions on $[0, 1]$ having derivatives of all orders | |
| $f'_-(x), f'_+(x)$ | Left, right hand derivatives of $f : I \to \mathbf{R}$ at $x$ | Section 11 |
| $f'(\mathbf{x}_0; \mathbf{v})$ | Directional derivative of $f : \mathbf{L} \to \mathbf{R}$ at $\mathbf{x}_0$ in the direction $\mathbf{v}$ | Section 23 |
| $f'(\mathbf{x})$ | The linear transformation $f'(\mathbf{x}) : \mathbf{L} \to \mathbf{R}$ that is the (Fréchet) derivative of $f$ at $\mathbf{x}$ | Section 23 |

# 1

## Convex Functions on the Real Line

> But definite concrete problems were first conquered in their undivided complexity, singlehanded by brute force, so to speak. Only afterwards the axiomaticians came along and stated: Instead of breaking in the door with all your might and bruising your hands, you should have constructed such and such a key of skill, and by it you would have been able to open the door quite smoothly. But they can construct the key only because they are able, after the breaking in was successful, to study the lock from within and without. Before you generalize, formalize, and axiomatize, there must be mathematical substance.
>
> HERMANN WEYL

## 10. Introduction

Historically, logically, and pedagogically, the study of convex functions begins in the context of real-valued functions of a real variable. Displayed in this setting where graphic representation guides our intuition, we find a rich diversity of theorems having an elegance that is rooted in the very simplicity of their proofs. Yet the results are not trivial. They have important applications, and at the same time they give rise to a wide variety of generalizations.

We take our functions $f: I \to \mathbf{R}$ to be defined on some interval of the real line $\mathbf{R}$. We mean to allow $I$ to be open, half-open, or closed, finite or infinite, and for technical reasons that appear in Section 15, we even allow the possibility that $I$ may be a point. A function $f: I \to \mathbf{R}$ is called **convex** if

$$f[\lambda x + (1 - \lambda)y] \leq \lambda f(x) + (1 - \lambda) f(y) \tag{1}$$

for all $x, y \in I$ and $\lambda$ in the open interval $(0, 1)$. (One could equivalently take $\lambda$ to be in the closed interval $[0, 1]$.) It is called **strictly convex** provided that the inequality (1) is strict for $x \neq y$. Geometrically, (1) means that if $P, Q,$ and $R$ are any three points on the graph of $f$ with $Q$ between $P$ and $R$, then $Q$ is on or below chord $PR$ (Fig. 10.1). In terms of slopes, it is equivalent to

$$\text{slope } PQ \leq \text{slope } PR \leq \text{slope } QR \tag{2}$$

with strict inequalities when $f$ is strictly convex.

Simple examples of convex functions are $f(x) = x^2$ on $(-\infty, \infty)$, $g(x) = \sin x$ on $[-\pi, 0]$, and $h(x) = |x|$ on $(-\infty, \infty)$. The first two are in fact strictly convex, the third is not. If $-f: I \to \mathbf{R}$ is convex, then we say that $f: I \to \mathbf{R}$ is **concave**. The theory of concave functions may therefore be subsumed under that of convex functions and we shall concentrate our attention on the latter. We say that $f: \mathbf{R} \to \mathbf{R}$ is **linear** if $f(\alpha x + \beta y) = \alpha f(x) + \beta f(y)$ for all $\alpha, \beta, x, y \in \mathbf{R}$. It is known and is easy to show that $f$ is linear if and only if $f(x) = mx$ for some constant $m$. We say that $f: I \to \mathbf{R}$ is **affine** if it is of the form $f(x) = mx + b$ on $I$. It is clear that any affine function is convex, but not strictly convex.

We begin in Section 11 with a careful look at the regularity properties of convex functions. It is perhaps not surprising that a convex function is continuous save at the endpoints of its domain, but it seems quite unexpected that it is, with the possible exception of a countable number of points, differentiable.

The class of convex functions can be characterized in a variety of ways

## 11. Continuity and Differentiability

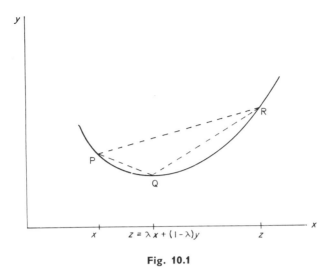

Fig. 10.1

introduced in Section 12. It is in this connection that we first encounter the important idea of a support function.

In Section 13 we consider a variety of operations (addition, taking limits, etc.) under which the class of convex functions remains closed. Subtraction is missing from this list of operations, but the class of functions representable as the difference of two convex functions is closed under all the standard functional operations and is interesting in its own right. It is the subject of our attention in Section 14. Finally in Section 15, we explore in its most elementary setting the notion of conjugate convex functions. Neither of these last two sections is essential to understanding the rest of the book.

## 11. Continuity and Differentiability

A function convex and finite on a closed interval $[a, b]$ is bounded from above by $M = \max\{f(a), f(b)\}$, since for any $z = \lambda a + (1 - \lambda)b$ in the interval,

$$f(z) \leqslant \lambda f(a) + (1 - \lambda) f(b) \leqslant \lambda M + (1 - \lambda)M = M$$

It is also bounded from below as we see by writing an arbitrary point in the form $(a + b)/2 + t$. Then

$$f\left(\frac{a+b}{2}\right) \leqslant \frac{1}{2} f\left(\frac{a+b}{2} + t\right) + \frac{1}{2} f\left(\frac{a+b}{2} - t\right)$$

or

$$f\left(\frac{a+b}{2}+t\right) \geq 2f\left(\frac{a+b}{2}\right) - f\left(\frac{a+b}{2}-t\right)$$

Using $M$ as the upper bound, $-f[(a+b)/2 - t] \geq -M$, so

$$f\left(\frac{a+b}{2}+t\right) \geq 2f\left(\frac{a+b}{2}\right) - M = m$$

It is easily seen that a convex function may not be continuous at the boundary points of its domain. It may, in fact, have upward jumps there. On the interior, however, it is not only continuous, but it satisfies a stronger condition. We will prove that for any closed subinterval $[a, b]$ of the interior of the domain, there is a constant $K$ so that for any two points $x, y \in [a, b]$,

$$|f(x) - f(y)| \leq K |x - y| \tag{1}$$

A function that satisfies (1) for some $K$ and all $x$ and $y$ in an interval is said to satisfy a **Lipschitz condition** (or to be Lipschitz) on the interval.

**Theorem A.** If $f: I \to \mathbf{R}$ is convex, then $f$ satisfies a Lipschitz condition on any closed interval $[a, b]$ contained in the interior $I^0$ of $I$. Consequently, $f$ is absolutely continuous on $[a, b]$ and continuous on $I^0$.

**Proof.** Choose $\varepsilon > 0$ so that $a - \varepsilon$ and $b + \varepsilon$ belong to $I$, and let $m$ and $M$ be the lower and upper bounds for $f$ on $[a - \varepsilon, b + \varepsilon]$. If $x$ and $y$ are distinct points of $[a, b]$, set

$$z = y + \frac{\varepsilon}{|y - x|}(y - x), \qquad \lambda = \frac{|y - x|}{\varepsilon + |y - x|}$$

Then $z \in [a - \varepsilon, b + \varepsilon]$, $y = \lambda z + (1 - \lambda)x$, and we have

$$f(y) \leq \lambda f(z) + (1 - \lambda) f(x) = \lambda [f(z) - f(x)] + f(x)$$

$$f(y) - f(x) \leq \lambda(M - m) < \frac{|y - x|}{\varepsilon}(M - m) = K|y - x|$$

where $K = (M - m)/\varepsilon$. Since this is true for any $x, y \in [a, b]$, we conclude that $|f(y) - f(x)| \leq K|y - x|$ as desired.

Next we recall that $f$ is **absolutely continuous** on $[a, b]$ if, corresponding to any $\varepsilon > 0$, we can produce a $\delta > 0$ such that for any

## 11. Continuity and Differentiability

collection $\{(a_i, b_i)\}_1^n$ of disjoint open subintervals of $[a, b]$ with $\sum_1^n (b_i - a_i) < \delta$, $\sum_1^n |f(b_i) - f(a_i)| < \varepsilon$. Clearly the choice $\delta = \varepsilon/K$ meets this requirement.

Finally the continuity of $f$ on $I^0$ is a consequence of the arbitrariness of $[a, b]$. ●

The derivative of a convex function is best studied in terms of the **left** and **right derivatives** defined by

$$f_-'(x) = \lim_{y \uparrow x} \frac{f(y) - f(x)}{y - x}$$

$$f_+'(x) = \lim_{y \downarrow x} \frac{f(y) - f(x)}{y - x}$$

**Theorem B.** If $f: I \to \mathbf{R}$ is convex [strictly convex], then $f_-'(x)$ and $f_+'(x)$ exist and are increasing [strictly increasing] on $I^0$.

**Proof.** Consider four points $w < x < y < z$ in $I^0$ with $P, Q, R$, and $S$ the corresponding points on the graph of $f$ (Fig. 11.1). Inequality (10.2) extended to four points gives

$$\text{slope } PQ \leq \text{slope } PR \leq \text{slope } QR \leq \text{slope } QS \leq \text{slope } RS \qquad (2)$$

with strict inequalities if $f$ is strictly convex. Now since slope

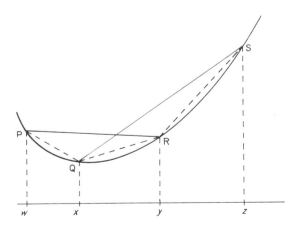

**Fig. 11.1**

$PR \leqslant$ slope $QR$, it is clear that slope $QR$ increases as $x \uparrow y$ and similarly that slope $RS$ decreases as $z \downarrow y$. Thus the left side of the inequality

$$\frac{f(x) - f(y)}{x - y} \leqslant \frac{f(z) - f(y)}{z - y}$$

increases as $x \uparrow y$ and the right side decreases as $z \downarrow y$. These facts guarantee that $f_-'(y)$ and $f_+'(y)$ exist and satisfy

$$f_-'(y) \leqslant f_+'(y) \tag{3}$$

a result that holds for all $y \in I^0$. Moreover, using (2) again, we see that

$$f_+'(w) \leqslant \frac{f(x) - f(w)}{x - w} \leqslant \frac{f(y) - f(x)}{y - x} \leqslant f_-'(y) \tag{4}$$

with strict inequalities prevailing if $f$ is strictly convex. This combined with (3) yields

$$f_-'(w) \leqslant f_+'(w) \leqslant f_-'(y) \leqslant f_+'(y)$$

establishing the monotone nature of $f_-'$ and $f_+'$. ●

Actually the results of Theorem B (appropriately interpreted) are valid for all of $I$, not just its interior. For example, if $I = (a, b]$, then $f_-'(b)$ exists at least in the infinite sense and $f_-'$ is increasing on $(a, b)$. The correct statement of Theorem B for the case $I = [a, b]$ is given in Problem C.

There are a number of other important facts having to do with the continuity properties of $f_+'$ and $f_-'$. The monotone character of $f_+'$ means that the limit of $f_+'(x)$ exists as $x \downarrow w$. From the inequality

$$f_+'(x) \leqslant \frac{f(y) - f(x)}{y - x}$$

and the continuity of $f$ it follows that

$$\lim_{x \downarrow w} f_+'(x) \leqslant \lim_{x \downarrow w} \frac{f(y) - f(x)}{y - x} = \frac{f(y) - f(w)}{y - w}$$

From this in turn we see that

$$\lim_{x \downarrow w} f_+'(x) \leqslant \lim_{y \downarrow w} \frac{f(y) - f(w)}{y - w} = f_+'(w)$$

## 11. Continuity and Differentiability

On the other hand, since $x > w$, monotonicity of $f_+'$ implies $f_+'(x) \geq f_+'(w)$. Thus

$$\lim_{x \downarrow w} f_+'(x) = f_+'(w) \tag{5}$$

Similar arguments show that

$$\lim_{x \uparrow w} f_+'(x) = f_-'(w) \tag{6}$$

We point out that (5) and (6) are also valid at the left and right endpoints of $I$, respectively, provided that $f$ is defined and continuous there. Finally we remark that statements analogous to (5) and (6) hold for the left and right limits of $f_-'(x)$.

> **Theorem C.** If $f: I \to \mathbf{R}$ is convex on the open interval $I$, then the set $E$ where $f'$ fails to exist is countable. Moreover, $f'$ is continuous on $I \setminus E$.

**Proof.** From (5) and (6), we conclude that $f_+'(w) = f_-'(w)$ if and only if $f_+'$ is continuous at $w$. Thus $E$ consists precisely of the discontinuities of the increasing function $f_+'$ and is therefore countable [Natanson I, 1961, p. 205]. On $I \setminus E$, $f_+'$ is continuous and so $f'$, which agrees with $f_+'$ on $I \setminus E$, is also continuous there. ●

### PROBLEMS AND REMARKS

A. Recall that $f$ is affine on $I$ if $f$ has the form $f(x) = mx + b$. Show

(1) $f$ is affine on $\mathbf{R} \Leftrightarrow f(\lambda x + (1 - \lambda)y) = \lambda f(x) + (1 - \lambda)f(y)$ for all $\lambda \in \mathbf{R}$ and $x, y \in \mathbf{R}$;
(2) $f$ is affine on $I \Leftrightarrow f$ and $-f$ are convex on $I$;
(3) if $f: [a, b] \to \mathbf{R}$ is convex and there is a single value of $\lambda \in (0, 1)$ for which $f(\lambda a + (1 - \lambda)b) = \lambda f(a) + (1 - \lambda)f(b)$, then $f$ is affine on $[a, b]$;
(4) if $f$ is convex on $I$, then it is strictly convex there $\Leftrightarrow$ there is no subinterval of $I$ on which $f$ is affine.

B. Let $f(x) = x^2$ if $x \in (-1, 1)$, $f(-1) = f(1) = 2$. Note that $f$ is convex on $[-1, 1]$. Show that if a convex function is not continuous at an endpoint of its domain, it has an upward jump there.

C. Theorem B can be extended to describe the behavior of $f$ at the endpoints when $I = [a, b]$. In this case, $f_+'(a)$ and $f_-'(b)$ exist at least in the infinite sense, $f_+'$ is increasing on $[a, b)$ and $f_-'$ is increasing on $(a, b]$.

D. A convex function $f: [a, b] \to \mathbf{R}$ is Lipschitz on $[a, b] \Leftrightarrow f_+'(a)$ and $f_-'(b)$ are finite. A convex function is Lipschitz on any compact (that is, closed and bounded) subset of the interior of its domain.

**\*E.** If $f$ has finite one-sided derivatives at all points of an interval $(a, b)$, then $f$ is continuous there and is differentiable except possibly at countably many points. If these one-sided derivatives are increasing, then $f$ is convex [Artin, 1964, p. 4].

**\*F.** Let $D$ be any countable subset of an interval $I$. There is a convex function $f: I \to \mathbf{R}$ such that $f'$ fails to exist on $D$. (This problem, stated here because of its relation to Theorem C, is more easily solved after Theorem 12A is available as a tool).

**G.** If $f: [0, 2\pi] \to \mathbf{R}$ is convex, then for $k \geqslant 1$,

$$a_k = \frac{1}{\pi} \int_0^{2\pi} f(x) \cos kx \, dx \geqslant 0.$$

**H.** If $f: (-\infty, \infty) \to \mathbf{R}$ is convex and bounded above, then it is constant.

**I.** The function $f: I \to \mathbf{R}$ is convex $\Leftrightarrow$ for all sets of distinct $x_1, x_2, x_3 \in I$,

$$\psi(x_1, x_2, x_3) = \frac{(x_3 - x_2) f(x_1) + (x_1 - x_3) f(x_2) + (x_2 - x_1) f(x_3)}{(x_1 - x_2)(x_2 - x_3)(x_3 - x_1)}$$

is nonnegative. Artin [1964] bases an elegant discussion of convex functions on this definition.

## BIBLIOGRAPHIC NOTES

The recognition of convex functions as a class of functions to be studied is generally traced to Jensen, but as is usually the case, earlier work can be cited that anticipated what was to come. Hölder proved that if $f''(x) \geqslant 0$, then $f$ satisfied what later came to be known as Jensen's inequality. Stolz proved that if $f$ is continuous on $[a, b]$ and satisfies

$$f\left(\frac{x+y}{2}\right) \leqslant \tfrac{1}{2}[f(x) + f(y)] \tag{7}$$

then $f$ has left and right derivatives at each point of $(a, b)$. Hadamard obtained a basic integral inequality for functions having an increasing derivative on $[a, b]$. Jensen used (7) to define convex functions and gave the first in a long series of results which together with (7) imply the continuity of $f$. We discuss this further in Chapter VII under the title of midconvex functions. The foundation work in convex functions to which we have referred is found in the following papers.

1889, H. Brunn, "Über Kurven ohne Wendepunkte," p. 74. München.
1889, O. Hölder, Über einen Mittelwertsatz. *Nachr. Ges. Wiss. Goettingen* pp. 38–47.
1893, J. Hadamard, Étude sur les propriétés des fonctions entières et en particulier d'une fonction considérée par Riemann. *J. Math. Pures Appl.* **58**, 171–215.
1893, O. Stolz, "Grundzüge der Differential und Integralrechung," Vol. I. Teubner, Leipzig.
1905, J. L. W. V. Jensen, Om konvexe Funktioner og Uligheder mellem Middelvaerdier. *Nyt Tidsskr. Math.* **16B**, 49–69.
1906, J. L. W. V. Jensen, Sur les fonctions convexes et les inegalités entre les valeurs moyennes. *Acta Math.* **30**, 175–193.

## 12. Characterizations

Certain results in this section and certain special extensions of these results suited to particular purposes are often developed in texts as preludes to something else. The following texts might be mentioned as having more than just a passing reference to convex functions of a single real variable. We also include as appropriate to this list several expository articles about convex functions and an unpublished set of notes by Beckenbach.

E. Artin (1964). "The Gamma Function," Chapter 1. Holt, New York.
E. F. Beckenbach (1948b). Convex functions. *Bull. Amer. Math. Soc.* **54**, 439–460.
E. F. Beckenbach (1953). Convexity (unpublished).
W. Fenchel (1953). Convex Cones, Sets, and Functions (mimeographed lecture notes). Princeton Univ., Princeton, New Jersey.
J. W. Green (1954). Recent applications of convex functions. *Amer. Math. Monthly* **61**, 449–454.
G. H. Hardy, J. E. Littlewood, and G. Polya (1952). "Inequalities," 2nd ed., pp. 70–101. Cambridge Univ. Press, London and New York.
M. A. Krasnosel'skii and Y. B. Rutickii (1961). "Convex Functions and Orlicz Spaces." Noordhoff, Groningen.
I. P. Natanson (1961). "Theory of Functions of a Real Variable," Vol. II, pp. 36–47, 230–234. Ungar, New York.
M. M. Peixoto (1948b). "Convexity of Curves." *Notas Mat.* **No. 6**, Livraria Boffoni, Rio de Janeiro.
T. Popoviciu (1944). "Les Fonctions Convexes." Hermann, Paris.
J. F. Randolph (1968). "Basic Real and Abstract Analysis." Academic Press, New York.
P. Sengenhorst (1952). Über konvexe Funktionen. *Math.-Phys. Semesterber.* **2**, 217–230.
A. Zygmund (1968). "Trigonometric Series," Vol. I, pp. 21–26. Cambridge Univ. Press, London and New York.

## 12. Characterizations

Legal descriptions of one's home are necessary and serve useful purposes, but people seldom describe their home in this way. They are more likely to give a physical description, a street address, and perhaps other instructions on how to find it. In like manner, the definition of a convex function serves useful purposes, but mathematicians often recognize or think about convex functions in other ways: by an integral representation, by properties of the derivatives, by geometric properties of the graph. All of these characterizations and others are considered in this section and in the exercises included at the end.

We begin by representing a convex function as an integral which, fortunately, can be taken in the sense of either Riemann or Lebesgue.

**Theorem A.** $f: (a, b) \to \mathbf{R}$ is convex [strictly convex] if and only if there is an increasing [strictly increasing] function

$g\colon (a, b) \to \mathbf{R}$ and a point $c \in (a, b)$ such that for all $x \in (a, b)$,

$$f(x) - f(c) = \int_c^x g(t)\, dt \tag{1}$$

**Proof.** We suppose first that $f$ is convex. Choose $g = f_+'$ which exists and is increasing (Theorem 11B) and let $c$ be any point in $(a, b)$. By Theorem 11A, $f$ is absolutely continuous on $[c, x]$. By an elementary argument for Riemann integrals (Problem A) or by a classical theorem for Lebesgue integrals [Natanson I, 1961, p. 255],

$$f(x) - f(c) = \int_c^x f_+'(t)\, dt = \int_c^x g(t)\, dt$$

Moreover, if $f$ is strictly convex, $g = f_+'$ will be strictly increasing (Theorem 11B).

Conversely, suppose that (1) holds with $g$ increasing. Let $\alpha, \beta$ be positive with $\alpha + \beta = 1$. Then for $x < y$ in $(a, b)$,

$$\alpha f(x) + \beta f(y) - (\alpha + \beta) f(\alpha x + \beta y) = \beta \int_{\alpha x + \beta y}^{y} g(t)\, dt - \alpha \int_x^{\alpha x + \beta y} g(t)\, dt$$

To bound this expression below, we replace both integrands by the constant $g(\alpha x + \beta y)$, this being the smallest value of the first integrand and the largest of the second. We obtain on the right-hand side

$$\beta g(\alpha x + \beta y)[y - (\alpha x + \beta y)] - \alpha g(\alpha x + \beta y)[\alpha x + \beta y - x]$$

which simplifies to 0. Thus,

$$\alpha f(x) + \beta f(y) - f(\alpha x + \beta y) \geq 0$$

which is equivalent to the inequality that defines convexity. Finally, we note that the estimate made above is a strict one when $g$ is strictly increasing. ∎

Theorem B of the previous section showed us that, for a differentiable function, convexity implies an increasing derivative. This also is a two-way street.

**Theorem B.** Suppose $f$ is differentiable on $(a, b)$. Then $f$ is convex [strictly convex] if and only if $f'$ is increasing [strictly increasing].

## 12. Characterizations

**Proof.** Having already established half of the theorem, let us suppose $f'$ is increasing [strictly increasing]. Then the fundamental theorem of calculus assures us that

$$f(x) - f(c) = \int_c^x f'(t)\, dt$$

for any $c \in (a, b)$. That $f$ is convex [strictly convex] now follows from Theorem A. ●

**Theorem C.** Suppose $f''$ exists on $(a, b)$. Then $f$ is convex if and only if $f''(x) \geq 0$. And if $f''(x) > 0$ on $(a, b)$, then $f$ is strictly convex on the interval.

**Proof.** Under the given assumption, $f'$ is increasing if and only if $f''$ is nonnegative and $f'$ is strictly increasing when $f''$ is positive. This combined with Theorem B gives us our result. ●

The last statement of Theorem C is not reversible. Consider, for example, $f(x) = x^4$ on $(-1, 1)$, or for more dramatic evidence see Problem G.

Our next characterization depends on the geometrically evident idea that through any point on the graph of a convex function, there is a line which lies on or below the graph (Fig. 12.1). More formally, we say that

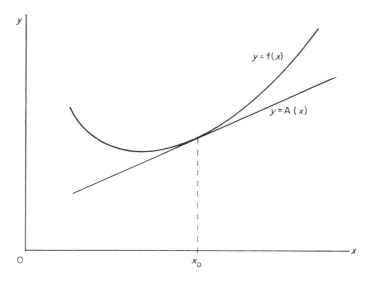

**Fig. 12.1**

a function $f$ defined on $I$ has **support** at $x_0 \in I$ if there exists an affine function $A(x) = f(x_0) + m(x - x_0)$ such that $A(x) \leq f(x)$ for every $x \in I$. The graph of the support function $A$ is called a **line of support** for $f$ at $x_0$.

> **Theorem D.** $f: (a, b) \to \mathbf{R}$ is convex if and only if there is at least one line of support for $f$ at each $x_0 \in (a, b)$.

**Proof.** If $f$ is convex and $x_0 \in (a, b)$, choose $m \in [f_-'(x_0), f_+'(x_0)]$. Then as we saw in Section 11,

$$\frac{f(x) - f(x_0)}{x - x_0} \geq m \quad \text{or} \quad \leq m$$

according as $x > x_0$ or $x < x_0$. In either case, $f(x) - f(x_0) \geq m(x - x_0)$; that is, $f(x) \geq f(x_0) + m(x - x_0)$.

Conversely, suppose that $f$ has a line of support at each point of $(a, b)$. Let $x, y \in (a, b)$. If $x_0 = \lambda x + (1 - \lambda)y$, $\lambda \in [0, 1]$, let $A(x) = f(x_0) + m(x - x_0)$ be the support function for $f$ at $x_0$. Then

$$f(x_0) = A(x_0) = \lambda A(x) + (1 - \lambda) A(y) \leq \lambda f(x) + (1 - \lambda) f(y)$$

as desired. ∎

While our next result is not a characterization of convex functions, we state it now because of its relation to Theorem D.

> **Theorem E.** Let $f: (a, b) \to \mathbf{R}$ be convex. Then $f$ is differentiable at $x_0$ if and only if the line of support for $f$ at $x_0$ is unique. And in this case, $A(x) = f(x_0) + f'(x_0)(x - x_0)$ provides this unique support.

**Proof.** It is clear from the proof of Theorem D that corresponding to each $m \in [f_-'(x_0), f_+'(x_0)]$, there is a line of support for $f$ at $x_0$. Uniqueness of the line therefore means $f_-'(x_0) = f_+'(x_0)$; that is, $f'(x_0)$ exists.

On the other hand, suppose $f'(x_0)$ exists. Any line of support $A(x) = f(x_0) + m(x - x_0)$ gives us $f(x) - f(x_0) \geq m(x - x_0)$. For $x_1 < x_0 < x_2$, we have

$$\frac{f(x_1) - f(x_0)}{x_1 - x_0} \leq m \leq \frac{f(x_2) - f(x_0)}{x_2 - x_0}$$

Taking limits as $x_1 \uparrow x_0$ and $x_2 \downarrow x_0$ gives $f_-'(x_0) \leq m \leq f_+'(x_0)$, so differentiability at $x_0$ implies uniqueness of $m$, hence of the support $A$ at $x_0$. ∎

## 12. Characterizations

## PROBLEMS AND REMARKS

**A.** An elementary argument involving only the concepts of the Riemann integral can be used to show that if $f$ is convex on $(a, b)$, then for all $c, x \in (a, b)$

$$f(x) - f(c) = \int_c^x f_-'(t)\, dt = \int_c^x f_+'(t)\, dt$$

(*Hint*: Let $\{c = x_0 < x_1 < \cdots < x_n = x\}$ be a partition of $[c, x]$ and note that

$$f_-'(x_{k-1}) \leq f_+'(x_{k-1}) \leq \frac{f(x_k) - f(x_{k-1})}{x_k - x_{k-1}} \leq f_-'(x_k) \leq f_+'(x_k)$$

Then use $f(x) - f(c) = \sum_1^n [f(x_k) - f(x_{k-1})]$.)

**B.** Theorem A can be proved in a form that applies to intervals of all kinds, not just open intervals. If $g: I \to \mathbf{R}$ is increasing and $c \in I$, then $f(x) = \int_c^x g(t)\, dt$ is convex and continuous on $I$. Conversely, if $f$ is convex and continuous on $I$, then for any $c$ and $x \in I$,

$$f(x) - f(c) = \int_c^x g(t)\, dt$$

for some increasing function $g$ defined at least on $I^0$. In fact $g$ may be taken as $f_+'$ or $f_-'$.

**C.** $f: [a, b] \to \mathbf{R}$ is convex and $f_+'(a)$ and $f_-'(b)$ are finite $\Leftrightarrow f(x) - f(a) = \int_a^x g(t)\, dt$ for some increasing function $g: [a, b] \to \mathbf{R}$.

**D.** The following are strictly convex:

(1) $e^x$ on $(-\infty, \infty)$,
(2) $-\log x$ on $(0, \infty)$,
(3) $x^p$ on $[0, \infty)$ if $p > 1$,
(4) $-x^p$ on $[0, \infty)$ if $0 < p < 1$,
(5) $x \log x$ on $(0, \infty)$.

**E.** Every convex polynomial of degree at least 2 is strictly convex. No polynomial of odd degree at least 3 is convex. What fourth-degree polynomials are convex?

**\*F.** There is a convex function whose first derivative fails to exist on a dense set. Such a function may be constructed on $[0, 1]$ using an enumeration $r_1, r_2, \ldots$ of the rationals on this interval. Let $A(t) = \{k: r_k < t\}$ and define

$$g(t) = \sum_{k \in A(t)} \frac{1}{2^k}$$

Then use Theorem $A$ to define the desired convex function $f$. Note that this example gives us a function $f$ which does not have an ordinary second derivative anywhere. However, if $h$ is any function satisfying $f_-'(x) \leq h(x) \leq f_+'(x)$ on $(0, 1)$, then $h$ is differentiable almost everywhere.

**\*G.** There is a strictly convex function whose second derivative is zero on a dense set or even almost everywhere. Such a function may be constructed using a function $g$

which is strictly increasing while $g'(t) = 0$ almost everywhere [Riesz and Sz.-Nagy, 1955, p. 48]. The desired convex function is then constructed using Theorem A.

*H. Suppose that $f''(x)$ exists on $(a, b)$ and $f''(x) > 0$ except on $D$. If $D$ is finite, then $f$ is strictly convex. What if $D$ is countable? a set of measure zero? the complement of a dense set?

I. Convex functions behave nicely with respect to maxima and minima.

(1) If $f: [a, b] \to \mathbf{R}$ is convex, then $f$ attains its maximum at $a$ or $b$.

(2) If $f: I \to \mathbf{R}$ is convex and attains a local minimum at $x_0 \in I$, then this minimum is the minimum of $f$ on all of $I$ (a global minimum). If $x_0 \in I^0$, then $f_-'(x_0) \leqslant 0 \leqslant f_+'(x_0)$.

(3) $f: (a, b) \to \mathbf{R}$ is convex $\Leftrightarrow$ for any closed subinterval $I$ of $(a, b)$ and any real number $\alpha$, $f(x) + \alpha x$ attains its maximum on $I$ at an endpoint of $I$.

(4) Let $f: I \to \mathbf{R}$ be continuous. Then $f$ is convex $\Leftrightarrow f(x) + \alpha x + \beta$ does not have an isolated maximum on the interior $I^0$ of $I$ for any real numbers $\alpha$ and $\beta$.

J. $f: (a, b) \to \mathbf{R}$ is convex $\Leftrightarrow$ given any point $P$ below the graph of $f$, there is a unique point on the graph nearest to $P$.

K. It is a bit complicated to state the support properties of a convex function at the endpoints of its domain $I$. Let us agree to say $f_-'(a) = -\infty$ if $I$ contains its left endpoint $a$ and $f_+'(b) = \infty$ if $I$ contains its right endpoint $b$. Then $f$ has support at $x_0 \in I \Leftrightarrow$ there is a real number $m$ such that $f_-'(x_0) \leqslant m \leqslant f_+'(x_0)$. Moreover, if $m$ is any such number, $A(x) = f(x_0) + m(x - x_0)$ supports $f$ at $x_0$. In particular, if $I = [a, b)$ or $I = [a, b]$ and $m \leqslant f_+'(a)$, then $A(x) = f(a) + m(x - a)$ supports $f$ at $a$.

L. $f: (a, b) \to \mathbf{R}$ is strictly convex $\Leftrightarrow$ at each $x_0 \in (a, b)$ there is a support function $A$ such that $f(x) > A(x)$ for all $x \in (a, b)$, $x \neq x_0$.

*M. Let $f: [a, b] \to \mathbf{R}$ be convex, and for each $t \in (a, b)$ let $A_t(x)$ be a support function at $t$. Then the minimum of $\int_a^b [f(x) - A_t(x)]\, dx$ occurs at $t = (a + b)/2$ [Miles, 1969].

*N. Problem F indicates some of the difficulties in using the ordinary second derivative to identify a convex function. However, a number of substitutes for the ordinary second derivative are available. Let $M$ be a class of real-valued functions on $(a, b)$ closed under addition with an operator $D$ satisfying the following.

(1) For $f \in M$, $Df$ is a function on $(a, b)$ with values in $[-\infty, \infty]$.

(2) $D(f + g) = Df + Dg$ provided that the right side makes sense; that is, does not involve $\infty - \infty$.

(3) If $f$ has a continuous ordinary second derivative $f''$ on $(a, b)$ then $f \in M$ and $Df \geqslant f''$.

(4) If $f \in M$ and $f$ attains its maximum at $x_0$, then $Df(x_0) \leqslant 0$.

We then assert the following.

(5) If $f$ is continuous, $f \in M$, and $Df \geqslant 0$ on $(a, b)$, then $f$ is convex [Hint: First show that $g(x) = f(x) + \varepsilon x^2$ is convex for $\varepsilon > 0$ and then let $\varepsilon \downarrow 0$].

(6) The class $M_1$ with operator $D_1$ and the class $M_2$ with operator $D_2$ both satisfy (1) through (4). $M_1$ is the class of all real-valued functions on $(a, b)$.

$$D_1 f(x) = \liminf_{h \downarrow 0} \frac{f(x + h) + f(x - h) - 2f(x)}{h^2}$$

## 13. Closure under Functional Operations

$M_2$ is the class of all integrable functions on $(a, b)$.

$$D_2 f(x) = \liminf_{h \downarrow 0} \frac{3 \int_{-h}^{h} f(x+t)\, dt - 6hf(x)}{h^3}$$

(7) If $f$ is continuous on $(a, b)$, then $f$ is convex $\Leftrightarrow D_1 f \geq 0 \Leftrightarrow D_2 f \geq 0$.
(8) If $f$ is continuous on $(a, b)$, then $f$ is convex $\Leftrightarrow$

$$f(x) \leq \frac{1}{2h} \int_{-h}^{h} f(x+t)\, dt$$

for every interval $[x - h, x + h]$ in $(a, b)$.

These and related results are found in the work of Beckenbach [1953].

*O. If $f: (a, b) \to \mathbf{R}$ is convex, must

$$\lim_{h \downarrow 0} \frac{f(x+h) + f(x-h) - 2f(x)}{h^2}$$

exist almost everywhere?

P. $f: I \to \mathbf{R}$ is convex $\Leftrightarrow x_1 < x_2 < x_3$ implies

$$\det \begin{bmatrix} 1 & x_1 & f(x_1) \\ 1 & x_2 & f(x_2) \\ 1 & x_3 & f(x_3) \end{bmatrix} \geq 0$$

Interpret this result geometrically.

Q. Let $f: (a, b) \to \mathbf{R}$ be continuous. Then $f$ is convex $\Leftrightarrow$

$$\int_{s}^{t} f(x)\, dx / (t - s) \leq \tfrac{1}{2}[f(s) + f(t)]$$

for all $a < s < t < b$. For generalizations, see Rado [1935] and Hartman [1972].

## 13. Closure under Functional Operations

The previous section offered a number of ways of identifying a convex function. Often, however, convex functions are most easily recognized by noting that they are built up from other functions known to be convex. For example, $f(x) = 3|x|^3 + 2e^{|x|}$ is readily seen to be convex by observing that $|x|$ is a convex function and then using Theorems B and A below.

**Theorem A.** If $f: I \to \mathbf{R}$ and $g: I \to \mathbf{R}$ are convex and $\alpha \geq 0$, then $f + g$ and $\alpha f$ are convex on $I$.

**Proof.** Check that $f+g$ and $\alpha f$ satisfy (10.1). ●

**Theorem B.** Let $f: I \to \mathbf{R}$ and $g: J \to \mathbf{R}$ where range$(f) \subseteq J$. If $f$ and $g$ are convex and $g$ is increasing, then the composite function $g \circ f$ is convex on $I$.

**Proof.** For $x, y \in I$ and $\lambda \in (0, 1)$
$$g[f(\lambda x + (1-\lambda)y)] \leq g[\lambda f(x) + (1-\lambda)f(y)]$$
$$\leq \lambda g[f(x)] + (1-\lambda)g[f(y)] \quad \bullet$$

**Theorem C.** If $f: I \to \mathbf{R}$ and $g: I \to \mathbf{R}$ are both non-negative, decreasing [increasing], and convex, then $h(x) = f(x) g(x)$ also exhibits these three properties.

**Proof.** Only the convexity is nontrivial. We begin by noting that for $x < y$,
$$[f(x) - f(y)][g(y) - g(x)] \leq 0$$
which implies that
$$f(x)g(y) + f(y)g(x) \leq f(x)g(x) + f(y)g(y)$$
an inequality we use below. Now if $\alpha > 0$, $\beta > 0$, and $\alpha + \beta = 1$,
$$f(\alpha x + \beta y) g(\alpha x + \beta y)$$
$$\leq [\alpha f(x) + \beta f(y)][\alpha g(x) + \beta g(y)]$$
$$= \alpha^2 f(x) g(x) + \alpha\beta[f(x)g(y) + f(y)g(x)] + \beta^2 f(y) g(y)$$
$$\leq \alpha^2 f(x) g(x) + \alpha\beta[f(x)g(x) + f(y)g(y)] + \beta^2 f(y) g(y)$$
$$= \alpha f(x) g(x) + \beta f(y) g(y) \quad \bullet$$

**Theorem D.** Let $f_\alpha: I \to \mathbf{R}$ be an arbitrary family of convex functions and let $f(x) = \sup_\alpha f_\alpha(x)$. If $J = \{x \in I : f(x) < \infty\}$ is nonempty, then $J$ is an interval and $f$ is convex on $J$.

**Proof.** If $\lambda \in (0, 1)$ and $x, y \in J$, then
$$f(\lambda x + (1-\lambda)y) = \sup_\alpha f_\alpha(\lambda x + (1-\lambda)y)$$
$$\leq \sup_\alpha [\lambda f_\alpha(x) + (1-\lambda) f_\alpha(y)]$$
$$\leq \lambda \sup_\alpha f_\alpha(x) + (1-\lambda) \sup_\alpha f_\alpha(y)$$
$$= \lambda f(x) + (1-\lambda) f(y)$$

## 13. Closure under Functional Operations

This shows simultaneously that $J$ is an interval (since it contains every point between any two of its points) and that $f$ is convex on it. ●

**Theorem E.** *If $f_n: I \to \mathbf{R}$ is a sequence of convex functions converging to a finite limit function $f$ on $I$, then $f$ is convex. Moreover, the convergence is uniform on any closed subinterval of $I^0$, the interior of $I$.*

**Proof.** If $\lambda \in (0, 1)$, $x, y \in I$,

$$f(\lambda x + (1-\lambda)y) = \lim_{n \to \infty} f_n(\lambda x + (1-\lambda)y)$$
$$\leqslant \lim_{n \to \infty} (\lambda f_n(x) + (1-\lambda) f_n(y))$$
$$= \lambda f(x) + (1-\lambda) f(y)$$

from which it follows that $f$ is convex.

Let $a < c < b$ be any three points of $I^0$ and let $\alpha = \sup f_n(a)$, $\gamma = \inf f_n(c)$, $\beta = \sup f_n(b)$. Further, let $L_1$, $L_2$, and $L_3$ be the three affine functions that satisfy $L_1(a) = \alpha$, $L_1(b) = \beta$; $L_2(c) = \gamma$, $L_2(b) = \beta$; $L_3(a) = \alpha$, $L_3(c) = \gamma$. We shall show that the sequence $\{f_n\}$ is uniformly bounded by these affine functions as illustrated in Fig. 13.1.

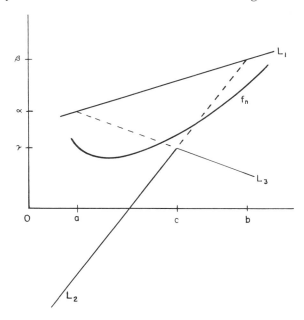

Fig. 13.1

If $x = \lambda a + (1 - \lambda)b$ is any point in $[a, b]$, then for arbitrary $n$,

$$f_n(x) \leq \lambda f_n(a) + (1 - \lambda) f_n(b) \leq \lambda L_1(a) + (1 - \lambda) L_1(b) = L_1(x)$$

On the other hand, if $x$ is any point of $[a, c]$, we may write $c = \lambda x + (1 - \lambda)b$ where $\lambda \in (0, 1]$. Then $x = (1/\lambda)c + [(\lambda - 1)/\lambda]b$, and

$$L_2(c) \leq f_n(c) \leq \lambda f_n(x) + (1 - \lambda) f_n(b) \leq \lambda f_n(x) + (1 - \lambda) L_2(b)$$

from which we conclude that

$$f_n(x) \geq \frac{1}{\lambda} L_2(c) + \frac{\lambda - 1}{\lambda} L_2(b) = L_2(x)$$

Similarly, if $x \in [c, b]$, we may show that $f_n(x) \geq L_3(x)$.

According to Theorem 11A, there is a number $K$ such that

$$|f_n(y) - f_n(x)| \leq K |y - x|$$

Moreover, since we have obtained upper and lower bounds on $f_n$ that are independent of $n$, it is clear from the proof of Theorem 11A that $K$ is independent of $n$; that is, our inequality is valid for all $n$ as well as for all $x, y \in [a, b]$. Choose a finite subset $E$ of $[a, b]$ such that each point of $[a, b]$ is within a distance $\varepsilon/3K$ of at least one point of $E$, $\varepsilon$ being an arbitrary positive number. Since $E$ is finite, there is an $N$ for which $m, n \geq N$ implies

$$|f_n(z) - f_m(z)| \leq \varepsilon/3$$

for all $z \in E$. Thus, if $x \in [a, b]$, $z \in E$, $|z - x| < \varepsilon/3K$, and $m, n \geq N$,

$$|f_n(x) - f_m(x)| \leq |f_n(x) - f_n(z)| + |f_n(z) - f_m(z)| + |f_m(z) - f_m(x)|$$
$$\leq K |x - z| + \varepsilon/3 + K |z - x| \leq \varepsilon$$

This, however, is just the Cauchy condition for uniform convergence on $[a, b]$. ●

We shall say that $f$ is **log-convex** on an interval $I$ if $f$ is positive and $\log f$ is convex on $I$. This is equivalent to requiring that $f$ be positive and satisfy

$$f(\alpha x + \beta y) \leq f^\alpha(x) f^\beta(y)$$

for $x, y \in I$, $\alpha > 0$, $\beta > 0$, $\alpha + \beta = 1$, which explains why some authors use the term **multiplicatively convex** to describe the functions we have called log-convex. Since $f(x) = \exp[\log f(x)]$, it follows from

## 13. Closure under Functional Operations

Theorem B that a log-convex function is convex. The class of log-convex functions has nice closure properties that we will need later.

**Theorem F.** *The class of log-convex functions on an interval $I$ is closed under addition, multiplication, and taking of limits, provided that the limit exists and is positive.*

**Proof.** Closure under multiplication and taking of limits follows from the identities $\log fg = \log f + \log g$ and $\log(\lim f_n) = \lim(\log f_n)$ combined with Theorems A and E. Addition is more difficult.

Let us note that if $a$, $b$, $c$, $d$, $\alpha$, $\beta$ are all positive numbers with $\alpha + \beta = 1$, then since $e^x$ is convex,

$$a^\alpha b^\beta = \exp(\alpha \log a + \beta \log b) \leq \alpha \exp(\log a) + \beta \exp(\log b) = \alpha a + \beta b$$

Thus

$$\frac{a^\alpha b^\beta + c^\alpha d^\beta}{(a+c)^\alpha (b+d)^\beta} = \left(\frac{a}{a+c}\right)^\alpha \left(\frac{b}{b+d}\right)^\beta + \left(\frac{c}{a+c}\right)^\alpha \left(\frac{d}{b+d}\right)^\beta$$

$$\leq \alpha \left(\frac{a}{a+c}\right) + \beta \left(\frac{b}{b+d}\right) + \alpha \left(\frac{c}{a+c}\right) + \beta \left(\frac{d}{b+d}\right)$$

$$= \alpha + \beta = 1$$

We have proved

$$a^\alpha b^\beta + c^\alpha d^\beta \leq (a+c)^\alpha (b+d)^\beta$$

Now choose $x, y \in I$ and use first the equivalent formulation of log-convexity and then the inequality above to obtain

$$f(\alpha x + \beta y) + g(\alpha x + \beta y)$$
$$\leq f^\alpha(x) f^\beta(y) + g^\alpha(x) g^\beta(y)$$
$$\leq [f(x) + g(x)]^\alpha [f(y) + g(y)]^\beta$$

We conclude that $f + g$ is log-convex, as desired. ●

### PROBLEMS AND REMARKS

**A.** If $f$ is convex on $I$, so are $f^+(x) = \sup(f(x), 0)$, $g(x) = f(x) + mx + b$, and $h(x) = f(x) + |mx + b|$.

**B.** The following functions are convex on $(-\infty, \infty)$:

(1) $f(x) = |x + a|$;
(2) $g(x) = |x + a|^p, p \geq 1$;

(3) $h(x) = 3|x + 1|^3 + 2|x - 1| - 4$;
(4) $k(x) = \sum_1^n c_j |x - a_j|^{b_j}$, $b_j \geq 1$, $c_j \geq 0$;
(5) $m(x) = (a + bx^2)^{1/2}$, $a \geq 0$, $b \geq 0$.

**C.** Let $\{f_n\}$ be a sequence of convex functions on $I$.

(1) If for each $n$, $f_{n+1}(x) \leq f_n(x)$ on $I$, and if there is a point $x_0 \in I^0$ where $\{f_n(x_0)\}$ is bounded below, then $\{f_n\}$ converges to a function $f$ convex on $I$.

(2) If for each $n$, $f_{n+1}(x) \geq f_n(x)$ on $I$, and if for two points $a, b \in I$ the sequences $\{f_n(a)\}$ and $\{f_n(b)\}$ are bounded above, then $\{f_n\}$ converges to a function $f$ convex on $[a, b]$.

(3) If $\sum_1^\infty f_n(x)$ converges to $f(x)$, then $f$ is convex.

(4) If each $f_n$ is positive and decreasing [increasing], and if $\prod_1^\infty f_n(x)$ converges to $f(x)$, then $f$ is convex.

(5) $\lim \sup_{n \to \infty} f_n(x)$ is convex on the set where it is finite.

(6) If $\{f_n\}$ is pointwise bounded, then $\{f_n\}$ is uniformly bounded on any compact subset of $I$.

*(7) If $\{f_n\}$ converges to a finite limit on a dense subset of $I$, then it converges for all $x \in I^0$ to a convex function and the convergence is uniform on each compact subset of $I^0$.

*(8) If $\{f_n\}$ is pointwise bounded, then there is a subsequence which converges uniformly on compact subsets of $I^0$.

*(9) If $\{f_n\}$ converges to a finite limit function $f$ everywhere on $I$, then $\{f_n'\}$ converges to $f'$ except possibly at countably many points of $I$.

*D. [Beckenbach, 1953]. Let $\{c_n\}$ be a sequence of positive numbers such that $\sum_1^\infty c_j < \infty$, and let $\{x_n\}$ be a bounded sequence of distinct real numbers. Then

(1) $f(x) = \sum_1^\infty c_n |x - x_n|$ converges on $(-\infty, \infty)$,
(2) $f$ is convex,
(3) $f'$ exists except at $x_1, x_2, \ldots$,
(4) $f_+'(x_n) - f_-'(x_n) = 2c_n$.

**E.** [Beckenbach, 1953]. Let $f: I \to \mathbf{R}$ and $g: J \to \mathbf{R}$ where range $f \subseteq J$. Conclusions about the composite function $g \circ f$ may be drawn as shown in the accompanying tabulation.

| $g : J \to \mathbf{R}$ | $f : I \to \mathbf{R}$ | $g \circ f : I \to \mathbf{R}$ |
|---|---|---|
| Convex, increasing | Convex | Convex |
| Convex, decreasing | Concave | Convex |
| Concave, increasing | Concave | Concave |
| Concave, decreasing | Convex | Concave |

**F.** Let $f: I \to \mathbf{R}$ be a positive concave function. Then the reciprocal $1/f$ is convex. If $f$ is not constant on any subinterval of $I$, then $1/f$ is strictly convex.

*G. Consider a function $f: I \times T \to \mathbf{R}$ of two variables and suppose that for each $t \in T$, $f(x, t)$ is convex on the open interval $I$.

(1) If $T$ is compact and $f$ is continuous in $t$ for each $x \in I$, then $f$ is continuous on $I \times T$.

(2) If $f$ is Riemann integrable on the interval $T$ for each $x \in I$, then $F(x) = \int_T f(x, t)\, dt$ is convex on $I$. Can Riemann be replaced by Lebesgue?

## 13. Closure under Functional Operations

(3) If $f$ is Riemann integrable on the interval $T$ for each $x \in I$ and is log-convex on $I$ for each $t \in T$, then $F(x) = \int_T f(x, t) \, dt$ is log-convex on $I$ [Artin, 1964, pp. 7–10].

(4) If $\phi$ is a positive continuous function on $(a, b)$, then $F(x) = \int_a^b \phi(t) t^{x-1} \, dt$ is log-convex on any interval where $F(x)$ exists as a (proper or improper) Riemann integral.

\*H.   Artin [1964] has drawn upon the facts of Problem G to give an elegant treatment of the **gamma** function

$$\Gamma(x) = \int_0^\infty e^{-t} t^{x-1} \, dt$$

(1) The integral converges for $x > 0$.
(2) $\Gamma(x + 1) = x\Gamma(x)$.
(3) $\Gamma(1) = 1$ and $\Gamma(n) = (n - 1)!$.
(4) $\Gamma$ is log-convex on $(0, \infty)$.
(5) The only function on $(0, \infty)$ which satisfies (2), (3), and (4) is the gamma function.

I.   If $f > 0$ and $f''$ exists on $I$, then $f$ is log-convex $\Leftrightarrow f \cdot f'' - (f')^2 \geq 0$.

J.   We say $f: I \to \mathbf{R}$ majorizes the affine function $A$ on $I$ if $A(x) \leq f(x)$ on $I$. For $f$ which majorizes at least one affine function, define

$$g(x) = \textbf{envelope } f(x) = \sup A(x)$$

where the supremum is taken over all affine functions majorized by $f$.

(1) $g$ is convex on $I$.
(2) If $h$ is convex and is majorized by $f$ on $I$, then $h(x) \leq g(x)$ for all $x \in I$.
(3) If $f$ is continuous on $I$, so is $g$.
(4) $g(x) = \inf\{\lambda_1 f(x_1) + \cdots + \lambda_n f(x_n): \lambda_1 x_1 + \cdots + \lambda_n x_n = x\}$ where the infimum is taken over all expressions of $x$ as a convex combination of points in $I$ [Rockafellar, 1970a, p. 36].
(5) If $f$ is convex, $f = g$.

K.   Let $f: [a, b] \to \mathbf{R}$ be convex and define

$$f^*(m) = \sup_{x \in [a,b]} \{mx - f(x)\}$$

Then $f^*$ is convex on $(-\infty, \infty)$; $f^*$ is called the **conjugate** of $f$ and is the subject of further study in Section 15.

L.   [Anderson, 1968].   Let $M$ be the class of nonnegative convex functions on $[0, 1]$ satisfying $f(0) = 0$.

(1) If $f \in M$, then $f$ is increasing.
(2) $M$ is closed under multiplication.
(3) If $f_*(x) = 2x \int_0^1 f(t) \, dt$, then $f \in M \Rightarrow f_* \in M$.
(4) $f, g \in M \Rightarrow \int_0^1 g(t) f(t) \, dt \geq \int_0^1 g(t) f_*(t) \, dt$.
(5) $f_1, \ldots, f_n \in M \Rightarrow$

$$\int_0^1 [f_1(t) \cdots f_n(t)] \, dt \geq \int_0^1 f_{1*}(t) \cdots f_{n*}(t) \, dt$$

$$= \frac{2^n}{n+1} \int_0^1 f_1(t) \, dt \cdots \int_0^1 f_n(t) \, dt$$

(6) There is equality throughout (5) if $f_i(x) = 2x\alpha_i$ where $\alpha_i$ is a constant.

\*M. [Besicovitch and Davies, 1965]. Let $f$ be a real-valued, nonnegative, continuous, monotone function defined on $[0, 1]$. Then there exist two convex functions $g_1$ and $g_2$ on $[0, 1]$ such that $0 \leq g_1 \leq f \leq g_2$ and

$$2 \int_0^1 g_1(x) \, dx \geq \int_0^1 f(x) \, dx \geq \tfrac{1}{2} \int_0^1 g_2(x) \, dx$$

Furthermore, the constants 2 and $\tfrac{1}{2}$ are the best possible. For generalizations of this result, see Problem 41H and [Nishiura and Schnitzer, 1972].

N. [Bruckner and Ostrow, 1962]. Let $f: [0, \infty) \to \mathbf{R}$ be convex, nonnegative, and $f(0) = 0$. Then a convex function on $[0, \infty)$ is defined by

$$F(x) = \frac{1}{x} \int_0^x f(t) \, dt, \quad F(0) = 0$$

\*O. If $f: (a, b) \to \mathbf{R}$ is strictly increasing and has a continuous second derivative, then $f$ has a representation $f(x) = g(h(x))$ where $h$ is increasing and convex on $(a, b)$ and $g$ is increasing and concave on an interval containing the range of $h$. (See Szekeres [1956], Marcus [1959a], Zamfirescu [1965], and Smajdor [1966] for this and related results.)

## 14. Differences of Convex Functions

The class of convex functions on an interval $I$ is closed under addition, but it is not closed under scalar multiplication or subtraction. We may, of course, consider the class of functions representable as the difference of two convex functions. This larger class is closed under all three operations, thus forming a so-called linear space (formally defined in Section 21). We wish to study this space; however, certain endpoint anomalies make it advisable to restrict it slightly. Moreover, it is convenient to work with closed intervals, though most of the results can be extended to open intervals with appropriate modifications.

Let $BC[a, b]$ be the class of functions $f : [a, b] \to \mathbf{R}$ representable in the form $f = g - h$ where $g$ and $h$ are convex and $g_+{'}(a)$, $g_-{'}(b)$, $h_+{'}(a)$, and $h_-{'}(b)$ are all finite. Then $BC[a, b]$ is a linear space and all its elements are actually Lipschitz (Problem A). Moreover, it is easily characterized in terms of $BV[a, b]$, the space of functions of bounded variation.

## 14. Differences of Convex Functions

**Theorem A.** $f \in BC[a, b]$ if and only if

$$f(x) - f(a) = \int_a^x r(t)\, dt \tag{1}$$

for some function $r \in BV[a, b]$.

**Proof.** If $f \in BC[a, b]$, then $f = g - h$ where $g$ and $h$ are convex and have finite endpoint derivatives. By a slight extension of Theorem 12A (Problem 12C),

$$g(x) - g(a) = \int_a^x p(t)\, dt, \qquad h(x) - h(a) = \int_a^x q(t)\, dt$$

for some increasing $p$ and $q$. Thus

$$f(x) - f(a) = \int_a^x [p(t) - q(t)]\, dt$$

where $(p - q) \in BV[a, b]$.

Conversely, if $f$ has an integral representation (1), then the fundamental characterization of functions of bounded variation [Natanson I, 1961, p. 218] enables us to write $r = p - q$ where $p$ and $q$ are increasing on $[a, b]$. Then

$$f(x) = f(a) + \int_a^x p(t)\, dt - \int_a^x q(t)\, dt$$

and it is therefore the difference of two convex functions by Theorem 12A. The endpoint conditions are easily established. ●

Functions in $BV[a, b]$ are characterized by the fact that

$$V_a^b(f) = \sup V(f, P) = \sup \sum_1^n |f(x_j) - f(x_{j-1})| < \infty$$

where the supremum is taken over all partitions

$$P = \{a = x_0 < x_1 < \cdots < x_n = b\}.$$

We seek similarly to characterize $BC[a, b]$ in terms of $K_a^b(f)$, defined by

$$K_a^b(f) = \sup K(f, P) = \sup \sum_1^{n-1} |\Box f_{j+1} - \Box f_j|$$

where
$$\Box f_j = \frac{f(x_j) - f(x_{j-1})}{x_j - x_{j-1}}$$

We note immediately that for a convex function $f$, $K_a^b(f) = f_-'(b) - f_+'(a)$. More generally, if $f = g - h$ where $g$ and $h$ are convex, then

$$K_a^b(f) \leq K_a^b(g) + K_a^b(h) = g_-'(b) - g_+'(a) + h_-'(b) - h_+'(a)$$

Thus $f \in BC[a, b]$ implies $K_a^b(f) < \infty$. To get the reverse implication is considerably more work. We begin with two preliminary results that are important in their own right.

**Theorem B.** If $K_a^b(f) < \infty$, then $f_+'$ exists on $[a, b)$ and $f_-'$ exists on $(a, b]$.

**Proof.** Let us first show that partition refinement serves if anything to increase the size of $K(f, P)$. We need only consider the consequence of inserting one point, say $\bar{x}$, between $x_k$ and $x_{k+1}$. Let

$$\Box f_t = \frac{f(x_{k+1}) - f(\bar{x})}{x_{k+1} - \bar{x}}, \qquad \Box f_s = \frac{f(\bar{x}) - f(x_k)}{\bar{x} - x_k}$$

so that $\Box f_{k+1} = \alpha \Box f_t + \beta \Box f_s$ where

$$\alpha = \frac{x_{k+1} - \bar{x}}{x_{k+1} - x_k}, \qquad \beta = \frac{\bar{x} - x_k}{x_{k+1} - x_k}$$

Note that $\alpha > 0$, $\beta > 0$, and $\alpha + \beta = 1$. We consider the case where $1 \leq k \leq n - 2$. (The argument is in fact simplified if $k = 0$ or $k = n - 1$.) Then

$$|\Box f_{k+1} - \Box f_k| + |\Box f_{k+2} - \Box f_{k+1}|$$
$$= |\alpha(\Box f_t - \Box f_s) + \Box f_s - \Box f_k| + |\Box f_{k+2} - \Box f_t + \beta(\Box f_t - \Box f_s)|$$
$$\leq |\Box f_s - \Box f_k| + |\Box f_t - \Box f_s| + |\Box f_{k+2} - \Box f_t|$$

which is the desired result.

This fact about partition refinement has two important consequences. First, it ensures the existence of a sequence of partitions $\{P_i\}$ such that

$$K_a^b(f) = \lim_{i \to \infty} K(f, P_i) \tag{2}$$

Second, it guarantees that $K_a^x(f)$ is increasing as a function of $x$.

## 14. Differences of Convex Functions

Now let $a = x_0 \leqslant x_1 < x_2 < x_3 < x_4 \leqslant b$. From the inequalities

$$|\Box f_2| - |\Box f_3| + |\Box f_3| - |\Box f_4| \leqslant |\Box f_3 - \Box f_2| + |\Box f_4 - \Box f_3|$$
$$\leqslant K_a^b(f)$$

we have

$$\frac{f(x_2) - f(x_1)}{x_2 - x_1} \leqslant K_a^b(f) + \frac{f(x_4) - f(x_3)}{x_4 - x_3} \tag{3}$$

If in (3) we let $x_2 \downarrow x_1$, we see that both

$$D^+ f(x_1) = \limsup_{x_2 \downarrow x_1} \frac{f(x_2) - f(x_1)}{x_2 - x_1}$$

and

$$D_+ f(x_1) = \liminf_{x_2 \downarrow x_1} \frac{f(x_2) - f(x_1)}{x_2 - x_1}$$

are finite. Clearly $D_+ f(x_1) \leqslant D^+ f(x_1)$. Hoping for a contradiction, let us suppose $D_+ f(x_1) < D^+ f(x_1)$. Set $\varepsilon = [D^+ f(x_1) - D_+ f(x_1)]/4$. We may then choose a sequence $t_1 > t_2 > \cdots > x_1$ such that

$$\frac{f(t_j) - f(x_1)}{t_j - x_1} \begin{cases} > D^+ f(x_1) - \varepsilon, & j \text{ odd} \\ < D_+ f(x_1) + \varepsilon, & j \text{ even} \end{cases}$$

Thus (Fig. 14.1) for the sequence $\{t_j\}$,

$$|\Box f_{j+1} - \Box f_j| > (D^+ f(x_1) - \varepsilon) - (D_+ f(x_1) + \varepsilon) = 2\varepsilon > 0$$

Hence, by taking a partition $P$ of $[a, b]$ containing enough points $\{t_j\}$, we may make $K(f, P)$ as large as we wish, thereby contradicting the finiteness of $K_a^b(f)$. We conclude that $f_+'(x)$ exists on $[a, b)$. A similar argument establishes the corresponding conclusion for $f_-'(x)$ on $(a, b]$. ●

**Theorem C.** If $K_a^b(f) < \infty$, then $f$ is Lipschitz and consequently absolutely continuous on $[a, b]$.

**Proof.** The existence of one-sided derivatives (Theorem B) ensures that $f$ is continuous on $[a, b]$. This together with the existence of $f_-'(b)$ allows us to choose $M$ so that

$$\left| \frac{f(b) - f(x_3)}{b - x_3} \right| \leqslant M \tag{4}$$

for all $x_3 \in [a, b)$. Referring to (3) with $x_4 = b$, we conclude that

$$|f(x_2) - f(x_1)| \leqslant [K_a^b(f) + M] |x_2 - x_1|$$

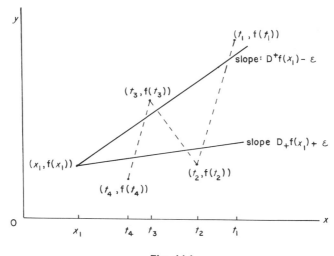

Fig. 14.1

for all $x_1$, $x_2 \in [a, b)$, a result that immediately extends to include $b$ by (4). Thus $f$ is Lipschitz on $[a, b]$. The absolute continuity of $f$ now follows from a simple verification of the definition. ●

**Theorem D.** $f \in BC[a, b]$ if and only if $K_a^b(f) < \infty$.

**Proof.** One-half of this theorem was proved in introducing Theorem B. For the other half, suppose $K_a^b(f) < \infty$ and define $\hat{f}$ by

$$\hat{f}(x) = f_-'(x) - f_+'(a) \quad \text{if} \quad x \in (a, b]$$

and $\hat{f}(a) = 0$. Let $a < u < v \leq b$ and choose a sequence of partitions

$$P_i = \{a = x_0 < x_1 < \cdots < x_m = u\} \cup \{u = x_m < x_{m+1} < \cdots < x_n = v\}$$
$$= P_i' \cup P_i''$$

such that $K(f, P_i') \to K_a^u(f)$, $K(f, P_i) \to K_a^v(f)$ with the length of the maximum subinterval of $P_i$ going to zero as $i \to \infty$, all of which can be done according to (2). Then

$$K(f, P_i) - K(f, P_i') = \sum_{j=m}^{n-1} |\Box f_{j+1} - \Box f_j|$$

$$\geq \left| \sum_{j=m}^{n-1} (\Box f_{j+1} - \Box f_j) \right| = |\Box f_n - \Box f_m|$$

## 14. Differences of Convex Functions

Taking limits on both sides as $i \to \infty$ gives us

$$K_a^v(f) - K_a^u(f) \geq |f_-'(v) - f_-'(u)|$$

or equivalently

$$K_a^v(f) - K_a^u(f) \geq |\hat{f}(v) - \hat{f}(u)| \tag{5}$$

This inequality, proved for $a < u < v \leq b$, is also valid if $u = a$ since

$$K_a^v(f) = \lim_{i \to \infty} K(f, P_i) \geq \lim_{i \to \infty} |\Box f_n - \Box f_1| = |\hat{f}(v) - \hat{f}(a)|$$

Now (5) means that for $v > u$, $K_a^v(f) - K_a^u(f)$ is greater than or equal to both $\hat{f}(v) - \hat{f}(u)$ and $\hat{f}(u) - \hat{f}(v)$. It follows that both $K_a^x(f) - \hat{f}(x)$ and $K_a^x(f) + \hat{f}(x)$ are increasing as functions of $x$, as are

$$n(x) = \tfrac{1}{2}[K_a^x(f) - \hat{f}(x)], \qquad p(x) = \tfrac{1}{2}[K_a^x(f) + \hat{f}(x)]$$

Thus $\hat{f}(x) = p(x) - n(x)$ is of bounded variation; and since $f$ is absolutely continuous (Theorem C), we have [Natanson I, 1961, p. 255]

$$f(x) - f(a) = \int_a^x f'(t)\, dt = \int_a^x [\hat{f}(t) + f_+'(a)]\, dt$$

This combined with Theorem A gives the desired conclusion. ●

### PROBLEMS AND REMARKS

**A.** Here are some further facts about $BC[a, b]$.

(1) If $f \in BC[a, b]$, then $f$ is Lipschitz, but not conversely (cf. Problem 11D).
(2) $BC[a, b]$ is closed under scalar multiplication, addition, and multiplication.
*(3) $BC[a, b]$ is closed under $\wedge$ and $\vee$. Here $f \vee g = \max(f, g)$ and $f \wedge g = \min(f, g)$.
*(4) Let $f$ be absolutely continuous on $[a, b]$. Then $f \in BC[a, b] \Leftrightarrow |f| \in BC[a, b]$. If $f(a) = f(b) = 0$, then $K_a^b(f) \leq K_a^b(|f|) \leq 2K_a^b(f)$.

**B.** $K_a^b(f) = 0 \Leftrightarrow f(x) = \alpha x + \beta$ for appropriate constants $\alpha$ and $\beta$. If $f(x) = g(x) + \alpha x + \beta$, then $K_a^b(f) = K_a^b(g)$.

**C.** Let $a < c < b$. Then $K_a^b(f) < \infty \Leftrightarrow K_a^c(f)$ and $K_c^b(f)$ are finite, and in this case

$$K_a^b(f) = K_a^c(f) + K_c^b(f) + |f_+'(c) - f_-'(c)|$$

**D.** $K_a^b(f)$ and $\hat{f}'$ are related as follows.

(1) If $K_a^b(f) < \infty$, then $f'(x)$ exists except for at most countably many points of $[a, b]$.
(2) If $f'$ exists on $[a, b]$, then $K_a^b(f) = V_a^b(f')$.

(3) If $f'$ exists and is absolutely continuous on $[a, b]$, then $K_a^b(f) = \int_a^b |f''(t)|\,dt$. (See Riesz and Sz.-Nagy [1955, p. 48]).

(4) If $K_a^b(f) < \infty$, then

$$\frac{d}{dx} K_a^x(f) = \left| \frac{d}{dx} f_-'(x) \right|$$

almost everywhere [Riesz and Sz.-Nagy, 1955, p. 15].

*E.  Let $K_a^b(f) < \infty$. Then $K_a^b(f) = K_a^b(P) + K_a^b(N) = V_a^b(\hat{f})$ where

$$P(x) = \int_a^x p(t)\,dt, \qquad N(x) = \int_a^x n(t)\,dt$$

The functions $p$, $n$, and $\hat{f}$ are those defined in Section 14.

*F.  Find good conditions on $f$ and $g$ which ensure that $f \circ g \in BC[a, b]$. (*Example*: If $f, g \in BC[a, b]$, $f \circ g$ is absolutely continuous, and $g$ is monotone, then $f \circ g \in BC[a, b]$. Of course this statement assumes that $g[a, b] \subseteq [a, b]$.)

*G.  $BC[a, b]$ may be characterized as the class of absolutely continuous functions whose derivatives are essentially of bounded variation (that is, equal almost everywhere to a function of bounded variation).

★ ★ ★ ★ ★

The material in this section seems not to be well known, though it has a long history. The equivalent of $K_a^b(f)$ first appears in the work of de la Vallée Poussin [1908] and was used by Riesz [1911] in one of his proofs of the theorem on representation of linear functionals. Popoviciu [1944] greatly extended the whole idea, replacing  by divided differences of higher order, a development discussed in Section 83. Alexandroff [1949, 1950], Landis [1951], Arsove [1953], Hartman [1959], and Zalgaller [1963] have considered generalizations on $R^n$. Other recent related papers are by Roberts and Varberg [1969], Russell [1970], Crownover and Simmons [1970], and Huggins [1972].

## 15. Conjugate Convex Functions

The concept of duality pervades much of mathematics and will appear several times in this book. When faced with a problem, a mathematician often finds another problem which, although it appears to be quite different, mirrors all aspects of the original and is perhaps easier to analyze. Thus, for example, it may be easier to decide if a certain set is closed (contains all its limit points) by investigating whether its complement is open. The two ideas are dual to each other. In its fullest development, duality involves two mathematical theories in which each theorem in one has its counterpart theorem in the other. Thus, in the elementary theory of sets, there is for each theorem expressed in terms of unions, intersections, and complements a dual theorem obtained by

## 15. Conjugate Convex Functions

interchanging unions with intersections and the empty set with the universal set (as in the two De Morgan laws). In plane projective geometry, any theorem concerning incidence of points and lines gives rise to a dual theorem in which point and line are interchanged. In developing any mathematical theory, one should always try to find a dual theory which may be lurking in the background. For as Valentine [1967, p. 492] has remarked, "... since it is a rare coincidence for the proofs of a theorem and its dual to be of equal difficulty, there is a double reason to investigate the dual. One may gain either a simpler proof or a less obvious theorem." We would add that it is often the interplay between dual theories that enlivens and clarifies both of them.

In the present section the notion of duality enters via our ability to associate with each convex function another one called the conjugate convex function. The relationship between a convex function and its conjugate is at the heart of much recent research. Our aim is to give an introduction to the subject in the most simple context possible, convex functions on the real line.

Let $g: [0, \infty) \to [0, \infty)$ be strictly increasing and continuous with $g(0) = 0$ and $g(x) \to \infty$ as $x \to \infty$. Then $g^{-1}$ exists and has the same properties as $g$. Moreover, if we let

$$f(x) = \int_0^x g(s)\, ds, \qquad f^*(y) = \int_0^y g^{-1}(t)\, dt$$

then $f$ and $f^*$ are both convex functions on $[0, \infty)$.

In Fig. 15.1 we have sketched the graph of $t = g(s)$. It is also, of

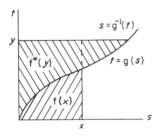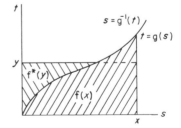

Fig. 15.1

course, the graph of $s = g^{-1}(t)$; one merely looks at it in a mirror held along the $t$ axis. A number of results follow from the integral representations above or can be seen directly from the figure. The first is known as **Young's inequality**.

(C1) $xy \leqslant f(x) + f^*(y)$ for all $x \geqslant 0$, $y \geqslant 0$.
(C2) $xy = f(x) + f^*(y)$ if and only if $y = g(x) = f'(x)$.
(C3) $(f^*)' = (f')^{-1}$.
(C4) $f^{**} = f$.
(C5) $f^*(y) = \sup_{x \geqslant 0} [xy - f(x)]$.

The last result follows immediately from (C1) and (C2).

We have now defined the conjugate for those convex functions $f$ representable as the integral of a function $g$ having special properties. Even in this limited context, the study of conjugate functions is of considerable importance in the theory of Birnbaum–Orlicz spaces [Hewitt and Stromberg 1965, pp. 203–204; Krasnosel'skiĭ and Rutickiĭ, 1961]. Our aim, however, is to extend the notion of conjugacy to all convex functions while preserving in some sense each of the properties (C1) through (C5). To that end, we take (C5) as a definition. More precisely, if $f: I \to \mathbf{R}$ is a convex function defined on an interval $I$, then $f^*: I^* \to \mathbf{R}$ will denote the **conjugate** function defined by

$$f^*(y) = \sup_{x \in I} [xy - f(x)] \qquad (1)$$

with domain $I^* = \{y \in \mathbf{R}: f^*(y) < \infty\}$. Then $f^*$ is convex (as is shown presently), but it also has another property that will be important in the sequel; it is closed. We say that a convex function $g: I \to \mathbf{R}$ is **closed** if the **level set** $L_\alpha = \{x \in I: g(x) \leqslant \alpha\}$ is a closed subset of $\mathbf{R}$ for each real $\alpha$.

**Theorem A.** If $f: I \to \mathbf{R}$ is convex, then its conjugate $f^*: I^* \to \mathbf{R}$ is convex and closed.

**Proof.** We first note that $I^* \neq \varnothing$. For if $I$ is the single point $x_0$, then $A(x) = f(x_0) + y(x - x_0)$ supports $f$ for each $y \in \mathbf{R}$. Otherwise we pick any interior point $x_0$, choose $y \in [f_-'(x_0), f_+'(x_0)]$, and again $A(x)$ will support $f$ (Theorem 12D). In either case then we are able to choose $y$ so that $f(x) \geqslant A(x)$; that is, we choose $y$ so that $xy - f(x) \leqslant x_0 y - f(x_0)$ for all $x \in I$. For this choice of $y$, $f^*(y) < \infty$ and $I^* \neq \varnothing$ as claimed. Secondly, we note that $f^*$ is the supremum of the convex (actually affine) functions $g_x: \mathbf{R} \to \mathbf{R}$ defined by $g_x(y) = xy - f(x)$, so by Theorem 13D, $I^*$ is an interval and $f^*$ is convex on it. To see that the level set $L_\alpha$ of $f^*$ is closed, suppose $\{y_n\}$ is a sequence of points of $L_\alpha$ converging, say to $\bar{y}$, as $n \to \infty$. Then $xy_n - f(x) \leqslant f^*(y_n) \leqslant \alpha$ for all $x \in I$. If we let $n \to \infty$, we obtain $x\bar{y} - f(x) \leqslant \alpha$ which in turn implies that $f^*(\bar{y}) \leqslant \alpha$. We conclude that $L_\alpha$, containing all its limit points, is closed. ●

## 15. Conjugate Convex Functions

It is evident now, since $f^{**}$ is closed, that we cannot possibly preserve (C4), $f^{**} = f$, unless we begin with a closed convex function $f$. If we impose this condition on $f$, we could prove directly that $f^{**} = f$, but we prefer to postpone this to Theorem D where it will come out with a number of other results. Our more immediate concern is to ask how restrictive a condition it is to require that $f$ be closed, since we are being forced to abandon our goal of preserving conditions (C1) to (C5) while defining the conjugate of any convex function.

In the top half of Fig. 15.2 we have sketched three convex functions.

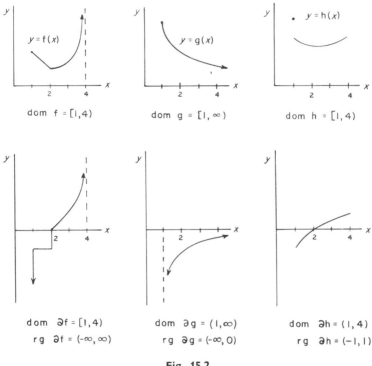

Fig. 15.2

The first two are closed; the third is not. A little thought convinces us that the property of being closed has only to do with the behavior of a convex function at the boundary of its domain $I$ since such a function is continuous on $I^0$. In fact the convex function $f: I \to \mathbf{R}$ is closed if and only if it is continuous at each endpoint of $I$ that is in $I$ and $f(x) \to \infty$ as $x$ approaches any finite endpoint not in $I$. This characterization should be kept in mind, especially in connection with Theorem B.

We now turn our attention to derivatives. As we know from Section 11, a convex function need not have a derivative everywhere, but it does have left and right derivatives everywhere. This last fact enables us to define the **generalized derivative** or **subdifferential**. For convex $f: I \to \mathbf{R}$, we define

$$\partial f(x) = \{y \in \mathbf{R}; \ y \text{ is the slope of a support line for } f \text{ at } x\} \tag{2}$$

Thus $\partial f$ is a set-valued function that is single valued and agrees with the ordinary two-sided derivative $f'$ wherever the latter exists (Theorem 12E). The domain of $\partial f$ (dom $\partial f$) is the set of $x$'s in $I$ where $f$ has support; its range ($rg\ \partial f$) is the set of support slopes. Note that $I^0 \subseteq \mathrm{dom}\ \partial f \subseteq I$.

In the bottom half of Fig. 15.2 we have sketched the graphs of the generalized derivative for each of the functions in the top half. In each case the graph is a continuous increasing curve with perhaps vertical as well as horizontal segments. In the case of $\partial f$ and $\partial g$, both ends of the curve recede infinitely far from the origin; for $\partial h$ they do not. These observations are related to the fact that $f$ and $g$ are closed while $h$ is not.

In order to make the descriptions of the previous paragraph precise, we introduce the notion of a maximal monotone increasing relation. A subset $\Gamma$ of $\mathbf{R} \times \mathbf{R}$ is called **monotone increasing** if $(x_1, y_1)$ and $(x_2, y_2)$ in $\Gamma$ imply $(x_2 - x_1)(y_2 - y_1) \geq 0$, and it is a **maximal monotone increasing set** if it is a monotone increasing set that is not a proper subset of any other monotone increasing set. We observe that the set $\Gamma^{-1}$ defined by $\Gamma^{-1} = \{(y, x): (x, y) \in \Gamma\}$ is a maximal monotone increasing set whenever $\Gamma$ is. We use this fact later on.

**Theorem B.** *If $f: I \to \mathbf{R}$ is convex and closed, then the graph of $\partial f$ is maximal monotone increasing.*

**Proof.** We recall from Section 11 that for $x_1, x_2 \in I$,

$$x_1 < x_2 \quad \text{implies} \quad f_-'(x_1) \leq f_+'(x_1) \leq f_-'(x_2) \leq f_+'(x_2) \tag{3}$$

On the other hand, from the support properties of $f$ that were enunciated in Theorem 12D and Problem 12K, we see that

$$\partial f(x) = \{y \in \mathbf{R}: f_-'(x) \leq y \leq f_+'(x)\} \tag{4}$$

Here as in (3) we must interpret $f_-'(a) = -\infty$ if $I$ contains its left endpoint $a$ and $f_+'(b) = \infty$ if $I$ contains its right endpoint $b$. Thus $\partial f(x)$ is always an interval (usually a point) unless of course it is empty.

## 15. Conjugate Convex Functions

Now let $(x_1, y_1)$ and $(x_2, y_2)$ belong to $\partial f$, meaning $y_1 \in \partial f(x_1)$, $y_2 \in \partial f(x_2)$. If $x_1 < x_2$, then by (4) and (3), $y_2 \geq f_-'(x_2) \geq f_+'(x_1) \geq y_1$ and so $(x_2 - x_1)(y_2 - y_1) \geq 0$. This inequality is obvious if $x_1 = x_2$, and a symmetric argument establishes it for $x_2 < x_1$. Hence, $\partial f$ is monotone increasing.

Our next task is to show maximality, and here it is important that $f$ is closed. Using the characterization given after Theorem A together with (4), we see that if $I$ has a finite right endpoint, then $rg\ \partial f$ extends to $\infty$; while if its left endpoint is finite, then $rg\ \partial f$ extends to $-\infty$. To show maximality, it is sufficient to demonstrate

$$(x_1, y_1) \notin \partial f \quad \text{implies} \quad (x - x_1)(y - y_1) < 0 \quad \text{for some} \quad (x, y) \in \partial f \quad (5)$$

Actually, replacing $f$ by $h(x) = f(x + x_1) - xy_1$, again a closed convex function, we see that we can assume $x_1 = 0 = y_1$. Our problem then is to show

$$(0, 0) \notin \partial f \quad \text{implies} \quad xy < 0 \quad \text{for some} \quad (x, y) \in \partial f \quad (6)$$

Now if $x < 0$ for all $x \in \text{dom}\ \partial f$, then $I$ is bounded above and $rg\ \partial f$ extends to $\infty$ and there surely is an $(x, y) \in \partial f$ for which $xy < 0$. Similar arguments prevail if $x > 0$ for all $x \in \text{dom}\ \partial f$. If $0 \in \text{dom}\ \partial f$, then $(0, 0) \notin \partial f$ means $0 \notin \partial f(0)$. But then, since $f(0)$ cannot be the minimum of $f$ in this case, there is an $x_2 \in I$ such that $f(x_2) < f(0)$. If $x_2 < 0$, then we can find $x_3 \in [x_2, 0)$ for which $f_+'(x_3) > 0$. Otherwise (see Problem 12B)

$$f(x_2) - f(0) = \int_0^{x_2} f_+'(t)\, dt \geq 0$$

violating the way $x_2$ was chosen. With this choice of $x_3$, it follows from (4) that $(x_3, f_+'(x_3)) \in \partial f$, and since $x_3 f_+'(x_3) < 0$, we have established (6). A similar argument works if $x_2 > 0$, in which case we can find $x_3 > 0$ such that $f_-'(x_3) < 0$ and again (6) is established. ●

We want to attach meaning to $\int_c^x \partial f(s)\, ds$. We do this by defining

$$\int_c^x \partial f(s)\, ds = \int_c^x f_+'(s)\, ds \quad (7)$$

for all $c, x \in I$. Note that any function $g$ such that $f_-'(x) \leq g(x) \leq f_+'(x)$ would have worked as well as $f_+'$ since $\int f_-' = \int f_+'$.

**Theorem C.** If $f: I \to \mathbf{R}$ is convex and closed, then

$$f(x) - f(c) = \int_c^x \partial f(s) \, ds$$

for all $x, c \in I$.

**Proof.** Since $f$ is closed, it is continuous. Hence the result follows from Theorem 12A as extended by Problem 12B. ●

Now we can prove the main theorem of this section.

**Theorem D.** Let $f: I \to \mathbf{R}$ be convex and closed. Then $f^*: I^* \to \mathbf{R}$ is also convex and closed and

(a) $xy \leq f(x) + f^*(y)$ for all $x \in I$, $y \in I^*$,
(b) $xy = f(x) + f^*(y)$ if and only if $y \in \partial f(x)$,
(c) $\partial(f^*) = (\partial f)^{-1}$,
(d) $f^{**} = f$.

**Proof.** We know from Theorem A that $f^*$ is closed and convex, and (a) is an immediate consequence of the definition of $f^*$. To prove (b) we observe that a convex function $g: I \to \mathbf{R}$ achieves a minimum at $x \in I$ if and only if $0 \in \partial g(x)$. Now

$$-f^*(y) = -\sup_{x \in I} [xy - f(x)] = \inf_{x \in I} [f(x) - xy]$$

But as noted above, $g(x) = f(x) - xy$ (being convex) achieves its infimum at $x$ if and only if $0 \in \partial g(x)$, or equivalently, if and only if $y \in \partial f(x)$. Thus,

$$-f^*(y) = f(x) - xy \quad \text{if and only if} \quad y \in \partial f(x) \tag{8}$$

which is (b).

On the other hand, for each (fixed) $x \in I$, we have from the definition of $f^*$,

$$f^*(y) - xy \geq -f(x) \quad \text{for all} \quad y \in I^* \tag{9}$$

Thus $f^*(y) - xy$ is minimized when there is equality in (9) which by (8) happens when $y \in \partial f(x)$. In other words, $y \in \partial f(x)$ implies $h(z) = f^*(z) - xz$ is minimized at $z = y$. But $h$, being convex, is minimized precisely when $0 \in \partial h(y)$, that is, when $x \in \partial(f^*)(y)$. We conclude after taking inverses that

$$x \in (\partial f)^{-1}(y) \quad \text{implies} \quad x \in \partial(f^*)(y)$$

## 15. Conjugate Convex Functions

Therefore $\partial(f^*)$ is an extension of $(\partial f)^{-1}$. The latter, however, is maximal monotone increasing (being the inverse of a maximal monotone increasing set) and so $\partial(f^*) = (\partial f)^{-1}$, which is (c).

Finally, if we apply (c) to $f^*$, we get

$$\partial(f^{**}) = (\partial(f^*))^{-1} = ((\partial f)^{-1})^{-1} = \partial f$$

and so by Theorem C

$$f(x) - f(c) = \int_c^x \partial f(s)\, ds = f^{**}(x) - f^{**}(c)$$

for all $x, c \in I$ and all $x, c \in I^{**}$. Consequently $I = I^{**}$ and we will be finished if we can find one $c$ for which $f(c) = f^{**}(c)$. Choose $c_0 \in I$ and $y_0 \in I^*$ so that $y_0 \in \partial f(c_0)$, which by (c) means also that $c_0 \in (\partial f)^{-1}(y_0) = \partial(f^*)(y_0)$. Applying (b) successively to $f$ and $f^*$, we get

$$c_0 y_0 = f(c_0) + f^*(y_0) = f^*(y_0) + f^{**}(c_0)$$

which means $f(c_0) = f^{**}(c_0)$. ●

### PROBLEMS AND REMARKS

**A.** If $f: I \to \mathbf{R}$ is convex, then $\{x \in I: f(x) \leq \alpha\}$ is an interval.

**B.** Given $f$ and $I$ as indicated, the other equalities may be verified.

(1) $f(x) = x^2, I = \mathbf{R}$

$\partial f(x) = 2x, x \in \mathbf{R}$

$\partial(f^*)(y) = y/2, y \in \mathbf{R}$

$f^*(y) = y^2/4, I^* = \mathbf{R}$.

(2) $f(x) = |x|, I = \mathbf{R}$

$\partial f(x) = \begin{cases} -1, & x < 0 \\ [-1, 1], & x = 0 \\ 1, & x > 0 \end{cases}$

$\partial(f^*)(y) = \begin{cases} (-\infty, 0], & y = -1 \\ 0, & y \in (-1, 1) \\ [0, \infty), & y = 1 \end{cases}$

$f^*(y) = 0, I^* = [-1, 1]$.

(3) $f(x) = e^x, I = \mathbf{R}$
$\partial f(x) = e^x, x \in \mathbf{R}$
$\partial(f^*)(y) = \log y, y > 0$
$f^*(y) = \begin{cases} y \log y - y, & y > 0 \\ 0, & y = 0 \end{cases}$
$I^* = [0, \infty)$.

(4) $f(x) = k, I = \{c\}$
$\partial f(c) = \mathbf{R}$
$\partial(f^*)(y) = c, y \in \mathbf{R}$
$f^*(y) = cy - k, I^* = \mathbf{R}$.

**C.** Let $f(x) = x^2, I = [0, 1)$. Is $f$ closed? Find $f^{**}$ and $I^{**}$.

**D.** Find $\partial f, \partial(f^*), f^*, I^*, f^{**}, I^{**}$ for

(1) $f(x) = x^2, I = [0, \infty)$,
(2) $f(x) = x^2/2, I = \mathbf{R}$,

(3) $f(x) = (1/p)|x|^p$, $p > 1$, $I = \mathbf{R}$,
(4) $f(x) = -(1 - x^2)^{1/2}$, $I = [-1, 1]$.

**E.** If $f: I \to \mathbf{R}$ is convex, then $I^0 \subseteq \mathrm{dom}\ \partial f \subseteq I$, but neither inclusion need be equality. If in addition $f$ is closed and has finite one-sided derivatives at any endpoints included in $I$ [so that, for example, $f_+'(a) < \infty$ if $I = [a, b]$], then $\mathrm{dom}\ \partial f = I$.

**F.** If $f: I \to \mathbf{R}$ is convex and closed and $I$ is bounded, then $I^* = \mathbf{R}$.

*__G.__ [Minty, 1962]. If $\Gamma$ is a maximal monotone increasing relation in $\mathbf{R} \times \mathbf{R}$, then $h: \Gamma \to \mathbf{R}$ defined by $h(x, y) = x + y$ is a homeomorphism of $\Gamma$ onto $\mathbf{R}$.

**H.** [Rockafellar, 1967]. Let $\Gamma$ be a maximal monotone increasing relation in $\mathbf{R} \times \mathbf{R}$. If $\gamma$ is a function representing $\Gamma$ [so, $(s, \gamma(s)) \in \Gamma$], we define

$$\int_c^x \Gamma(s)\ ds = \int_c^x \gamma(s)\ ds$$

for all $c, x \in \mathrm{dom}\ \Gamma = \{s \in \mathbf{R}: (s, t) \in \Gamma\}$. Then

(1) $\mathrm{dom}\ \Gamma$ is an interval,
(2) $\int_c^x \Gamma(s)\ ds$ is unambiguously defined; that is, its value does not depend on the choice of the representative of $\Gamma$,
(3) $f(x) = \int_c^x \Gamma(s)\ ds$ is a convex function on $\mathrm{dom}\ \Gamma$,
(4) the graph of $\partial f$ is $\Gamma$.

**I.** If $\Gamma_1$ and $\Gamma_2$ are maximal monotone increasing relations, so are $\Gamma_1 + \Gamma_2 = \{(x, y_1 + y_2): (x, y_1) \in \Gamma_1, (x, y_2) \in \Gamma_2\}$ and $\Gamma_1 \square \Gamma_2 = \{(x_1 + x_2, y): (x_1, y) \in \Gamma_1, (x_2, y) \in \Gamma_2\}$ provided in the first case $(\mathrm{dom}\ \Gamma_1) \cap (\mathrm{dom}\ \Gamma_2) \neq \varnothing$ and in the second that $(\mathrm{rg}\ \Gamma_1) \cap (\mathrm{rg}\ \Gamma_2) \neq \varnothing$. In fact $\Gamma_1 \square \Gamma_2 = (\Gamma_1^{-1} + \Gamma_2^{-1})^{-1}$ and $\Gamma_1 + \Gamma_2 = (\Gamma_1^{-1} \square \Gamma_2^{-1})^{-1}$.

★ ★ ★ ★ ★

The topic of conjugate convex functions really originated in a paper of Young [1912]. However, except for some applications in connection with Birnbaum–Orlicz spaces [Birnbaum and Orlicz, 1931; Krasnosel'skiĭ and Rutickiĭ, 1961], the subject attracted little interest until after the work of Fenchel [1949, 1953] who greatly generalized the whole idea and applied it to the programming problem. The bare outline of this development is suggested in Section 43 and in Problems 43C, 43D, and 54E. A complete modern treatment for convex functions on $\mathbf{R}^n$ together with an extensive bibliography is given in the book by Rockafellar [1970a]. Our introduction follows an earlier paper by the same author [Rockafellar, 1967]. Readers interested in infinite-dimensional generalizations may consult Brøndsted [1964], Moreau [1962], and Ioffe and Tikhomirov [1968], the latter with a bibliography of 108 items. Applications of monotone increasing curves in connection with flows and electrical networks may be found in the work of Berge and Ghouila–Houri [1965].

# 11

# Normed Linear Spaces

> The unity which it [axiomatic method] gives to mathematics is not the armor of formal logic, the unity of a lifeless skeleton; it is the nutritive fluid of an organism at the height of its development, the supple and fertile research instrument to which all the great mathematical thinkers since Gauss have contributed, all those who in the words of Lejeune-Dirichlet, have always labored to substitute "ideas for calculations."
>
> N. BOURBAKI

> The little excursion that we propose to make into functional analysis may seem a bit tough on a beginner. However, this is partly a matter of preconceived ideas. There is some similarity to mountain climbing: some beginners take to it like a duck takes to water, while others are inclined to get dizzy. In this case there is nothing to get dizzy about; the work is exactly the same as in Euclidean space for a good part of the way, and the rest can be omitted until the reader has more confidence, or it can be taken at the "slow pace of men of the hills."
>
> L. C. YOUNG

## 20. Introduction

We have studied convex functions defined on the real line. It is both natural and correct to suppose that most of the results of Chapter I are valid for convex functions defined on $n$-dimensional Euclidean space $\mathbf{R}^n$. And historically this was the next step in the development. But $\mathbf{R}^n$ is in many important ways just a prototype for other finite-dimensional spaces, and even the finite dimensionality is of no consequence for many of the theorems. What we need is a setting appropriate to the level of our study and yet general enough to include many of the interesting special cases.

The class of normed linear spaces fits our needs admirably. By restricting attention to functions defined on a normed linear space, we avoid certain technical difficulties that would be encountered in more general linear topological spaces. At the same time, the development of normed linear spaces illustrates one of the major themes in modern mathematics. When observation indicates the existence of common structural features in a variety of contexts, one should isolate the most relevant of these features and make them the object of study. Thereby we achieve great economy of thought. Moreover, the absence of rich but irrelevant undergrowth may actually make it easier to find our way across the landscape.

Our object, then, in this chapter is to develop the basic theory of a normed linear space $\mathbf{L}$. The definition of $\mathbf{L}$ and a number of basic results are given in Section 21. In keeping with our primary aim, we turn in Section 22 to the study of functions, especially linear and affine functions, defined on $\mathbf{L}$. Finally in Section 23, we introduce several notions of differentiation, putting emphasis on the Fréchet derivative as the concept most natural to the setting of normed linear spaces. All of our discussion is pointed toward Chapter IV where we shall investigate convex functions on a normed linear space.

## 21. Normed Linear Spaces

Many of the definitions commonly used in the study of Euclidean $n$-space have very natural generalizations that we intend to describe in this section. In order to review the essential ideas as well as to fix terminology and notation, we begin with some definitions commonly met in either advanced calculus or linear algebra.

The elements of Euclidean $n$-space $\mathbf{R}^n$, called either points or vectors,

## 21. Normed Linear Spaces

are ordered $n$-tuples $(x_1, ..., x_n)$ of real numbers. They may be indicated either by displaying the $n$-tuple as we have already done, or by using a boldface letter. Thus,

$$\mathbf{x} = (x_1, ..., x_n), \qquad \mathbf{y} = (y_1, ..., y_n)$$

These elements can be added according to the rule

$$\mathbf{x} + \mathbf{y} = (x_1 + y_1, ..., x_n + y_n)$$

They can also be multiplied by a real number $\alpha$, an operation called **scalar multiplication**, and defined by

$$\alpha \mathbf{x} = (\alpha x_1, ..., \alpha x_n)$$

The special element $\mathbf{O} = (0, ..., 0)$ satisfies the relation

$$\mathbf{x} + \mathbf{O} = \mathbf{x}$$

for all $\mathbf{x}$.

A sum of the form

$$\alpha_1 \mathbf{x}_1 + \cdots + \alpha_k \mathbf{x}_k, \qquad \alpha_i \text{ real}, \qquad \mathbf{x}_i \in \mathbf{R}^n \qquad (1)$$

is said to be a **linear combination** of the vectors $\{\mathbf{x}_1, ..., \mathbf{x}_k\} \subseteq \mathbf{R}^n$. The set of vectors $\{\mathbf{x}_1, ..., \mathbf{x}_k\}$ is said to be **linearly independent** if, whenever the sum (1) is $\mathbf{O}$, it follows that $\alpha_1 = \cdots = \alpha_k = 0$; otherwise it is **linearly dependent**. The set $\{\mathbf{x}_1, ..., \mathbf{x}_k\}$ is called a **basis** for $\mathbf{R}^n$ if

(F1) the set is linearly independent,
(F2) given any $\mathbf{x} \in \mathbf{R}^n$, it is possible to find $k$ numbers $\alpha_1, ..., \alpha_k$ so that $\mathbf{x} = \alpha_1 \mathbf{x}_1 + \cdots + \alpha_k \mathbf{x}_k$.

It is well known that any basis of $\mathbf{R}^n$ consists of $n$ vectors. The most obvious basis, henceforth referred to as the **standard basis**, is the set

$$\mathbf{e}_1 = (1, 0, ..., 0)$$
$$\vdots$$
$$\mathbf{e}_n = (0, ..., 0, 1)$$

where the only nonzero entry in $\mathbf{e}_i$ is a 1 in the $i$th position.

So far, only algebraic ideas have entered our discussion. To introduce topological notions, we first define the **length** or the **Euclidean norm** of a vector $\mathbf{x} = (x_1, ..., x_n)$ to be

$$\| \mathbf{x} \| = (x_1^2 + \cdots + x_n^2)^{1/2}$$

and we define the distance between two vectors to be

$$\text{dist}(\mathbf{x}, \mathbf{y}) = \| \mathbf{x} - \mathbf{y} \|$$

The ε-**neighborhood** of $\mathbf{x}$ is the set

$$N_\varepsilon(\mathbf{x}) = \{\mathbf{y} \in \mathbf{R}^n : \| \mathbf{y} - \mathbf{x} \| < \varepsilon\}$$

The point $\mathbf{x}$ is **interior** to a set $U$ if there is an ε-neighborhood of $\mathbf{x}$ containing only points of $U$; it is **exterior** to $U$ if there is an ε-neighborhood of $\mathbf{x}$ containing no points of $U$. A point $\mathbf{x}$ that is not interior or exterior to $U$, meaning that every neighborhood of $\mathbf{x}$ contains at least one point in $U$ and one point not in $U$, is called a **boundary** point.

A sequence of points $\{\mathbf{x}_j\}$ in $\mathbf{R}^n$ is said to **converge** to $\mathbf{y}$ if for each $\varepsilon > 0$ there is a corresponding integer $N$ such that $j \geqslant N$ implies $\| \mathbf{x}_j - \mathbf{y} \| < \varepsilon$. In this case, we write $\lim_{j \to \infty} \mathbf{x}_j = \mathbf{y}$. The point $\mathbf{y}$ is a **limit point** of the set $U$ if there is a sequence of points $\{\mathbf{x}_j\}$, $\mathbf{x}_j \in U$, $\mathbf{x}_j \neq \mathbf{y}$, that converges to $\mathbf{y}$.

A set $U$ is **open** if every point of $U$ is an interior point; it is **closed** if its **complement** (the set $U'$ of all points not in $U$) is open. A set is closed if and only if it contains all its limit points. For a given set $U$, not necessarily open or closed, the set of interior points is designated by $U^0$, and the set $U$ taken together with its limit points is $\overline{U}$, the **closure** of $U$. The entire space $\mathbf{R}^n$ and the empty set $\varnothing$ are the only sets that are both open and closed. The set $U$ is said to be **bounded** if there is a number $B$ such that for all $\mathbf{x} \in U$, $\| \mathbf{x} \| \leqslant B$.

A collection of open sets $U_\alpha$ is said to be an **open cover** of the set $D$ if $\bigcup_\alpha U_\alpha \supseteq D$. The set $D$ is **compact** if every open cover of $D$ has a finite subcollection that still covers $D$. In laying the foundations for real analysis, it is common to take as axiomatic that every set of real numbers that has an upper bound has a least upper bound, also called a **supremum** (sup). From this one proves the **Heine–Borel** theorem; any closed bounded set in $\mathbf{R}$ is compact. It is possible to prove the same result in $\mathbf{R}^n$, paving the way for the following useful characterization of compact sets in $\mathbf{R}^n$.

> **Theorem A.** A set $D \subseteq \mathbf{R}^n$ is compact if and only if $D$ is closed and bounded.

Using this theorem, it is easy to prove the **Bolzano–Weierstrass** theorem.

> **Theorem B.** Every bounded infinite set in $\mathbf{R}^n$ has a limit point.

## 21. Normed Linear Spaces

This in turn enables us to obtain important results about infinite sequences in $\mathbf{R}^n$. For example, any bounded sequence must have a convergent subsequence. More important for our purposes is a fact about Cauchy sequences. A **Cauchy sequence** is a sequence $\{\mathbf{x}_j\}$ having the property that for any $\varepsilon > 0$, there exists an $N$ such that if $i, j \geqslant N$, then $\|\mathbf{x}_i - \mathbf{x}_j\| < \varepsilon$.

**Theorem C.** Every Cauchy sequence in $\mathbf{R}^n$ converges to a point in $\mathbf{R}^n$.

The proofs of these three theorems are outlined in Problem D. The reader wishing to see further examples or more details regarding the definitions and theorems mentioned here is referred to Buck [1965, pp. 24–48].

There is one more property to which we wish to draw attention in our review of the structure of $\mathbf{R}^n$. The **inner product** of $\mathbf{x} = (x_1, \ldots, x_n)$ and $\mathbf{y} = (y_1, \ldots, y_n)$ is defined by

$$\langle \mathbf{x}, \mathbf{y} \rangle = x_1 y_1 + \cdots + x_n y_n$$

Clearly, $\langle \mathbf{x}, \mathbf{x} \rangle = \|\mathbf{x}\|^2$. It is an easy exercise using the law of cosines from trigonometry to show in $\mathbf{R}^2$ and $\mathbf{R}^3$ that

$$\langle \mathbf{x}, \mathbf{y} \rangle = \|\mathbf{x}\| \|\mathbf{y}\| \cos \theta$$

where $\theta$ is the angle between $\mathbf{x}$ and $\mathbf{y}$. For this reason, two vectors are said to be **orthogonal** if $\langle \mathbf{x}, \mathbf{y} \rangle = 0$. Since $|\cos \theta| \leqslant 1$, we see immediately that for $\mathbf{R}^2$ and $\mathbf{R}^3$,

$$|\langle \mathbf{x}, \mathbf{y} \rangle| \leqslant \|\mathbf{x}\| \|\mathbf{y}\|$$

This inequality, which holds in $\mathbf{R}^n$ and in many more general settings, is variously attributed to Cauchy, Bunyakovskiĭ, and/or Schwarz. We will meet it again, referring to it as the **CBS inequality**.

With our review completed, we turn to the generalizations of these ideas that will be of interest to us. A **real linear space** $\mathbf{L}$, alternately called a **real vector space**, is a collection of elements (points, vectors) together with two operations called addition and scalar multiplication that satisfy the following axioms. Denote the members of $\mathbf{L}$ by $\mathbf{x}, \mathbf{y}, \mathbf{z}$, and real numbers by $\alpha, \beta$. Then

(V1) $(\mathbf{x} + \mathbf{y}) \in \mathbf{L}$,
(V2) $(\mathbf{x} + \mathbf{y}) + \mathbf{z} = \mathbf{x} + (\mathbf{y} + \mathbf{z})$,
(V3) there is an element $\mathbf{O} \in \mathbf{L}$ so that for any $\mathbf{x}$, $\mathbf{O} + \mathbf{x} = \mathbf{x}$,

(V4)  for each **x**, there is an **x̄** so that **x** + **x̄** = **O**,
(V5)  **x** + **y** = **y** + **x**,
(V6)  α**x** ∈ **L**,
(V7)  α(**x** + **y**) = α**x** + α**y**,
(V8)  (α + β)**x** = α**x** + β**x**,
(V9)  (αβ)**x** = α(β**x**),
(V10) 1**x** = **x**.

It is easily verified that **x̄** = (−1)**x**, so we normally write **x̄** = −**x**. Also note that 0**x** = **O**.

A linear combination of a set of vectors $\{\mathbf{x}_1, ..., \mathbf{x}_k\}$ and the linear independence of such a set are defined exactly as in $\mathbf{R}^n$. A set $B$, possibly infinite, is called a **basis** for **L** if

(B1)  any finite subset of $B$ is linearly independent,
(B2)  given any **x** ∈ **L**, it is possible to represent **x** as a linear combination of some finite collection of $\mathbf{x}_i \in B$.

Every linear space has such a basis [Day, 1962, p. 2]. An important theorem [Halmos, 1958, p. 13] says that if one basis of a space contains $n$ elements, every basis will contain $n$ elements. In this case the space is said to be **finite dimensional** with dimension $n$, and the conditions (B1) and (B2) are equivalent to conditions (F1) and (F2) used to define a basis of the space $\mathbf{R}^n$. Any linear space not finite dimensional is said to be **infinite dimensional**. It is easy to prove that for a fixed basis, the representation of an arbitrary **x** as a linear combination of basis vectors is unique.

**Example A.**  The set $\mathbf{P}_n$ of all polynomials

$$p(x) = a_1 + a_2 + \cdots + a_n x^{n-1}$$

of degree $n - 1$ or less, with addition and scalar multiplication defined as usual, forms an $n$-dimensional linear space with basis $\{1, x, ..., x^{n-1}\}$.

**Example B.**  The set $\mathbf{C}_n$ of all functions on $[0, \pi]$ of the form

$$f(x) = b_1 \cos x + b_2 \cos 2x + \cdots + b_n \cos nx$$

with the usual addition and scalar multiplication forms an $n$-dimensional linear space with basis $\{\cos x, ..., \cos nx\}$.

There is a sense in which $\mathbf{P}_n$ and $\mathbf{C}_n$ are alike. As linear spaces they are indistinguishable; addition and scalar multiplication behave in exactly the same way in the two spaces. This phenomenon is charac-

## 21. Normed Linear Spaces

teristic of all $n$-dimensional linear spaces. They all have the same linear structure, the structure we described for the addition and scalar multiplication of points in $\mathbf{R}^n$. We now introduce language to state this more precisely.

Let $\{\mathbf{u}_1, \ldots, \mathbf{u}_n\}$ be a basis for an $n$-dimensional space $\mathbf{L}$. Then for any $\mathbf{x} \in \mathbf{L}$, we may write

$$\mathbf{x} = x_1 \mathbf{u}_1 + \cdots + x_n \mathbf{u}_n$$

and since this representation in terms of $\{\mathbf{u}_1, \ldots, \mathbf{u}_n\}$ is unique, there is a one-to-one correspondence between $\mathbf{L}$ and $\mathbf{R}^n$ described by

$$\mathbf{x} \leftrightarrow (x_1, \ldots, x_n)$$

Moreover, addition and scalar multiplication are preserved by this correspondence; for if $\mathbf{y} \leftrightarrow (y_1, \ldots, y_n)$, then

$$\mathbf{x} + \mathbf{y} = (x_1 + y_1)\mathbf{u}_1 + \cdots + (x_n + y_n)\mathbf{u}_n$$
$$\leftrightarrow (x_1 + y_1, \ldots, x_n + y_n) = (x_1, \ldots, x_n) + (y_1, \ldots, y_n)$$

and

$$\alpha \mathbf{x} = \alpha x_1 \mathbf{u}_1 + \cdots + \alpha x_n \mathbf{u}_n$$
$$\leftrightarrow (\alpha x_1, \ldots, \alpha x_n) = \alpha(x_1, \ldots, x_n)$$

The correspondence $\leftrightarrow$ is called an isomorphism. More generally, a mapping (function) $\phi: \mathbf{L} \to \mathbf{M}$ from a linear space $\mathbf{L}$ onto a linear space $\mathbf{M}$ is an **isomorphism** if $\phi$ is one to one and satisfies

$$\phi(\mathbf{x} + \mathbf{y}) = \phi(\mathbf{x}) + \phi(\mathbf{y}), \qquad \phi(\alpha \mathbf{x}) = \alpha \phi(\mathbf{x})$$

for all $\mathbf{x}, \mathbf{y} \in \mathbf{L}$ and $\alpha \in \mathbf{R}$; $\mathbf{L}$ and $\mathbf{M}$ are then said to be **isomorphic** linear spaces. In this language we have demonstrated the following result.

> **Theorem D.** Every $n$-dimensional linear space $\mathbf{L}$ is isomorphic to $\mathbf{R}^n$.

A word of caution is in order. We have not said that $\mathbf{L}$ is like $\mathbf{R}^n$ in all respects. For $\mathbf{R}^n$ has structure (inner product, length) that $\mathbf{L}$ may not have, and even if $\mathbf{L}$ has these notions, we have no reason to think that $\phi$ will preserve them. We elaborate on these remarks later.

**Example C.** The set of all real $m \times n$ matrices forms a linear space using the customary rules for addition and scalar multiplication of matrices. The space is finite dimensional, one basis being the $mn$ distinct matrices having a single entry of 1 with all other entries 0.

We now give some examples of infinite-dimensional spaces.

**Example D.** The set of all real-valued functions defined and continuous on $[0, 1]$, with $(f + g)(x) = f(x) + g(x)$ and $(\alpha f)(x) = \alpha f(x)$, forms an infinite-dimensional linear space. Call it $C[0, 1]$.

Verification that $C[0, 1]$ is a linear space involves checking the 10 axioms listed. Though the verification amounts to nothing more than a recitation of the familiar properties of continuous functions [axiom (V1), for instance, depends on the fact that the sum of two continuous functions is again a continuous function], it should be carried out in monotonous detail by anyone who has not previously encountered it. Similar comments pertain to the other examples of this chapter since we typically limit our comments to verifications involving some particular difficulty.

**Example E.** The set **P** of all polynomials defined on $[0, 1]$ forms a linear space of infinite dimension having as basis the polynomials $p_0(x) = 1$ (the constant function), $p_1(x) = x$, $p_2(x) = x^2, \ldots$.

Note that the space **P** of Example E is contained in the space $C[0, 1]$ described in Example D. It is for this reason called a subspace. More generally, if **M** is a subset of **L** which is itself a linear space under the operations of **L**, then **M** is called a **subspace** of **L**. Our examples show that an infinite-dimensional space may properly contain a subspace that is also infinite dimensional. The set $\mathbf{P}_n$ of all polynomials on $[0, 1]$ of degree less than or equal to $n - 1$ (Example A) is a finite-dimensional subspace of both $C[0, 1]$ and **P**. To verify that a subset **M** of a linear space **L** is a subspace, it is sufficient (Problem H) to show that if $\mathbf{x}, \mathbf{y} \in \mathbf{M}$, then $(\mathbf{x} + \alpha \mathbf{y}) \in \mathbf{M}$ for all real $\alpha$.

**Example F.** The set **S** of all infinite sequences $\{x_i\}$ of real numbers forms an infinite-dimensional linear space.

We move on now to the generalization of length. A real-valued function $N: \mathbf{L} \to \mathbf{R}$, often written $N(\mathbf{x}) = \|\mathbf{x}\|$, is called a **norm** on **L** if it satisfies for all $\mathbf{x}, \mathbf{y} \in \mathbf{L}$ and $\alpha \in \mathbf{R}$

(N1) $\|\mathbf{x}\| \geqslant 0$ with equality if and only if $\mathbf{x} = \mathbf{O}$,
(N2) $\|\alpha \mathbf{x}\| = |\alpha| \|\mathbf{x}\|$,
(N3) $\|\mathbf{x} + \mathbf{y}\| \leqslant \|\mathbf{x}\| + \|\mathbf{y}\|$ (the triangle inequality).

A space with a norm defined on it is called a **normed linear space**.

## 21. Normed Linear Spaces

It is important to remember that one must know both the space and the norm, since different norms may be defined on the same space. For example, in the plane where points are located by giving coordinates with respect to two orthogonal axes, some common norms for $\mathbf{x} = (x_1, x_2)$ are

$$\|\mathbf{x}\| = |x_1| + |x_2|$$
$$\|\mathbf{x}\| = \max\{|x_1|, |x_2|\}$$
$$\|\mathbf{x}\| = (x_1^2 + x_2^2)^{1/2}$$

When a concept of length has been introduced, one can speak about **unit vectors**, meaning vectors having length (norm) equal to one, and both the **unit sphere** $S$ and the **unit ball** $B$

$$S = \{\mathbf{x} \in \mathbf{L}: \|\mathbf{x}\| = 1\}, \qquad B = \{\mathbf{x} \in \mathbf{L}: \|\mathbf{x}\| < 1\}$$

Corresponding to the three norms just defined for points in the plane, we get the three unit spheres indicated in Fig. 21.1. In only one case does the unit sphere look completely smooth. In the study of normed linear spaces, there is considerable interest in the smoothness of the unit sphere, a topic to which we shall return (Problem 44E).

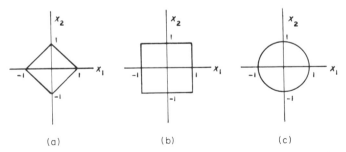

**Fig. 21.1.** (a) $|x_1| + |x_2| = 1$. (b) $\max\{|x_1|, |x_2|\} = 1$. (c) $(x_1^2 + x_2^2)^{1/2} = 1$.

We return to some of the examples already introduced, this time defining some of the norms commonly associated with the respective spaces.

**Example D'.** For $f \in C[0, 1]$, both $\|f\| = \max |f(x)|$ over $[0, 1]$ and $|f| = [\int_0^1 f^2(x)\, dx]^{1/2}$ define norms.

Verification that $\|f\|$ actually defines a norm proceeds smoothly enough for (N1) and (N2), and (N3) follows from

$$\max |f(x) + g(x)| \leq \max[|f(x)| + |g(x)|] \leq \max |f(x)| + \max |g(x)|$$

all maximums taken over [0, 1]. In verifying that $|f|$ defines a norm, (N1) explains the reason for using $\int f^2$ instead of $\int f$, (N2) explains the reason for taking the square root of the integral, and the justification of (N3) requires some thought, not an unusual feature of this most troublesome of the three properties.

The same norms may be used on the spaces $\mathbf{P}_n$ and $\mathbf{P}$ defined in Examples A and E, once again underlining the necessity of knowing both the space and the norm in talking about a normed linear space. In our next example we obtain many classical normed linear spaces by specifying a particular subspace of $\mathbf{S}$, the space of all infinite sequences (Example F), and a particular norm to be used in that subspace.

**Example F'.** We consider the subspaces of $\mathbf{S}$ shown in the accompanying table, defining norms for $\mathbf{x} = \{x_i\}$ and naming the resulting spaces as they are commonly named in mathematical literature.

| Name | Subspace | Norm |
|---|---|---|
| $c$ | Convergent sequences | $\|\mathbf{x}\| = \max |x_i|$ |
| $l^1$ | $\sum_1^\infty |x_i|$ converges | $\|\mathbf{x}\| = \sum_1^\infty |x_i|$ |
| $l^2$ | $\sum_1^\infty x_i^2$ converges | $\|\mathbf{x}\| = (\sum_1^\infty x_i^2)^{1/2}$ |
| $l^\infty$ | Bounded sequences | $\|\mathbf{x}\| = \sup |x_i|$ |

Verification that the indicated sets are normed linear spaces is left as Problem H. While many of the required steps are tedious, we would warn the novice that some of them raise questions of mathematical substance. In the case of $l^2$, for example, the first step in showing that we even have a linear space involves showing that for $\mathbf{x}, \mathbf{y} \in l^2$ $\sum_1^\infty (x_i + y_i)^2$ converges, and as is often the case, (N3) poses the greatest challenge.

As in $\mathbf{R}^n$, so in any normed linear space, the notion of length (norm) enables us to define the distance between points, writing dist$(\mathbf{x}, \mathbf{y}) = \|\mathbf{x} - \mathbf{y}\|$. The accomplishment is not to be regarded lightly as we have now said, to look again at Example D', what we shall mean by the distance between two functions continuous on [0, 1]. In fact we have two ways, $\|f - g\|$ and $|f - g|$, to measure the "distance" between $f$ and $g$.

With the distance between two points defined, it again makes sense to speak about the $\varepsilon$-neighborhood of $\mathbf{x}$;

$$N_\varepsilon(\mathbf{x}) = \{\mathbf{y} \in \mathbf{L}: \ \|\mathbf{y} - \mathbf{x}\| < \varepsilon\}$$

## 21. Normed Linear Spaces

Once neighborhoods of points have been defined, we may go on to describe relative to any set $U \subseteq \mathbf{L}$ what we shall mean by interior points, exterior points, or boundary points; the descriptions of sets as open, closed, bounded, or compact make sense; sequences of points may be described as convergent in the same way. In summary, all the topological notions defined for $\mathbf{R}^n$ may be carried to the new setting.

While there are obvious advantages to retaining all our familiar terminology, one must take particular heed to the warning that not all the familiar relationships involving these terms can be expected to hold in arbitrary normed linear spaces. Stated another way, we cannot prove for any space $\mathbf{L}$ all of the theorems that we can prove for $\mathbf{R}^n$. For example, a closed bounded set in an arbitrary normed linear space $\mathbf{L}$ need not be compact; in fact, if $\mathbf{L}$ is infinite dimensional, then the unit sphere is not compact (Problem F). In an infinite-dimensional space $\mathbf{L}$, we can always construct a sequence $\{\mathbf{x}_i\}$ having the properties that $\|\mathbf{x}_i\| = 1$ for all $i$, $\|\mathbf{x}_i - \mathbf{x}_j\| > \frac{1}{2}$ for $i \neq j$ (Problem F2); hence there is no hope of extending the Bolzano–Weierstrass theorem to an arbitrary space $\mathbf{L}$. A Cauchy sequence may or may not converge to a point in the space.

The last-mentioned difficulty deserves special attention. Consider the space $\mathbf{P}$ of all polynomials on the interval $[0, 1]$ with the norm $\|p\| = \max |p(x)|$ for $x \in [0, 1]$. The Weierstrass approximation theorem [Buck, 1965, p. 66] assures us that there is a sequence $\{p_j\}$ in $\mathbf{P}$ converging to $S(x) = \sin x$ in the norm we are using. We note that $S \notin \mathbf{P}$. On the other hand,

$$\|p_i - p_j\| \leqslant \|p_i - S\| + \|p_j - S\|$$

so $\{p_j\}$ is a Cauchy sequence. For this reason, $\mathbf{P}$ is said to be incomplete.

A normed linear space $\mathbf{L}$ is said to be **complete** if every Cauchy sequence in $\mathbf{L}$ converges to an element in $\mathbf{L}$. A complete normed linear space is called a **Banach space**. Theorem C tells us that $\mathbf{R}^n$ is a Banach space.

All the topological notions we have introduced derive from the use of a norm to define an $\varepsilon$-neighborhood. We are aware from looking at the plane (Fig. 21.1) that different norms may lead to quite different looking neighborhoods. It is still possible (and in fact true in the plane) that these neighborhoods may determine the same open sets and give the same limits for convergent sequences. We now set about the business of making these general remarks more explicit.

We say that two norms $\| \ \|_1$ and $\| \ \|_2$ defined on the same linear space $\mathbf{L}$ are **topologically equivalent** if there are positive numbers $m_1$ and $m_2$ such that $\|\mathbf{x}\|_2 \leqslant m_1 \|\mathbf{x}\|_1$ and $\|\mathbf{x}\|_1 \leqslant m_2 \|\mathbf{x}\|_2$ for all $\mathbf{x} \in \mathbf{L}$. In this case, each $\varepsilon$-neighborhood of a point with respect to $\| \ \|_1$ will contain a

$\delta$-neighborhood of the point with respect to $\|\ \|_2$, and conversely. The reason for calling two such norms topologically equivalent is seen from the following easily established facts.

(E1) A point **x** is an interior, exterior, or boundary point of a set $U$ with respect to neighborhoods determined by $\|\ \|_1$ if and only if it is similarly related to $U$ with respect to neighborhoods determined by $\|\ \|_2$.

(E2) A set $U$ is open, closed, compact, or bounded with respect to the topology induced by $\|\ \|_1$ if and only if it is so classified with respect to the topology induced by $\|\ \|_2$.

(E3) A sequence $\{\mathbf{x}_j\}$ is Cauchy with respect to $\|\ \|_1$ if and only if it is Cauchy with respect to $\|\ \|_2$. It converges to **y** with respect to $\|\ \|_1$ if and only if it converges to **y** with respect to $\|\ \|_2$.

As an important illustration, consider the two norms defined on the space of $n$-tuples $\mathbf{x} = (x_1,\ldots, x_n)$ by

$$\|\mathbf{x}\| = (x_1^2 + \cdots + x_n^2)^{1/2}, \qquad |\mathbf{x}| = \max |x_i|, \qquad i = 1,\ldots, n$$

The elementary inequality

$$\max |x_i| \leqslant (x_1^2 + \cdots + x_n^2)^{1/2} \leqslant n \max |x_i|$$

implies that the norms are topologically equivalent. We can use this to obtain a useful fact about sequences in $\mathbf{R}^n$. Let $\mathbf{x}_k = (x_{1k},\ldots, x_{nk})$ be such a sequence. Using the two norms just discussed, we have for $\mathbf{x} = (x_1,\ldots, x_n)$

$$\|\mathbf{x}_k - \mathbf{x}\| = [(x_{1k} - x_1)^2 + \cdots + (x_{nk} - x_n)^2]^{1/2}$$

$$|\mathbf{x}_k - \mathbf{x}| = \max_{1 \leqslant i \leqslant n} |x_{ik} - x_i|$$

Equivalence of the norms together with (E3) shows that $\{\mathbf{x}_k\}$ converges to **x** in the Euclidean norm if and only if each of the coordinate sequences $\{x_{ik}\}_{k=1}^{\infty}$ converges to $x_i$. This is a fact we shall use repeatedly.

It is not difficult to give an example of two nonequivalent norms on a space **L** (Problem L), but our next theorem makes it clear that in such an example, the space **L** will have to be infinite dimensional.

**Theorem E.** All norms on a finite-dimensional linear space are topologically equivalent.

**Proof.** Let **L** be a finite-dimensional space with a basis $\{\mathbf{u}_1,\ldots, \mathbf{u}_n\}$. Any $\mathbf{x} \in \mathbf{L}$ may be represented in the form $\mathbf{x} = x_1\mathbf{u}_1 + \cdots + x_n\mathbf{u}_n$ and

## 21. Normed Linear Spaces

it is easy to verify that $\|\mathbf{x}\| = (x_1^2 + \cdots + x_n^2)^{1/2}$ is a norm on $\mathbf{L}$. Since topological equivalence is a transitive relationship, it will suffice to show that any other norm $|\ |$ for $\mathbf{L}$ is topologically equivalent to $\|\ \|$. To begin with, let $M = \max|\mathbf{u}_i|$, $i = 1,\ldots, n$. Then

$$|\mathbf{x}| = |x_1\mathbf{u}_1 + \cdots + x_n\mathbf{u}_n| \leqslant |x_1|\,|\mathbf{u}_1| + \cdots + |x_n|\,|\mathbf{u}_n|$$
$$\leqslant M(|x_1| + \cdots + |x_n|) \leqslant Mn \max|x_i| \leqslant Mn\|\mathbf{x}\|$$

It remains to show the existence of an $m$ such that $\|\mathbf{x}\| \leqslant m|\mathbf{x}|$. If no such $m$ exists, then corresponding to each positive integer $k$, there is a vector $\mathbf{y}_k$ such that $\|\mathbf{y}_k\| > k|\mathbf{y}_k|$. Set $\mathbf{z}_k = \mathbf{y}_k/\|\mathbf{y}_k\|$ so that $\|\mathbf{z}_k\| = 1$ and $|\mathbf{z}_k| = |\mathbf{y}_k|/\|\mathbf{y}_k\| < 1/k$. Let us write

$$\mathbf{z}_k = z_{1k}\mathbf{u}_1 + \cdots + z_{nk}\mathbf{u}_n$$

Since $\|\mathbf{z}_k\| = 1$, the sequence $\{(z_{1k},\ldots, z_{nk})\}_{k=1}^{\infty}$ is a bounded sequence in Euclidean space $\mathbf{R}^n$, and by the Bolzano–Weierstrass theorem it contains a subsequence converging say to $(z_1,\ldots, z_n)$. For ease of notation we suppose the subsequence to be the sequence itself. Thus if $\mathbf{z} = z_1\mathbf{u}_1 + \cdots + z_n\mathbf{u}_n$,

$$\lim_{k\to\infty}\|\mathbf{z}_k - \mathbf{z}\| = \lim_{k\to\infty}[(z_{1k} - z_1)^2 + \cdots + (z_{nk} - z_n)^2]^{1/2} = 0$$

Since by the triangle inequality,

$$\|\mathbf{z}_k\| - \|\mathbf{z}_k - \mathbf{z}\| \leqslant \|\mathbf{z}\| \leqslant \|\mathbf{z} - \mathbf{z}_k\| + \|\mathbf{z}_k\|$$

it follows that

$$\|\mathbf{z}\| = \lim_{k\to\infty}\|\mathbf{z}_k\| = 1$$

On the other hand,

$$|\mathbf{z}| \leqslant |\mathbf{z} - \mathbf{z}_k| + |\mathbf{z}_k|$$
$$\leqslant Mn\|\mathbf{z} - \mathbf{z}_k\| + 1/k$$

Since this is true for arbitrarily large $k$, it must be that $|\mathbf{z}| = 0$, which in turn means that $\mathbf{z} = \mathbf{O}$. This contradicts $\|\mathbf{z}\| = 1$, completing our proof. ●

We have actually proved more than we have claimed, as is made clear by viewing things in the following way. Beginning with $\mathbf{x} = x_1\mathbf{u}_1 + \cdots + x_n\mathbf{u}_n$ in $\mathbf{L}$, the mapping $\phi$ that sends $\mathbf{x}$ onto $(x_1,\ldots, x_n)$ in $\mathbf{R}^n$ is an isomorphism (Theorem D). The proof just given establishes a relationship between norms on two different spaces: $|\ |$ on $\mathbf{L}$ and $\|\ \|$ on $\mathbf{R}^n$. We showed, using $k = Mn$,

$$|\mathbf{x}| \leqslant k \|\phi(\mathbf{x})\| \quad \text{and} \quad \|\phi(\mathbf{x})\| \leqslant m |\mathbf{x}| \tag{2}$$

If $\phi$ is an isomorphism between two linear spaces $\mathbf{L}$ and $\mathbf{M}$ having the respective norms $|\ |$ and $\|\ \|$, and if the relations (2) hold (for some constants $k$ and $m$), we say the spaces are **topologically isomorphic**. The reason for the name is clear. Not only does $\phi$ preserve the linear structure of the two spaces, it also preserves the topological structure (technically, any concept that depends on the notion of an open set). In particular, the analogs of (E1) through (E3) are valid.

(T1) A point $\mathbf{x}$ is an interior, exterior, or boundary point of a set $U \subseteq \mathbf{L}$ with respect to neighborhoods determined by $|\ |$ if and only if $\phi(\mathbf{x})$ is similarly related to $\phi(U) \subseteq \mathbf{M}$ with respect to neighborhoods determined by $\|\ \|$.

(T2) A set $U \subseteq \mathbf{L}$ is open, closed, compact, or bounded if and only if $\phi(u) \subseteq \mathbf{M}$ is so classified.

(T3) A sequence $\{\mathbf{x}_j\}$ is Cauchy in $\mathbf{L}$ if and only if the sequence $\{\phi(\mathbf{x}_j)\}$ is Cauchy in $\mathbf{M}$. The sequence $\{\mathbf{x}_j\}$ converges to $\mathbf{y}$ in $\mathbf{L}$ if and only if $\{\phi(\mathbf{x}_j)\}$ converges to $\phi(\mathbf{y}) \in \mathbf{M}$.

We are now in a position to state precisely what was proved above.

**Theorem F.** Every finite-dimensional normed linear space of dimension $n$ is topologically isomorphic to $\mathbf{R}^n$.

On the strength of this theorem, we may state theorems about the topological properties mentioned in (T1) through (T3) for an arbitrary finite-dimensional normed linear space, and get by with proving them for $\mathbf{R}^n$. For instance, we now have the following result for any $n$-dimensional normed linear space $\mathbf{L}^n$.

**Theorem G.** In $\mathbf{L}^n$, every Cauchy sequence converges to a point in $\mathbf{L}^n$, every bounded infinite set has a limit point, and a set is compact if and only if it is closed and bounded.

**Proof.** This follows from Theorems A, B, and C. ●

We now come to our last generalization of a property of $\mathbf{R}^n$, this having to do with the inner product. A linear space $\mathbf{L}$ is said to be a **real inner product space** (or a **pre-Hilbert space**) if there is a

## 21. Normed Linear Spaces

real-valued function $g: \mathbf{L} \times \mathbf{L} \to \mathbf{R}$, usually written $g(\mathbf{x}, \mathbf{y}) = \langle \mathbf{x}, \mathbf{y} \rangle$, satisfying for all $\mathbf{x}, \mathbf{y} \in \mathbf{L}$, $\alpha \in \mathbf{R}$,

(H1) $\langle \mathbf{x}, \mathbf{x} \rangle \geqslant 0$ with equality if and only if $\mathbf{x} = \mathbf{O}$,
(H2) $\langle \alpha \mathbf{x}, \mathbf{y} \rangle = \alpha \langle \mathbf{x}, \mathbf{y} \rangle$,
(H3) $\langle \mathbf{x}, \mathbf{y} \rangle = \langle \mathbf{y}, \mathbf{x} \rangle$,
(H4) $\langle \mathbf{x} + \mathbf{y}, \mathbf{z} \rangle = \langle \mathbf{x}, \mathbf{z} \rangle + \langle \mathbf{y}, \mathbf{z} \rangle$.

Except for the usual difficulties with the triangle inequality, necessitating in this case an appeal to the CBS inequality (Problem J), it is easy to show that an inner product space can always be made into a normed linear space with norm defined by $\| \mathbf{x} \| = \langle \mathbf{x}, \mathbf{x} \rangle^{1/2}$.

The inner product in $\mathbf{R}^n$ satisfies axioms (H1) through (H4), meaning $\mathbf{R}^n$ is an inner product space. Here is a characteristic of $\mathbf{R}^n$ not preserved by a topological isomorphism; it does not follow from Theorem F that all finite-dimensional linear spaces are inner product spaces.

Though verification requires some effort, it is true that for two sequences $\mathbf{x} = \{x_i\}$, $\mathbf{y} = \{y_i\}$ in $l^2$,

$$\langle \mathbf{x}, \mathbf{y} \rangle = \sum_1^\infty x_i y_i \qquad (3)$$

defines an inner product; it requires even more effort to show that with the norm induced by this inner product, $l^2$ is complete. A complete inner product space is called a **Hilbert space**. In Problem K we shall see that $C[0, 1]$ with

$$\langle f, g \rangle = \int_0^1 f(x) g(x) \, dx \qquad (4)$$

gives us an example of an inner product (pre-Hilbert) space that is not complete (not a Hilbert space).

### PROBLEMS AND REMARKS

A. Prove the following about a set $U$ in a normed linear space $\mathbf{L}$.

(1) $U^0$ is open; $\bar{U}$ is closed.
(2) $U$ can contain a limit point that is not a boundary point, and a boundary point that is not a limit point.
(3) $U$ is closed if and only if it contains all of its limit points.
(4) $U$ is closed if and only if it contains all of its boundary points.

B. Frequent use is made of the preservation of certain topological properties under the set operations of $\cup$ and $\cap$ in a normed linear space $\mathbf{L}$.

(1) The union of any collection (finite or infinite) of open sets is open.

(2) The intersection of a finite collection of open sets is open, but the intersection of an infinite collection of open sets may not be open.

(3) If in parts (1) and (2) we replace the word open with closed and everywhere interchange the words union and intersection, the resulting (so-called dual) statements are true.

**C.** For two sets $U$ and $V$ in a normed linear space **L**, we define
$$U \setminus V = \{\mathbf{x} \in U : \mathbf{x} \notin V\}$$

(1) $U \setminus V = U \cap V'$ where $V'$ is the complement of $V$.

(2) If $U$ is closed and $V$ is open, then $U \setminus V$ is closed.

(3) The dual (defined in Problem B) of (2) is true.

**D.** Theorems A, B, and C of this section, together with some related properties of $\mathbf{R}^n$, can be derived from the axiomatic property of the real numbers **R** which says that every set of real numbers bounded from above has a least upper bound. We outline one procedure by which this may be accomplished.

(1) The least upper bound axiom for the real numbers can be used to show that **R** is complete.

(2) Using the equivalence of two norms on $\mathbf{R}^n$ as discussed in this section, completeness of $\mathbf{R}^n$ follows from the completeness of **R**.

(3) $\mathbf{R}^n$ has the **Lindelöf property**: any set covered by an arbitrary collection of open sets can be covered by a countable subcollection of these same open sets [Roydon, 1968, p. 130].

(4) We say a normed linear space **L** has the **nested set property** (NSP) if every nested sequence of closed, bounded, nonempty sets $A_1 \supseteq A_2 \supseteq \cdots$ with $\lim_{k \to \infty}(\text{diam } A_k) = 0$ has a nonempty intersection; $\bigcap_1^\infty A_k \neq \emptyset$. In $\mathbf{R}^n$, completeness $\Rightarrow$ NSP $\Rightarrow$ Heine–Borel $\Rightarrow$ Bolzano–Weierstrass $\Rightarrow$ completeness.

(5) A bounded sequence in $\mathbf{R}^n$ must have a convergent subsequence.

**E.** Not all the equivalent properties in $\mathbf{R}^n$ are equivalent in an arbitrary normed linear space **L**.

(1) The space $C[0, 1]$ is complete with respect to the norm $\| \ \|$ defined in Example D'.

(2) Let $f_k$ be the real-valued funtion defined on $[0, 1]$ so that its graph consists of two line segments joining $(0, 0)$, $(1 - 1/k, 0)$, and $(1, 1)$. Use this set $\{f_k\}$ to show that in $C[0, 1]$ with the norm $\| \ \|$ as in part (1), the Bolzano–Weierstrass theorem fails.

(3) A set $U \subseteq \mathbf{L}$ is **sequentially compact** if every sequence of points in $U$ has a subsequence converging to a point of $U$. A set in **L** is sequentially compact if and only if it is compact.

(4) The unit sphere in $C[0, 1]$ (with the same norm $\| \ \|$) is not compact.

(5) The space **L** is complete if and only if it has the NSP.

**F.** We saw in Problem E that the unit sphere in $C[0, 1]$ is not compact. The following sequence of ideas is used by several authors [Naimark, 1970, pp. 69–70; Taylor, 1958, pp. 96–97] to show that the unit sphere of an infinite-dimensional normed linear space **L** is never compact.

(1) Any finite-dimensional subspace of **L** is closed.

(2) Let **M** be a proper closed subspace of **L**. Then for each $\alpha \in (0, 1)$, there is a vector $\mathbf{x} \in \mathbf{L}$ such that $\|\mathbf{x}\| = 1$ and $\|\mathbf{y} - \mathbf{x}\| \geq \alpha$ for all $\mathbf{y} \in \mathbf{M}$ (**Riesz's lemma**).

(3) The unit sphere in **L** is not compact.

## 21. Normed Linear Spaces

**G.** For two points $\mathbf{x} = (x_1,\ldots, x_n)$, $\mathbf{y} = (y_1,\ldots, y_n)$ in $\mathbf{R}^n$, prove the CBS inequality. (*Hint*: Show that for any two real numbers $\alpha$ and $\beta$, $0 \leqslant \|\alpha\mathbf{x} - \beta\mathbf{y}\|^2 = \langle\alpha\mathbf{x} - \beta\mathbf{y}, \alpha\mathbf{x} - \beta\mathbf{y}\rangle = \alpha^2\|\mathbf{x}\|^2 - 2\alpha\beta\langle\mathbf{x},\mathbf{y}\rangle + \beta^2\|\mathbf{y}\|^2$. Choose $\alpha = \|\mathbf{x}\|$, $\beta = \|\mathbf{y}\|$.)

**H.** Suppose $M$ is a subset of the linear space $\mathbf{L}$. On the face of it, one needs to verify properties (V1) through (V10) in order to show that $M$ is a subspace. In fact, the work can be reduced.

(1) $M$ will be a subspace if for all $\mathbf{x}, \mathbf{y} \in M$ and for all real $\alpha$, $(\mathbf{x} + \alpha\mathbf{y}) \in M$.
(2) Use part (1) to reduce the work, and complete Example F'.

**I.** For two norms $\|\ \|_1$ and $\|\ \|_2$ on $\mathbf{L}$ that are topologically equivalent, verify (E1), (E2), and (E3).

**J.** Let $\mathbf{L}$ be an inner product space.

(1) For any two vectors $\mathbf{x}, \mathbf{y} \in \mathbf{L}$, we have the **CBS inequality**

$$|\langle\mathbf{x},\mathbf{y}\rangle| \leqslant \|\mathbf{x}\|\|\mathbf{y}\|$$

(2) $\|\mathbf{x}\| = \langle\mathbf{x},\mathbf{x}\rangle^{1/2}$ defines a norm on $\mathbf{L}$.
(3) A basis $B$ is called an **orthonormal basis** if for $\mathbf{x}_i, \mathbf{x}_j \in B$, $\langle\mathbf{x}_i, \mathbf{x}_j\rangle = 0$ for $i \neq j$, $\langle\mathbf{x}_i, \mathbf{x}_j\rangle = 1$ for $i = j$. Every inner product space has an orthonormal basis [Taylor, 1958, p. 114].
(4) Let $B$ be an orthonormal basis for $\mathbf{L}$. Given $\mathbf{x} \in \mathbf{L}$, we can find $k$ members of $B$ and $k$ real numbers $\alpha_i$ such that $\mathbf{x} = \alpha_1\mathbf{x}_1 + \cdots + \alpha_k\mathbf{x}_k$. Then $\alpha_i = \langle\mathbf{x}_i, \mathbf{x}\rangle$.
(5) For any two points $\mathbf{x}$ and $\mathbf{y} \in \mathbf{L}$,

$$\|\mathbf{x}+\mathbf{y}\|^2 + \|\mathbf{x}-\mathbf{y}\|^2 = 2\|\mathbf{x}\|^2 + 2\|\mathbf{y}\|^2$$

The latter property, called the **parallelogram law**, characterizes inner product spaces in the sense that if $\mathbf{L}$ is a normed linear space for which the parallelogram law holds for arbitrary elements $\mathbf{x}, \mathbf{y} \in \mathbf{L}$, then an inner product may be defined on $\mathbf{L}$ that induces the given norm [Day, 1962, p. 115].

**K.** We suggested two examples of inner product spaces in the text showing that an infinite-dimensional space may or may not be complete.

(1) Show that $l^2$ with the inner product (3) is complete.
(2) Show that $C[0, 1]$ with the inner product (4) is not complete. *Hint*: Define $f_k: [0, 1] \to \mathbf{R}$ to be the function whose graph consists of three line segments joining $(0, 0)$, $(1/2 - 1/2^{k+1}, 0)$, $(1/2 + 1/2^{k+1}, 1)$, and $(1, 1)$. To what does this Cauchy sequence converge with respect to the norm $|f| = \langle f,f\rangle^{1/2}$?

**L.** We give two examples of nonequivalent norms defined on a linear space $\mathbf{L}$.

(1) Consider the two norms defined on $C[0, 1]$ in Example D'. The results of Problems E1 and K2, combined with the conclusion to Problem I, show that these norms are not topologically equivalent.
(2) The elements of $l^2$ can also be normed by

$$|\mathbf{x}| = \sum_1^\infty \frac{1}{i}|x_i|$$

Now consider the sequence $\{\mathbf{e}_i\}$ where $\mathbf{e}_1 = \{1, 0, 0,\ldots\}$, $\mathbf{e}_2 = \{0, 1, 0,\ldots\}$, and $\mathbf{e}_i$ has a 1 in the $i$th position as its only nonzero entry. Conclude that the two norms are not topologically equivalent.

**M.** Let $\| \ \|_1$ and $\| \ \|_2$ be two norms on **L** which determine the same open sets. Then $\| \ \|_1$ and $\| \ \|_2$ are topologically equivalent.

**\*N.** The space $BV[a, b]$ of functions of bounded variation on $[a, b]$ is a linear space. Moreover, $\|f\| = V_a^b(f) + |f(a)|$ defines a norm on this space, with respect to which $BV[a, b]$ is a Banach space.

**\*O.** The space $BC[a, b]$ of Section 14 is a linear space, and $\|f\| = K_a^b(f) + |f_+'(a)| + |f(a)|$ defines a norm on this space, with respect to which $BC[a, b]$ is a Banach space.

★ ★ ★ ★ ★

A full treatment of the linear space structure of $\mathbf{R}^n$ and $\mathbf{L}^n$ is given by Halmos [1958]. The topological ideas summarized here for $\mathbf{R}^n$ are treated in greater detail in Buck [1965]. Corresponding material for the infinite-dimensional case is covered in Taylor [1958] and Roydon [1968]. Advanced treatments of normed linear spaces are given in Day [1962] and Naimark [1970].

## 22. Functions on Normed Linear Spaces

We are concerned here with functions having their domain $U$ in a normed linear space **L**. The range of $f$ will be assumed to lie in a (perhaps different) normed linear space **M**. We write

$$f: U \to \mathbf{M}$$

understanding the mapping is into **M**, not necessarily onto **M**. When $\mathbf{M} = \mathbf{R}$, $f$ is called a **functional**.

A function $f$ defined at $\mathbf{x} \in U$ is said to be **continuous** at $\mathbf{x}$ if for every $\varepsilon > 0$ we can find a $\delta > 0$ such that $\|f(\mathbf{y}) - f(\mathbf{x})\| < \varepsilon$ whenever $\mathbf{y} \in U$ and $\|\mathbf{y} - \mathbf{x}\| < \delta$. Though the two norms here may be quite different, being on different spaces, we commonly use the same symbol $\| \ \|$, letting the context indicate which norm is meant. Observe that if $\| \ \|_1$ and $\| \ \|_2$ are topologically equivalent norms on **L**, then $f$ is continuous at $\mathbf{x}$ with respect to $\| \ \|_1$ if and only if it is continuous at $\mathbf{x}$ with respect to $\| \ \|_2$.

Having used a definition of continuity that parallels the one used in elementary calculus, it is to be expected that the usual theorems regarding sums, products, quotients, and limits of continuous functions all remain valid. Similar remarks apply to theorems dealing with functions known to be continuous (or uniformly continuous, or Lipschitz) on a set $U$. A few specific results are indicated in the problems at the end of this section.

## 22. Functions on Normed Linear Spaces

In the remainder of this section we confine our attention to linear and affine functions. In deference to common practice, we call such functions **transformations** and we denote them with capital letters. A transformation $T: \mathbf{L} \to \mathbf{M}$ is called **linear** if for all $\mathbf{x}, \mathbf{y} \in \mathbf{L}$ and all real $\alpha$,

$$T(\mathbf{x} + \mathbf{y}) = T(\mathbf{x}) + T(\mathbf{y}) \quad \text{and} \quad T(\alpha \mathbf{x}) = \alpha T(\mathbf{x})$$

Closely related to linear transformations are affine transformations. The transformation $A: \mathbf{L} \to \mathbf{M}$ is **affine** if for every $\mathbf{x} \in \mathbf{L}$, $A(\mathbf{x}) = T(\mathbf{x}) + \mathbf{b}$ where $T$ is linear and $\mathbf{b}$ is a constant in $\mathbf{M}$. It is clear from the definition that if $T$ is linear, then $T(\mathbf{O}) = \mathbf{O}$, and if $A$ is affine, then $A(\mathbf{O}) = \mathbf{b}$. When $\mathbf{L} = \mathbf{M} = \mathbf{R}$, the linear functions are those functions described by $T(x) = mx$ and the affine functions are described by $A(x) = mx + b$.

**Theorem A.** $A: \mathbf{L} \to \mathbf{M}$ is affine if and only if

$$A\left(\sum_1^n \alpha_i \mathbf{x}_i\right) = \sum_1^n \alpha_i A(\mathbf{x}_i) \tag{1}$$

for all choices of $\mathbf{x}_i \in \mathbf{L}$ and real $\alpha_i$ such that $\sum_1^n \alpha_i = 1$.

**Proof.** If $A$ is affine, then there is a linear transformation $T$ and a constant $\mathbf{b}$ such that

$$A\left(\sum_1^n \alpha_i \mathbf{x}_i\right) = T\left(\sum_1^n \alpha_i \mathbf{x}_i\right) + \mathbf{b} = \sum_1^n \alpha_i [T(\mathbf{x}_i) + \mathbf{b}] = \sum_1^n \alpha_i A(\mathbf{x}_i)$$

If, on the other hand, $A$ is known to satisfy (1), we set $T(\mathbf{x}) = A(\mathbf{x}) - A(\mathbf{O})$. For any real $\alpha$,

$$A(\alpha \mathbf{x}) = A[\alpha \mathbf{x} + (1 - \alpha)\mathbf{O}] = \alpha A(\mathbf{x}) + (1 - \alpha) A(\mathbf{O})$$

so

$$T(\alpha \mathbf{x}) = \alpha A(\mathbf{x}) + (1 - \alpha) A(\mathbf{O}) - A(\mathbf{O}) = \alpha T(\mathbf{x})$$

Finally,

$$\begin{aligned}
T(\mathbf{x}_1 + \mathbf{x}_2) &= T[2(\tfrac{1}{2}\mathbf{x}_1 + \tfrac{1}{2}\mathbf{x}_2)] = 2T[\tfrac{1}{2}\mathbf{x}_1 + \tfrac{1}{2}\mathbf{x}_2] \\
&= 2[A(\tfrac{1}{2}\mathbf{x}_1 + \tfrac{1}{2}\mathbf{x}_2) - A(\mathbf{O})] \\
&= 2\{\tfrac{1}{2}[A(\mathbf{x}_1) - A(\mathbf{O})] + \tfrac{1}{2}[A(\mathbf{x}_2) - A(\mathbf{O})]\} \\
&= T(\mathbf{x}_1) + T(\mathbf{x}_2)
\end{aligned}$$

Thus $T$ is linear and $A$ is affine. ●

Contrary to the expectations of persons not accustomed to thinking in terms of infinite-dimensional spaces, a linear transformation $T$ may be discontinuous (Problem B). Since $\|T(\mathbf{y}) - T(\mathbf{x})\| = \|T(\mathbf{y} - \mathbf{x})\|$, however, it is clear that if $T$ is continuous at the origin, it is continuous at any $\mathbf{x} \in \mathbf{L}$, and conversely.

If $T: \mathbf{L} \to \mathbf{M}$ is linear and continuous, then continuity at the origin means we can always find a $\delta > 0$ so that $\|T(\mathbf{x})\| < 1$ whenever $\|\mathbf{x}\| \leq \delta$. Then for any $\mathbf{x} \in \mathbf{L}$ with $\|\mathbf{x}\| = 1$, $\|\delta\mathbf{x}\| = \delta$ and we have

$$\|T(\mathbf{x})\| = \left\|T\left(\frac{1}{\delta}\delta\mathbf{x}\right)\right\| = \frac{1}{\delta}\|T(\delta\mathbf{x})\| < \frac{1}{\delta}$$

This shows that the set $B$ of real numbers defined by

$$B = \{\|T(\mathbf{x})\|: \mathbf{x} \in \mathbf{L}, \quad \|\mathbf{x}\| = 1\} \tag{2}$$

has $1/\delta$ as an upper bound, hence that $B$ has a least upper bound. Linear transformations $T$ for which the set $B$ has an upper bound are called **bounded linear transformations**. (See Problem D for a word of caution about this terminology.) We have seen that continuous linear transformations are bounded. The converse is also true.

**Theorem B.** Let $T: \mathbf{L} \to \mathbf{M}$ be a linear transformation from one normed linear space to another. Then $T$ is continuous if and only if $T$ is bounded.

**Proof.** We only need to show that if $T$ is bounded, then $T$ is continuous. Suppose not. Then in particular, $T$ is not continuous at $\mathbf{O}$ so we can find a sequence $\mathbf{x}_i$ with $\|\mathbf{x}_i\| < 1/i$ and $\|T(\mathbf{x}_i)\| > \varepsilon$ for some $\varepsilon > 0$. It follows that for all $i$,

$$\left\|T\left(\frac{\mathbf{x}_i}{\|\mathbf{x}_i\|}\right)\right\| = \frac{1}{\|\mathbf{x}_i\|}\|T(\mathbf{x}_i)\| > i\varepsilon$$

contradicting the boundedness of $T$ on the unit sphere. ●

In alerting the reader to the possibility that a linear transformation might be discontinuous, we made specific reference to infinite-dimensional space. Our next theorem makes it clear that a discontinuous linear transformation must be defined on an infinite-dimensional space.

**Theorem C.** Let $T: \mathbf{L} \to \mathbf{M}$ be a linear transformation from one normed linear space to another. If $\mathbf{L}$ is finite dimensional, then $T$ is continuous.

## 22. Functions on Normed Linear Spaces

**Proof.** Appealing to Theorem 21F we need only prove the theorem for the case where $\mathbf{L} = \mathbf{R}^n$. Using the standard basis and letting $m = \max\| T(\mathbf{e}_i)\|$, $i = 1,\ldots, n$, we can write for any unit vector $\mathbf{x} \in \mathbf{R}^n$, $\mathbf{x} = \sum_1^n \alpha_i \mathbf{e}_i$,

$$\| T(\mathbf{x})\| = \left\| T\left(\sum_1^n \alpha_i \mathbf{e}_i\right) \right\| \leqslant \sum_1^n \| \alpha_i T(\mathbf{e}_i)\| \leqslant m \sum_1^n |\alpha_i|$$

Now $\alpha_i = \langle \mathbf{x}, \mathbf{e}_i \rangle$ (Problem 21J), so the CBS inequality gives $|\alpha_i| \leqslant \| \mathbf{x} \| \| \mathbf{e}_i \| = 1$ and we conclude that $\| T(\mathbf{x})\| \leqslant mn$. Then $T$ is a bounded, hence continuous linear transformation. ●

If $T: \mathbf{L} \to \mathbf{R}$ is linear, then the subset $\mathbf{N} \subseteq \mathbf{L}$ defined by

$$\mathbf{N} = \{\mathbf{x}: \quad T(\mathbf{x}) = 0\}$$

is a subspace of $\mathbf{L}$. It is called the **null space** of $T$. If $T$ is a **nontrivial** (that is, not identically 0) linear functional, then $\mathbf{N}$ is a **maximal proper subspace** of $\mathbf{L}$; that is, if $\mathbf{K}$ is a subspace of $\mathbf{L}$ such that $\mathbf{N} \subseteq \mathbf{K} \subseteq \mathbf{L}$, then either $\mathbf{N} = \mathbf{K}$ or $\mathbf{K} = \mathbf{L}$. In fact, maximal subspaces characterize the linear functionals on a space as indicated in Theorem D.

> **Theorem D.** $\mathbf{N}$ is a maximal proper subspace of $\mathbf{L}$ if and only if $\mathbf{N}$ is the null space of a nontrivial linear functional. The maximal proper subspace is closed if and only if $f$ is continuous.

**Proof.** Let $\mathbf{N}$ be a maximal proper subspace. There exists a $\mathbf{y} \notin \mathbf{N}$. Form

$$\mathbf{K} = \{\alpha \mathbf{y} + \mathbf{x}: \quad \mathbf{x} \in \mathbf{N}, \quad \alpha \text{ real}\}$$

It is easily verified that $\mathbf{K}$ is a subspace and (since we may take $\alpha = 0$) that it properly contains $\mathbf{N}$; $\mathbf{N} \subset \mathbf{K} \subseteq \mathbf{L}$, so $\mathbf{K} = \mathbf{L}$. We can now define $T: \mathbf{L} \to \mathbf{R}$ by

$$T(\alpha \mathbf{y} + \mathbf{x}) = \alpha$$

It is easily verified that $T$ is linear and that $\mathbf{N}$ is its null space.

Conversely, let $T$ be a nontrivial linear functional with null space $\mathbf{N}$. Suppose $\mathbf{K}$ is a subspace that properly contains $\mathbf{N}$; $\mathbf{N} \subset \mathbf{K} \subseteq \mathbf{L}$. Then there is a $\mathbf{y} \in \mathbf{K}$ such that $T(\mathbf{y}) = \alpha \neq 0$. Let $\mathbf{y}_0 = \mathbf{y}/\alpha$ so that $T(\mathbf{y}_0) = 1$. Since $\mathbf{K}$ is a subspace, $\mathbf{K}$ contains all elements of the form $\beta \mathbf{y}_0 + \mathbf{x}$, $\mathbf{x} \in \mathbf{N}$. Choose any $\mathbf{z} \in \mathbf{L}$. Setting

$$\mathbf{x} = \mathbf{z} - T(\mathbf{z}) \mathbf{y}_0$$

we see that
$$T(\mathbf{x}) = T(\mathbf{z}) - T(\mathbf{z}) T(\mathbf{y}_0) = 0$$
so $\mathbf{x} \in \mathbf{N}$. That is, any $\mathbf{z} \in \mathbf{L}$ can be written in the form $\mathbf{z} = T(\mathbf{z})\mathbf{y}_0 + \mathbf{x}$, which is a member of $\mathbf{K}$; $\mathbf{K} = \mathbf{L}$.

If a linear functional $T$ is known to be continuous on $\mathbf{L}$, then for any sequence $\{\mathbf{x}_i\}$ in $\mathbf{N}$ converging to $\mathbf{x}$, $\lim_{i \to \infty} T(\mathbf{x}_i) = T(\mathbf{x})$. Thus $T(\mathbf{x}) = 0$ and $\mathbf{N}$ must contain all its limit points; that is, $\mathbf{N}$ is closed.

Finally we suppose the null space $\mathbf{N}$ of a linear functional $T$ is closed. We have just showed that all $\mathbf{z} \in \mathbf{L}$ can be written in the form $\mathbf{z} = \alpha \mathbf{y}_0 + \mathbf{x}$ where $\mathbf{y}_0 \notin \mathbf{N}$, $\mathbf{x} \in \mathbf{N}$, and $\alpha = T(\mathbf{z})$. Since $\mathbf{N}$ is closed, we can find an $\varepsilon$-neighborhood of $\mathbf{y}_0$ disjoint from $\mathbf{N}$. Then for all $\mathbf{x} \in \mathbf{N}$, $\| \mathbf{x} - \mathbf{y}_0 \| > \varepsilon$, and since $-\mathbf{x}$ is also in $\mathbf{N}$, $\| -\mathbf{x} - \mathbf{y}_0 \| = \| \mathbf{x} + \mathbf{y}_0 \| > \varepsilon$. For any $\mathbf{z} \in \mathbf{L}$,

$$\| \mathbf{z} \| = \| \alpha \mathbf{y}_0 + \mathbf{x} \| = |\alpha| \left\| \mathbf{y}_0 + \frac{1}{\alpha} \mathbf{x} \right\| \geq |\alpha| \varepsilon = \varepsilon |T(\mathbf{z})|$$

This shows that $T$ is a bounded linear functional, hence continuous. •

When $\mathbf{L} = \mathbf{L}^n$ and $\mathbf{M} = \mathbf{L}^m$ are finite-dimensional spaces, we take the relationship between linear transformations $T: \mathbf{L}^n \to \mathbf{L}^m$ and $m \times n$ matrices to be well known. It is to be remembered that the matrix representation of a linear transformation depends on the choice of a basis in each space. When a basis is understood as chosen in each space, we will use $[T]$ to represent the matrix corresponding to $T$. In order that the image of $\mathbf{x} \in \mathbf{L}^n$ be obtained as an ordinary matrix product, it is convenient to think of $\mathbf{x}$ written as a column vector, a practice we shall follow whenever $\mathbf{x}$ enters into a matrix equation. Thus, $\mathbf{y} = T\mathbf{x}$ suggests the matrix equation

$$\begin{bmatrix} y_1 \\ \vdots \\ y_m \end{bmatrix} = \begin{bmatrix} a_{11} & \cdots & a_{1n} \\ \vdots & & \vdots \\ a_{m1} & \cdots & a_{mn} \end{bmatrix} \begin{bmatrix} x_1 \\ \vdots \\ x_n \end{bmatrix} \qquad (3)$$

and where the inner product of two vectors $\mathbf{x}, \mathbf{y} \in \mathbf{R}^n$ is to be thought of as a matrix product, we write $\langle \mathbf{x}, \mathbf{y} \rangle = \mathbf{x}^t \mathbf{y}$. Consistent use of $\mathbf{x}$ to represent a column vector would require writing $\mathbf{x}^t = (x_1, \ldots, x_n)$, but we shall avoid the resulting profusion of superscripts $t$ by writing $\mathbf{x} = (x_1, \ldots, x_n)$ when no confusion seems likely.

We have seen (Example 21C) that the set of $m \times n$ matrices forms a linear space. The correspondence between matrices and linear transformations means that the set of all linear transformations from $\mathbf{L}^n$ to $\mathbf{L}^m$ should also form a linear space. That they do is a special case of a more

## 22. Functions on Normed Linear Spaces

general result. For two normed linear spaces **L** and **M**, let $\mathscr{L}(\mathbf{L}, \mathbf{M})$ be the class of all continuous linear transformations $T \colon \mathbf{L} \to \mathbf{M}$. If we define addition and scalar multiplication as is usually done for functions, that is, if for $S, T \in \mathscr{L}(\mathbf{L}, \mathbf{M})$ and real $\alpha$

$$(S + T)(\mathbf{x}) = S(\mathbf{x}) + T(\mathbf{x})$$

$$(\alpha T)(\mathbf{x}) = \alpha T(\mathbf{x})$$

then $\mathscr{L}(\mathbf{L}, \mathbf{M})$ is a linear space. We omit verifications of properties (V1) through (V10), partly because the reader can easily supply them and partly because they are available in detail in books on the subject of linear transformations [Lorch, 1962, pp. 33–34].

We specified that members of $\mathscr{L}(\mathbf{L}, \mathbf{M})$ should be continuous. We know (Theorem B) that we could also have described $\mathscr{L}(\mathbf{L}, \mathbf{M})$ as the set of all bounded linear transformations. This is, in fact, the usual terminology since it suggests a natural way to define a norm on the space. Recall that $T$ is continuous if and only if the set $B$ defined in (2) has an upper bound. We define

$$\|T\| = \sup \|T(\mathbf{x})\| \quad \text{over} \quad \|\mathbf{x}\| = 1$$

Observe (Problem E) that $T \in \mathscr{L}(\mathbf{L}, \mathbf{M})$ if and only if there exists a number $k \geq 0$ such that for all $\mathbf{x} \in \mathbf{L}$, $\|T(\mathbf{x})\| \leq k\|\mathbf{x}\|$; and when $T \in \mathscr{L}(\mathbf{L}, \mathbf{M})$, $\|T\|$ is the smallest value of $k$ that works. Thus,

$$\|T(\mathbf{x})\| \leq \|T\| \|\mathbf{x}\|$$

Using this inequality, it is easily established that we have indeed defined a norm on $\mathscr{L}(\mathbf{L}, \mathbf{M})$.

**Theorem E.** *The class $\mathscr{L}(\mathbf{L}, \mathbf{M})$ is a normed linear space. If $\mathbf{M}$ is a Banach space, then $\mathscr{L}(\mathbf{L}, \mathbf{M})$ is a Banach space.*

**Proof.** Having already left the proof of the first statement as a straightforward verification to be either worked out or looked up, we turn our attention to the second assertion. Suppose $\{T_n\}$ is a Cauchy sequence in $\mathscr{L}(\mathbf{L}, \mathbf{M})$. For any $\mathbf{x} \in L$, $\{T_n(\mathbf{x})\}$ is a Cauchy sequence in **M** since

$$\|T_m(\mathbf{x}) - T_n(\mathbf{x})\| = \|(T_m - T_n)(\mathbf{x})\| \leq \|T_m - T_n\| \|\mathbf{x}\|$$

which goes to zero as $n, m \to \infty$. Since **M** is complete, $\{T_n(\mathbf{x})\}$ converges to some element in **M**, enabling us to define $T: \mathbf{L} \to \mathbf{M}$ by $T(\mathbf{x}) = \lim_{n\to\infty} T_n(\mathbf{x})$. For $\mathbf{x}, \mathbf{y} \in \mathbf{L}$ and real $\alpha$,

$$T(\mathbf{x} + \alpha\mathbf{y}) = \lim_{n\to\infty} T_n(\mathbf{x} + \alpha\mathbf{y}) = \lim_{n\to\infty} [T_n(\mathbf{x}) + \alpha T_n(\mathbf{y})]$$

by the linearity of $T_n$. It follows that $T$ is linear. Taking $k$ as a bound on $\| T_n \|$ (all Cauchy sequences must be bounded), and using the continuity of the norm function, we have for any $\mathbf{x} \in \mathbf{L}$,

$$\| T(\mathbf{x}) \| = \| \lim_{n\to\infty} T_n(\mathbf{x}) \| = \lim_{n\to\infty} \| T_n(\mathbf{x}) \|$$
$$\leqslant \lim_{n\to\infty} \| T_n \| \| \mathbf{x} \| \leqslant k \| \mathbf{x} \|$$

so $T$ is bounded. We now know $T \in \mathscr{L}(\mathbf{L}, \mathbf{M})$ and it only remains to show that $\{T_n\}$ actually converges to $T$. Since $\| T_n - T \| = \sup\|(T_n - T)\mathbf{x}\|$ over $\|\mathbf{x}\| = 1$, there is for any $\varepsilon > 0$ a unit vector $\mathbf{x}$ such that

$$\| T_n - T \| \leqslant \|(T_n - T)\mathbf{x}\| + \varepsilon/2 = \| T_n(\mathbf{x}) - T(\mathbf{x}) \| + \varepsilon/2$$

Since $T_n(\mathbf{x})$ converges to $T(\mathbf{x})$, the first term on the right may be made less than $\varepsilon/2$ by taking $n$ sufficiently large. For such $n$, $\| T_n - T \| < \varepsilon$ as desired. ●

The case in which $\mathbf{M} = \mathbf{R}$ is of special interest. $\mathscr{L}(\mathbf{L}, \mathbf{R})$ is called the **dual space** of **L**, often designated by $\mathbf{L}^*$. The theorem just proved shows that $\mathbf{L}^*$ is always a Banach space; its elements are linear functionals.

If **L** is an inner product space, then $\mathbf{L} \subseteq \mathbf{L}^*$ in the following sense. Corresponding to any $\mathbf{a} \in \mathbf{L}$, define $T: \mathbf{L} \to \mathbf{R}$ by

$$T(\mathbf{x}) = \langle \mathbf{a}, \mathbf{x} \rangle \tag{4}$$

By properties (H2) and (H4) of Section 21, it is clear that $T$ is linear, and by the CBS inequality,

$$\| T(\mathbf{x}) \| = |\langle \mathbf{a}, \mathbf{x} \rangle| \leqslant \| \mathbf{a} \| \| \mathbf{x} \|$$

so $T$ is bounded. In fact (Problem F), $\| T \| = \| \mathbf{a} \|$. We see therefore that for each $\mathbf{a} \in \mathbf{L}$, (4) defines a $T \in \mathbf{L}^*$, explaining the sense in which $\mathbf{L} \subseteq \mathbf{L}^*$. (More properly, we have exhibited a one-to-one mapping $\phi: \mathbf{L} \xrightarrow{\text{into}} \mathbf{L}^*$ that preserves length; $\phi$ is an isometric isomorphism.) When **L** is a Hilbert space, $\mathbf{L} = \mathbf{L}^*$ (the mapping $\phi$ is onto). In this case any $T \in \mathbf{L}^*$ may be described by (4) (Problem F).

Consider the dual space $\mathbf{R}^{n^*}$. Since $\mathbf{R}^n$ is a Hilbert space, $\mathbf{R}^{n^*} = \mathbf{R}^n$.

## 22. Functions on Normed Linear Spaces

A linear functional $T: \mathbf{R}^n \to \mathbf{R}$, in the notation of (3), corresponds to a $1 \times n$ matrix $[a_1 \cdots a_n] = [T]$ and the representation (4) of $T$ as an inner product gives

$$T(\mathbf{x}) = \langle \mathbf{a}, \mathbf{x} \rangle = \mathbf{a}^t \mathbf{x} = [T]\mathbf{x} \tag{5}$$

### PROBLEMS AND REMARKS

**A.** Let $\mathbf{L}, \mathbf{M}$ be normed linear spaces, and suppose that $f: D \to \mathbf{M}$ is continuous on a compact set $D \subseteq \mathbf{L}$.

(1) $f$ is uniformly continuous on $D$.
(2) $f(D) = \{\mathbf{y} \in \mathbf{M}: \mathbf{y} = f(\mathbf{x}) \text{ for some } \mathbf{x} \in D\}$ is compact.
(3) $\|f(\mathbf{x})\|$ assumes both a maximum and a minimum on $D$.

**B.** We may construct an example of a discontinuous linear functional as follows.

(1) $\mathbf{S} = \{\{x_i\}: \sum_1^\infty |x_i| < \infty\}$ is a linear space.
(2) $\|\{x_i\}\| = \sup |x_i|$ defines a norm on $\mathbf{S}$.
(3) $T(\{x_i\}) = \sum_1^\infty x_i$ defines a linear transformation $T: \mathbf{S} \to \mathbf{R}$.
(4) Let $\mathbf{s}_n = \{1, 1,\ldots, 1, 0,\ldots\}$ be the member of $\mathbf{S}$ having the first $n$ entries 1, the rest 0. Then $\|\mathbf{s}_n\| = 1$, $T(\mathbf{s}_n) = n$; thus $T$ is not continuous.

**C.** Let $N: \mathbf{L} \to \mathbf{R}$ be a norm; $N(\mathbf{x}) = \|\mathbf{x}\|$. Then

$$|\|\mathbf{x}\| - \|\mathbf{y}\|| \leq \|\mathbf{x} + \mathbf{y}\| \leq \|\mathbf{x}\| + \|\mathbf{y}\|$$

and $N$ is continuous on all of $\mathbf{L}$.

**D.** For a linear transformation, we must distinguish between saying it is a bounded function and saying it is a bounded linear transformation. A function $f: \mathbf{R} \to \mathbf{R}$ is said to be bounded if there is a number $k$ such that $|f(x)| < k$ for all $x \in \mathbf{R}$.

(1) The linear function $T: \mathbf{R} \to \mathbf{R}$, defined by $T(x) = mx$, is not bounded if $m \neq 0$.
(2) The linear function defined in part (1) is a bounded linear transformation with $\|T\| = m$.
(3) The linear transformation $T: \mathbf{R}^2 \to \mathbf{R}^2$ has matrix

$$[T] = \begin{bmatrix} a & b \\ -b & a \end{bmatrix}$$

with respect to the standard basis. Find $\|T\|$.

**E.** Let $T: \mathbf{L} \to \mathbf{M}$ be a linear transformation from one normed linear space to another.

(1) $T \in \mathscr{L}(\mathbf{L}, \mathbf{M})$ if and only if $\sup\|T(\mathbf{x})\| = k$, $\|\mathbf{x}\| \leq 1$, is finite, in which case $k = \|T\|$.
(2) $T \in \mathscr{L}(\mathbf{L}, \mathbf{M})$ if and only if there is a $k \geq 0$ such that for all $\mathbf{x} \in \mathbf{L}$, $\|T(\mathbf{x})\| \leq k\|\mathbf{x}\|$, in which case $\|T\|$ is the smallest value of $k$ that will work.
(3) $\|T\|$ defines a norm on $\mathscr{L}(\mathbf{L}, \mathbf{M})$.

**F.** We now investigate the conditions under which a linear functional $T: \mathbf{L} \to \mathbf{R}$

defined on an inner product space **L** may be represented, for some $\mathbf{y} \in \mathbf{L}$, in the form

$$T(\mathbf{x}) = \langle \mathbf{y}, \mathbf{x} \rangle$$

(1) When $T$ can be so represented, $\|T\| = \|\mathbf{y}\|$.

(2) If every $T$ with $\|T\| = 1$ can be represented in the desired form, then every linear functional can be so represented.

*(3) For any linear functional $T$ with $\|T\| = 1$, there is a sequence $\{\mathbf{y}_i\}$ in **L** such that $\|\mathbf{y}_i\| = 1$ and $|T(\mathbf{y}_i)| \to 1$. This sequence is Cauchy. If **L** is a Hilbert space, $\{\mathbf{y}_i\}$ converges to $\mathbf{y} \in \mathbf{L}$.

(4) Using **y** as determined in part (3), prove that $T$ may be represented in the desired form.

Conclude from parts (2) and (4) that every linear functional defined on a Hilbert space can be represented in the desired form. This is a fundamental result about Hilbert spaces and is proved in most books on functional analysis [Lorch, 1962, pp. 63-64].

*G. A linear functional $f: \mathbf{L} \to \mathbf{R}$ that fails to assume some finite value $\alpha \in \mathbf{R}$ on a nonempty open set $U \subseteq \mathbf{L}$ is necessarily continuous.

H. A normed linear space is finite dimensional $\Leftrightarrow$ all the linear functionals on it are continuous.

★ ★ ★ ★ ★

The basic work on functions defined on a normed linear space is by Banach [1932]. Of the many good expositions now available, we have used those by Lorch [1962] and Taylor [1958] because they seemed to be written at a level consistent with our own work. Among more advanced treatments, it is probably correct to say that the standard reference is by Dunford and Schwartz [1958].

## 23. Derivatives in a Normed Linear Space

In discussing the derivative of a function of one variable, we considered two limits

$$f_+'(x_0) = \lim_{t \downarrow 0} \frac{f(x_0 + t) - f(x_0)}{t}$$

$$f_-'(x_0) = \lim_{t \uparrow 0} \frac{f(x_0 + t) - f(x_0)}{t}$$

corresponding to the two directions from which $x_0$ may be approached. For a function defined in a neighborhood of $\mathbf{x}_0$ in a general normed linear space **L**, there are an infinite number of ways to approach $\mathbf{x}_0$. The approach along a line parallel to **v** gives rise to the (two-sided) **directional derivative**

$$f'(\mathbf{x}_0; \mathbf{v}) = \lim_{t \to 0} \frac{f(\mathbf{x}_0 + t\mathbf{v}) - f(\mathbf{x}_0)}{t}$$

## 23. Derivatives in a Normed Linear Space

Along $\mathbf{v}$ the two possibilities for $t$ going to zero correspond to the **one-sided directional derivatives** $f_+'(\mathbf{x}_0; \mathbf{v})$ and $f_-'(\mathbf{x}_0; \mathbf{v})$. Note that when $f'(\mathbf{x}_0; \mathbf{v})$ exists, then

$$f'(\mathbf{x}_0; -\mathbf{v}) = -f'(\mathbf{x}_0; \mathbf{v}) \tag{1}$$

When $\mathbf{L} = \mathbf{R}^n$ and $\mathbf{v}$ is taken to be one of the standard basis vectors $\mathbf{e}_i = (0, ..., 1, ..., 0)$, then the corresponding directional derivative is called the **ith partial derivative**, written variously as

$$\frac{\partial f}{\partial x_i}(\mathbf{x}_0) = f_i(\mathbf{x}_0) = f'(\mathbf{x}_0; \mathbf{e}_i)$$

In $\mathbf{R}^n$ we continue to denote points by $\mathbf{x} = (x_1, ..., x_n)$, though again in examples in $\mathbf{R}^2$ we write $\mathbf{x} = (r, s)$ to avoid the use of subscripts.

It is clear from the study of functions of one variable like $f(x) = |x|$ that existence of one-sided derivatives at a point does not tell us much about the "smoothness" of a function there. It is the existence of the (two-sided) derivative $f'(x_0) = m$ that tells us that the graph $y = f(x)$ is smooth at $x_0$ and that we can approximate $f(x)$ for values of $x$ near $x_0$ by the function $A(x) = f(x_0) + m(x - x_0)$.

Unfortunately, for functions of two variables, even the existence of the directional derivative from every direction is not enough to guarantee the smoothness of $f$ (Problem B). In order to relate differentiability to smoothness, we need a concept of differentiability that tells us, at least in the case of two variables where geometric intuition still operates, when there is a unique plane tangent to the graph of $f$. This leads us to the so-called **Fréchet derivative**, defined as follows.

Let $f$ be defined on an open set $U \subseteq \mathbf{L}$, taking values in a second normed linear space $\mathbf{M}$. Then $f$ is differentiable at $\mathbf{x}_0 \in U$ if there is a linear transformation $T: \mathbf{L} \to \mathbf{M}$ such that, for sufficiently small $\mathbf{h} \in \mathbf{L}$,

$$f(\mathbf{x}_0 + \mathbf{h}) = f(\mathbf{x}_0) + T(\mathbf{h}) + \|\mathbf{h}\| \varepsilon(\mathbf{x}_0, \mathbf{h})$$

where $\varepsilon(\mathbf{x}_0, \mathbf{h}) \in \mathbf{M}$ goes to zero as $\|\mathbf{h}\| \to 0$. The linear transformation $T$ is called the derivative and is denoted by $f'(\mathbf{x}_0)$.

We wish to examine this definition in the case where $\mathbf{L} = \mathbf{R}^n$, $\mathbf{M} = \mathbf{R}^m$. In this situation, using the standard bases for both spaces, we may associate with a linear transformation $T: \mathbf{R}^n \to \mathbf{R}^m$ a unique $m \times n$ matrix $[T]$. Thus, for example, a linear transformation

$T: \mathbf{R}^2 \to \mathbf{R}$ which takes the form $T(r, s) = a_1 r + a_2 s$ is determined by the matrix $[T] = [a_1 \ a_2]$. That is,

$$T(r, s) = a_1 r + a_2 s = [a_1 \ a_2] \begin{bmatrix} r \\ s \end{bmatrix}$$

Now suppose $w = f(r, s)$ is a function from $\mathbf{R}^2$ to $\mathbf{R}$ which has a derivative in the sense described above. Then at $\mathbf{x}_0 = (r_0, s_0)$, $f'(\mathbf{x}_0)$ may be represented by a matrix $[a_1 \ a_2]$. Moreover, for the particular choice $\mathbf{h} = t\mathbf{e}_1 = (t, 0)$,

$$f(\mathbf{x}_0 + \mathbf{h}) - f(\mathbf{x}_0) = f'(\mathbf{x}_0)\mathbf{h} + \|\mathbf{h}\| \varepsilon(\mathbf{x}_0, \mathbf{h})$$

becomes

$$f(\mathbf{x}_0 + t\mathbf{e}_1) - f(\mathbf{x}_0) = [a_1 \ a_2] \begin{bmatrix} t \\ 0 \end{bmatrix} + |t| \varepsilon(\mathbf{x}_0, t\mathbf{e}_1)$$

Subtracting $a_1 t$ from both sides and dividing by $t$, we see that

$$\lim_{t \to 0} \frac{f(r_0 + t, s_0) - f(r_0, s_0)}{t} - a_1 = 0$$

Thus $a_1 = (\partial f / \partial r)(r_0, s_0)$. Similar considerations using $\mathbf{h} = t\mathbf{e}_2$ show that $a_2 = (\partial f / \partial s)(r_0, s_0)$; hence that $[f'(\mathbf{x}_0)] = [f_1(\mathbf{x}_0) \ f_2(\mathbf{x}_0)]$. The Fréchet derivative turns out in this case to be the familiar gradient vector. It is known from elementary calculus, at least when the partial derivatives are continuous, that the gradient vector does determine a tangent plane to the graph of $f$, and in general the existence of the Fréchet derivative $f'(\mathbf{x}_0)$ is precisely what is needed to ensure the existence of a unique tangent plane to the graph of $w = f(r, s)$ at $\mathbf{x}_0$.

The methods just illustrated serve equally well to determine the matrix $[f'(\mathbf{x}_0)]$ when $f$ has its domain in $\mathbf{R}^n$, its range in $\mathbf{R}^m$. Such a transformation is determined by a set of coordinate functions

$$f: \begin{array}{c} y_1 = f^1(x_1, \ldots, x_n) \\ \vdots \\ y_m = f^m(x_1, \ldots, x_n) \end{array}$$

**Theorem A.** If $f: \mathbf{R}^n \to \mathbf{R}^m$ is differentiable at $\mathbf{x}$, then the partial derivatives of the coordinate functions all exist and

$$[f'(x)] = \begin{bmatrix} \dfrac{\partial f^1}{\partial x_1}(\mathbf{x}) & \cdots & \dfrac{\partial f^1}{\partial x_n}(\mathbf{x}) \\ \vdots & & \vdots \\ \dfrac{\partial f^m}{\partial x_1}(\mathbf{x}) & \cdots & \dfrac{\partial f^m}{\partial x_n}(\mathbf{x}) \end{bmatrix}$$

## 23. Derivatives in a Normed Linear Space

**Proof.** From the definition of the derivative, we know there is a linear transformation $T$, represented by a matrix $[a_{ij}]$ such that for $\mathbf{h} = t\mathbf{e}_j$,

$$\begin{bmatrix} f^1(\mathbf{x}+t\mathbf{e}_j) \\ f^2(\mathbf{x}+t\mathbf{e}_j) \\ \vdots \\ f^m(\mathbf{x}+t\mathbf{e}_j) \end{bmatrix} - \begin{bmatrix} f^1(\mathbf{x}) \\ f^2(\mathbf{x}) \\ \vdots \\ f^m(\mathbf{x}) \end{bmatrix} = \begin{bmatrix} a_{11} & a_{12} & \cdots & a_{1n} \\ a_{21} & a_{22} & \cdots & a_{2n} \\ \vdots & \vdots & & \vdots \\ a_{m1} & a_{m2} & \cdots & a_{mn} \end{bmatrix} \begin{bmatrix} 0 \\ 0 \\ \vdots \\ t \\ \vdots \\ 0 \end{bmatrix} + |t| \begin{bmatrix} \varepsilon^1(\mathbf{x},t\mathbf{e}_j) \\ \varepsilon^2(\mathbf{x},t\mathbf{e}_j) \\ \vdots \\ \varepsilon^m(\mathbf{x},t\mathbf{e}_j) \end{bmatrix}$$

Vectors are equal if and only if their entries are equal, so for any $i = 1, 2, \ldots, m$

$$f^i(\mathbf{x}+t\mathbf{e}_j) - f^i(\mathbf{x}) = ta_{ij} + |t|\,\varepsilon^i(\mathbf{x}, t\mathbf{e}_j)$$

It follows that for $\mathbf{x} = (x_1, \ldots, x_n)$,

$$\lim_{t \to 0} \frac{f^i(x_1, \ldots, x_j + t, \ldots, x_n) - f^i(x_1, \ldots, x_n)}{t} = \frac{\partial f^i}{\partial x_j}(\mathbf{x}) = a_{ij} \quad \bullet$$

In the case of a mapping of the plane into itself described by

$$f: \begin{array}{l} u = g(r,s) \\ v = h(r,s) \end{array} \tag{2}$$

the derivative $f'(\mathbf{x})$ is the linear transformation with the matrix representation

$$[f'(\mathbf{x})] = \begin{bmatrix} \dfrac{\partial g}{\partial r}(\mathbf{x}) & \dfrac{\partial g}{\partial s}(\mathbf{x}) \\ \dfrac{\partial h}{\partial r}(\mathbf{x}) & \dfrac{\partial h}{\partial s}(\mathbf{x}) \end{bmatrix}$$

It is clear from Theorem A that the existence of the derivative implies the existence of the partial derivatives of the coordinate functions. Problem A cautions us about thinking that these partial derivatives must be continuous, but we also note (Problem B) that if the partial derivatives exist in a neighborhood of $\mathbf{x}_0$ without being continuous there, we cannot conclude that $f'(\mathbf{x}_0)$ exists. Things fit together most nicely when the partials exist and are continuous.

**Theorem B.** *Let $f: U \to \mathbf{R}^m$ be described by coordinate functions having continuous partial derivatives throughout the open set $U \subseteq \mathbf{R}^n$. Then $f'(\mathbf{x})$ exists for each $\mathbf{x} \in U$.*

**Proof.** Except for the language of the Fréchet derivative, this is the theorem proved in advanced calculus [Buck, 1965, p. 264]. ●

If $U \subseteq \mathbf{L}$ and $f: U \to \mathbf{M}$ is differentiable throughout $U$, then $f'$ maps $U$ into the space of linear transformations from $\mathbf{L}$ to $\mathbf{M}$. We have not required in our definition of the derivative that $f'(\mathbf{x}_0)$ be a continuous linear transformation, but this is obviously the case of greatest interest. The linear transformation $f'(\mathbf{x}_0)$ is continuous if and only if $f$ is continuous at $\mathbf{x}_0$ (Problem E). If $f$ is continuous at $\mathbf{x}_0$ so that $f'(\mathbf{x}_0)$ is a continuous linear transformation, then of course $f'(\mathbf{x}_0)$ is a bounded linear transformation (Theorem 22B), a member of the normed linear space $\mathscr{L}(\mathbf{L}, \mathbf{M})$, and for any $\mathbf{h} \in \mathbf{L}$ we may write

$$\|f'(\mathbf{x}_0)h\| \leqslant \|f'(\mathbf{x}_0)\| \|\mathbf{h}\|$$

Care must be taken to distinguish between saying that $f'(\mathbf{x}_0)$, a member of $\mathscr{L}(\mathbf{L}, \mathbf{M})$, is continuous and saying that $f': U \to \mathscr{L}(\mathbf{L}, \mathbf{M})$ is continuous at $\mathbf{x}_0$.

A fundamental property of differentiation is the so-called **chain rule**, which we now prove.

**Theorem C.** Let $\mathbf{L}$, $\mathbf{M}$, and $\mathbf{N}$ be normed linear spaces, and let $U$ and $V$ be open sets in $\mathbf{L}$ and $\mathbf{M}$ respectively. Suppose that $f: U \to \mathbf{M}$ and $g: V \to \mathbf{N}$ are continuous with $f(U) \subseteq V$. If $\mathbf{y}_0 = f(\mathbf{x}_0)$, if $f'(\mathbf{x}_0)$ and $g'(\mathbf{y}_0)$ both exist, then $H = g \circ f$ is differentiable at $\mathbf{x}_0$ and

$$H'(\mathbf{x}_0) = g'(\mathbf{y}_0) \circ f'(\mathbf{x}_0).$$

**Proof.** The stated differentiability conditions enable us to write, for sufficiently small $\mathbf{h}$ and $\mathbf{k}$,

$$f(\mathbf{x}_0 + \mathbf{h}) = f(\mathbf{x}_0) + f'(\mathbf{x}_0)(\mathbf{h}) + \|\mathbf{h}\| \varepsilon_1(\mathbf{h})$$

$$g(\mathbf{y}_0 + \mathbf{k}) = g(\mathbf{y}_0) + g'(\mathbf{y}_0)(\mathbf{k}) + \|\mathbf{k}\| \varepsilon_2(\mathbf{k})$$

Then

$$H(\mathbf{x}_0 + \mathbf{h}) = g\{f(\mathbf{x}_0 + \mathbf{h})\} = g\{f(\mathbf{x}_0) + f'(\mathbf{x}_0)(\mathbf{h}) + \|\mathbf{h}\| \varepsilon_1(\mathbf{h})\}$$

If we set $\mathbf{k} = f'(\mathbf{x}_0)(\mathbf{h}) + \|\mathbf{h}\|\varepsilon_1(\mathbf{h})$, then $\|\mathbf{k}\| \leqslant \|\mathbf{h}\|\{\|f'(\mathbf{x}_0)\| + \|\varepsilon_1(\mathbf{h})\|\}$

## 23. Derivatives in a Normed Linear Space

so $\mathbf{k} \to \mathbf{O}$ as $\mathbf{h} \to \mathbf{O}$ and

$$H(\mathbf{x}_0 + \mathbf{h}) = g(\mathbf{y}_0 + \mathbf{k}) = g(\mathbf{y}_0) + g'(\mathbf{y}_0)(\mathbf{k}) + \|\mathbf{k}\| \varepsilon_2(\mathbf{k})$$
$$= g(\mathbf{y}_0) + g'(\mathbf{y}_0) \circ f'(\mathbf{x}_0)\mathbf{h} + \{g'(\mathbf{y}_0)(\|\mathbf{h}\| \varepsilon_1(\mathbf{h})) + \|\mathbf{k}\| \varepsilon_2(\mathbf{k})\}$$

Since the term in braces is in absolute value less than or equal to

$$\|\mathbf{h}\| \{ \| g'(\mathbf{y}_0) \varepsilon_1(\mathbf{h})\| + (\|f'(\mathbf{x}_0)\| + \|\varepsilon_1(\mathbf{h})\|) \|\varepsilon_2(\mathbf{k})\| \}$$

it follows that the linear transformation $g'(\mathbf{y}_0) \circ f'(\mathbf{x}_0) = H'(\mathbf{x}_0)$. ●

In the case where **L**, **M**, and **N** are finite dimensional so that $g'(\mathbf{y}_0)$ and $f'(\mathbf{x}_0)$ are represented by matrices, the chain rule gives the matrix representing $H'(\mathbf{x}_0)$ as a matrix product $[H'(\mathbf{x}_0)] = [g'(\mathbf{y}_0)][f'(\mathbf{x}_0)]$.

We also wish to discuss the form of the second derivative. Toward this end it will be helpful to have a specific example to consider. Let $f: \mathbf{R}^2 \to \mathbf{R}$ be described by

$$f(r, s) = r^2 + 3rs + 5s^2$$

Then

$$[f'(r, s)] = [2r + 3s \quad 3r + 10s]$$

We now see that $f'$ maps the vector $(r, s)$ onto the vector $(u, v)$ according to the rule

$$f': \begin{array}{l} u = 2r + 3s \\ v = 3r + 10s \end{array}$$

This is exactly the form of (2), however, and we therefore know how to find the derivative of $f'$. Naturally this is called the second derivative of $f$, written

$$[f''(r, s)] = \begin{bmatrix} 2 & 3 \\ 3 & 10 \end{bmatrix}$$

These computations illustrate the following summary.

$$\begin{array}{ll} f: & \mathbf{R}^2 \to \mathbf{R} \\ f'(\mathbf{x}): & \mathbf{R}^2 \to \mathbf{R} \quad \text{(linearly)} \\ f': & \mathbf{R}^2 \to \mathbf{R}^2 \\ f''(\mathbf{x}): & \mathbf{R}^2 \to \mathbf{R}^2 \quad \text{(linearly)} \end{array}$$

Note that $f''(\mathbf{x})(\mathbf{h})$ is an element of $\mathbf{R}^2$. If we think of this element as a matrix determining a linear transformation, then $[f''(\mathbf{x})(\mathbf{h})](\mathbf{k})$ makes sense for $\mathbf{k} \in \mathbf{R}^2$. The expression $[f''(\mathbf{x})(\mathbf{h})](\mathbf{k})$ is linear in both $\mathbf{h} = (h_1, h_2)$ and $\mathbf{k} = (k_1, k_2)$. Viewed this way, $f''(\mathbf{x})$ is a bilinear transformation from $\mathbf{R}^2 \times \mathbf{R}^2$ to $\mathbf{R}$. We often emphasize this by writing $f''(\mathbf{x})(\mathbf{h}, \mathbf{k})$.

All that we have said for functions defined on $U \subseteq \mathbf{R}^n$ can now be said for $U \subseteq \mathbf{L}$. To simplify the discussion somewhat, let us assume that $f$ is continuous on $U$, that it has a derivative throughout $U$, and that this derivative is continuous; that is, the mapping

$$f': U \to \mathscr{L}(\mathbf{L}, \mathbf{M})$$

is continuous. We say such functions are **continuously differentiable**. Then if in addition $f'$ is differentiable at $\mathbf{x}$, $f''(\mathbf{x})$ will be a continuous linear transformation from $\mathbf{L}$ to $\mathscr{L}(\mathbf{L}, \mathbf{M})$,

$$f''(\mathbf{x}): \mathbf{L} \to \mathscr{L}(\mathbf{L}, \mathbf{M})$$

Then, $f''(\mathbf{x})(\mathbf{h}) \in \mathscr{L}(\mathbf{L}, \mathbf{M})$ and $[f''(\mathbf{x})(\mathbf{h})](\mathbf{k})$ makes sense for $\mathbf{k} \in \mathbf{L}$. The expression $[f''(\mathbf{x})(\mathbf{h})](\mathbf{k})$ is again linear in both $\mathbf{h}$ and $\mathbf{k}$. Again $f''(\mathbf{x})$ is a bilinear transformation from $\mathbf{L} \times \mathbf{L}$ to $\mathbf{M}$ which we denote by $f''(\mathbf{x})(\mathbf{h}, \mathbf{k})$.

Having made this little flight into abstract normed linear space, we return once again to finite-dimensional spaces for insights into what else we may expect in the more general setting. Real-valued bilinear transformations on $\mathbf{R}^n \times \mathbf{R}^n$ are usually studied along with matrix theory. Just as linear transformations from $\mathbf{R}^n$ to $\mathbf{R}^n$ correspond to $n \times n$ matrices, so do bilinear transformations from $\mathbf{R}^n \times \mathbf{R}^n$ to $\mathbf{R}$. For example, using $\mathbf{h} = (h_1, h_2)$ and $\mathbf{k} = (k_1, k_2)$, the bilinear transformation on $\mathbf{R}^2 \times \mathbf{R}^2$ takes the form

$$B(\mathbf{h}, \mathbf{k}) = a_{11} k_1 h_1 + a_{21} k_2 h_1 + a_{12} k_1 h_2 + a_{22} k_2 h_2$$

which can be written

$$B(\mathbf{h}, \mathbf{k}) = \begin{bmatrix} k_1 & k_2 \end{bmatrix} \begin{bmatrix} a_{11} & a_{12} \\ a_{21} & a_{22} \end{bmatrix} \begin{bmatrix} h_1 \\ h_2 \end{bmatrix}$$

For a variety of reasons, one of which will soon appear to us, there is great interest in bilinear forms for which $B(\mathbf{h}, \mathbf{k}) = B(\mathbf{k}, \mathbf{h})$. These are called **symmetric bilinear transformations**. Happily, they correspond to symmetric matrices (those matrices $[a_{ij}]$ for which $a_{ij} = a_{ji}$), meaning in the two-dimensional case just illustrated that $B$ is symmetric if and only if $a_{12} = a_{21}$.

## 23. Derivatives in a Normed Linear Space

A symmetric bilinear transformation $B$ defined on $\mathbf{R}^n \times \mathbf{R}^n$ is called **positive definite** if for all choices of $\mathbf{h}$ different from $\mathbf{O}$, $B(\mathbf{h}, \mathbf{h}) > 0$. It is a fact proved in matrix algebra [Halmos, 1958, p. 153] that the symmetric bilinear transformation $B$ is positive definite if and only if the characteristic values (eigenvalues) of the corresponding matrix are positive real numbers. The transformation is called **nonnegative definite** if $B(\mathbf{h}, \mathbf{h}) \geqslant 0$ for all $\mathbf{h}$, a circumstance equivalent to the corresponding matrix having nonnegative characteristic values.

We illustrate these ideas by returning to our example.

$$f(r, s) = r^2 + 3rs + 5s^2$$
$$[f'(r, s)] = [2r + 3s \quad 3r + 10s]$$
$$[f''(r, s)] = \begin{bmatrix} 2 & 3 \\ 3 & 10 \end{bmatrix}$$

We see that $[f''(r, s)]$ is symmetric. Its characteristic values are the roots of the characteristic equation

$$\det \begin{bmatrix} 2 - \lambda & 3 \\ 3 & 10 - \lambda \end{bmatrix} = \lambda^2 - 12\lambda + 11 = 0$$

The roots, 11 and 1, are positive so $[f''(r, s)]$ is positive definite.

We have now defined for a bilinear transformation on $\mathbf{R}^n \times \mathbf{R}^n$ the concepts of symmetry and positive or nonnegative definiteness, and we have illustrated these ideas using a function defined on $\mathbf{R}^2$. The way is paved to return to the general situation. A bilinear transformation $B$ defined on $\mathbf{L} \times \mathbf{L}$ is **symmetric** if $B(\mathbf{h}, \mathbf{k}) = B(\mathbf{k}, \mathbf{h})$ for all $\mathbf{h}, \mathbf{k} \in \mathbf{L}$; and such a transformation is **positive definite** [**nonnegative definite**] if for every nonzero $\mathbf{h} \in \mathbf{L}$, $B(\mathbf{h}, \mathbf{h}) > 0$ $[B(\mathbf{h}, \mathbf{h}) \geqslant 0]$.

The symmetry of $f''(r, s)$ observed in our example above is not the consequence of having carefully selected our illustrative function. Rather it illustrates a basic fact about differentiation. The second derivative, when it exists in a region where the first derivative is continuous, will be symmetric. This fundamental fact, in the case where $\mathbf{L} = \mathbf{R}^n$, is equivalent to the statement that the mixed partial derivatives are equal. Thus, in the example, symmetry is equivalent to $\partial^2 f / \partial r \, \partial s = \partial^2 f / \partial s \, \partial r$.

**Theorem D.** Let $f: U \to \mathbf{M}$ be continuously differentiable on the open set $U \subseteq \mathbf{L}$. Then $f''(\mathbf{x})$ is symmetric wherever it exists.

**Proof.** A proof is outlined in Problem I for the case in which $f''$ exists and is continuous in $U$ and $\mathbf{M} = \mathbf{R}$. For the proof of the theorem as stated, see Dieudonné [1960, p. 175].  •

Equipped with the chain rule and our knowledge of functions of a real variable, we can carry many familiar theorems into a new setting. As an example of the general technique, we prove the following modest form of **Taylor's theorem** that we shall need in Section 42.

**Theorem E.** Let $f: U \to \mathbf{R}$ be continuously differentiable on the open convex set $U \subseteq \mathbf{L}$ and suppose $f''(\mathbf{x})$ exists throughout $U$. Then for any $\mathbf{x}, \mathbf{x}_0 \in U$, there is an $s \in (0, 1)$ such that

$$f(\mathbf{x}) = f(\mathbf{x}_0) + f'(\mathbf{x}_0)(\mathbf{h}) + \tfrac{1}{2}f''(\mathbf{x}_0 + s\mathbf{h})(\mathbf{h}, \mathbf{h})$$

where $\mathbf{h} = \mathbf{x} - \mathbf{x}_0$.

**Proof.** Given $\mathbf{x}, \mathbf{x}_0 \in U$, define $\phi: (a, b) \to \mathbf{R}$ on an interval containing $[0, 1]$ by $\phi(t) = f(\mathbf{x}_0 + t\mathbf{h})$. The chain rule applied first to $\phi(t)$ and then to $\theta(t) = f'(\mathbf{x}_0 + t\mathbf{h})(\mathbf{h})$ gives

$$\phi'(t) = f'(\mathbf{x}_0 + t\mathbf{h})(\mathbf{h})$$
$$\phi''(t) = f''(\mathbf{x}_0 + t\mathbf{h})(\mathbf{h}, \mathbf{h})$$

We know that for $t > 0$, there is an $s \in (0, t)$ such that

$$\phi(t) = \phi(0) + \phi'(0)t + \tfrac{1}{2}\phi''(s)\, t^2$$

so substitution gives

$$f(\mathbf{x}_0 + t\mathbf{h}) = f(\mathbf{x}_0) + f'(\mathbf{x}_0)\,t\mathbf{h} + \tfrac{1}{2}f''(\mathbf{x}_0 + s\mathbf{h})(t\mathbf{h}, t\mathbf{h})$$

Setting $t = 1$ gives the desired result.  •

### PROBLEMS AND REMARKS

**A.** Each of the functions below is Fréchet differentiable at $\mathbf{O} = (0, 0)$. The partial derivatives exist in a neighborhood of $\mathbf{O}$, but are discontinuous at $\mathbf{O}$.

(1) $f(r, s) = rs \sin(1/rs)$, $r \neq 0$, $s \neq 0$; $f(r, s) = 0$ otherwise.
(2) $g(r, s) = (r^2 + s^2) \sin[1/(r^2 + s^2)]$, $(r, s) \neq \mathbf{O}$; $f(\mathbf{O}) = 0$.

**B.** Let $f(r, s) = r^2 s/(r^2 + s^2)$, $(r, s) \neq \mathbf{O}$; $f(\mathbf{O}) = 0$. The directional derivative at $\mathbf{O}$ exists from every direction; in particular $(\partial f/\partial r)(\mathbf{O}) = (\partial f/\partial s)(\mathbf{O}) = 0$. Yet $f'(\mathbf{O})$ does not exist.

## 23. Derivatives in a Normed Linear Space

**C.** If $f$ is differentiable at $\mathbf{x}_0$, the derivative is unique.

**D.** If $f: \mathbf{M} \to \mathbf{N}$ is linear, then for every $\mathbf{x} \in \mathbf{M}$, $f'(\mathbf{x}) = f$.

**E.** From Problem 22B and Problem D above, it is clear that a differentiable function may be discontinuous. If $f$ is differentiable at $\mathbf{x}_0$, then $f$ is continuous at $\mathbf{x}_0 \Leftrightarrow f'(\mathbf{x}_0)$ is a continuous linear transformation. Relate this result to the theorem often proved in the elementary calculus course which says that if $f$ is differentiable at $\mathbf{x}_0$, then $f$ is continuous at $\mathbf{x}_0$. (Note Theorem 22C).

**F.** Suppose $B: \mathbf{M} \times \mathbf{N} \to \mathbf{R}$ is bilinear. At any $(\mathbf{x}_0, \mathbf{y}_0) \in \mathbf{M} \times \mathbf{N}$, $B'(\mathbf{x}_0, \mathbf{y}_0)(\mathbf{x}, \mathbf{y}) = B(\mathbf{x}_0, \mathbf{y}) + B(\mathbf{x}, \mathbf{y}_0)$.

**G.** For a function $f: \mathbf{R}^3 \to \mathbf{R}$, verify that symmetry of the matrix corresponding to $f''(\mathbf{x})$ is equivalent to the equality of the various mixed partial derivatives.

**H.** For $f: U \to \mathbf{M}$ differentiable on the open convex set $U \subseteq \mathbf{L}$ we have the following **mean value theorem**. Corresponding to $\mathbf{x}_0$ and $\mathbf{x} \in U$ there exists a real number $s \in (0, 1)$ such that $f(\mathbf{x}) - f(\mathbf{x}_0) = f'(\mathbf{x}_1)(\mathbf{x} - \mathbf{x}_0)$ where $\mathbf{x}_1 = \mathbf{x}_0 + s(\mathbf{x} - \mathbf{x}_0)$.

**I.** We outline a proof for Theorem D for the special case where $f''$ exists and is continuous in $U$ and $\mathbf{M} = \mathbf{R}$.

(1) Given $\mathbf{x}_0 \in U$, choose $\mathbf{h}$, $\mathbf{k}$ such that $\mathbf{x}_0 + \mathbf{h}$, $\mathbf{x}_0 + \mathbf{k}$, $\mathbf{x}_0 + \mathbf{h} + \mathbf{k}$ are all in $U$. Set $d = f(\mathbf{x}_0 + \mathbf{h} + \mathbf{k}) - f(\mathbf{x}_0 + \mathbf{h}) - f(\mathbf{x}_0 + \mathbf{k}) + f(\mathbf{x}_0)$.

(2) Define $G(\mathbf{x}) = f(\mathbf{x} + \mathbf{h}) - f(\mathbf{x})$; $H(\mathbf{x}) = f(\mathbf{x} + \mathbf{k}) - f(\mathbf{x})$. Notice $d = H(\mathbf{x}_0 + \mathbf{h}) - H(\mathbf{x}_0)$.

(3) Two appeals to the mean value theorem give
$$d = f''(\mathbf{x}_0 + t_1\mathbf{h} + t_2\mathbf{k})(\mathbf{k}, \mathbf{h}), \; t_1, t_2 \in (0, 1).$$

(4) Similarly prove, using $G$, that $d = f''(\mathbf{x}_0 + t_3\mathbf{h} + t_4\mathbf{k})(\mathbf{h}, \mathbf{k})$.

(5) For arbitrary $\mathbf{u}, \mathbf{v}$, choose $\alpha \in (0, 1)$ so that $\alpha\mathbf{u}$, $\alpha\mathbf{v}$ meet the restrictions on $\mathbf{h}$ and $\mathbf{k}$, respectively. Then
$$0 = f''(\mathbf{x}_0 + t_1\alpha\mathbf{u} + t_2\alpha\mathbf{v})(\alpha\mathbf{v}, \alpha\mathbf{u}) - f''(\mathbf{x}_0 + t_3\alpha\mathbf{u} + t_4\alpha\mathbf{v})(\alpha\mathbf{u}, \alpha\mathbf{v})$$

(6) Multiply by $1/\alpha^2$ and use the continuity of $f''$ to conclude that $0 = f''(\mathbf{x}_0)(\mathbf{v}, \mathbf{u}) - f''(\mathbf{x}_0)(\mathbf{u}, \mathbf{v})$.

We have already referred to Dieudonné [1960] as a place where Theorem D is proved as we have stated it. We might add that Chapter 8 of Dieudonné's book is the best source for further information on the topics of this section available in English. Readers of German may also consult Nevanlinna and Nevanlinna [1959].

# III

## *Convex Sets*

> Geometry, however, supplies sustenance and meaning to bare formulas. Geometry, remains the major source of rich and fruitful intuitions, which in turn supply creative power to mathematics. Most mathematicians think in terms of geometric schemes, even though they leave no trace of that scaffolding when they present the complicated analytical structures. One can still believe Plato's statement that "geometry draws the soul toward truth."
>
> MORRIS KLINE

> The study of convex sets is a branch of geometry, analysis, and linear algebra that has numerous connections with other areas of mathematics and serves to unify many apparently diverse mathematical phenomena.
>
> VICTOR KLEE

## 30. Introduction

The primary purpose of our work is to study convex functions. The natural domains for such functions, it turns out, are convex sets. Partly for this reason, and partly because they help to create for the uninitiated reader a feeling for the ideas, methods, and terminology typical of the study of convex geometry as well as convex functions, we consider in this chapter a few important facts about convex sets. These modest goals mean that we shall omit much that is known about convex sets, a topic developed very fully in a variety of good texts [Bonnesen and Fenchel, 1934; Eggleston, 1958; Yaglom and Boltyanskiĭ, 1961; Valentine, 1964; Grünbaum, 1967].

In Section 31 we introduce basic definitions and facts about convex sets, convex hulls, and affine sets, sometimes called **flats** in the literature. In Section 32 we prove three fundamental theorems about convex sets in $\mathbf{R}^n$ dealing with the topics of separation, support, and extreme points.

## 31. Convex Sets and Affine Sets

Let $U$ be a subset of a linear space $\mathbf{L}$. We say that $U$ is **convex** if $\mathbf{x}, \mathbf{y} \in U$ implies that $\mathbf{z} = [\lambda \mathbf{x} + (1 - \lambda)\mathbf{y}] \in U$ for all $\lambda \in [0, 1]$. Similarly, $U$ is **affine** if $\mathbf{z} \in U$ for all $\lambda \in \mathbf{R}$. Interpreted geometrically, we see that to be convex a set must contain the line segment connecting any two of its points; to be affine it must contain the whole line through any two of its points. It is clear that every affine set is convex, but not conversely. $\mathbf{L}$ itself, along with the trivial examples of the empty set, and sets consisting of one point are both affine and convex. Besides these, convex sets in $\mathbf{R}^2$ include line segments, interiors of triangles and ellipses, and hosts of other sets (Fig. 31.1). A nontrivial proper affine set

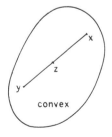
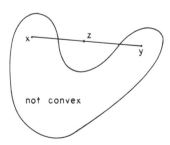

Fig. 31.1

in $\mathbf{R}^2$ is more easily described, being a straight line. Since the nontrivial proper subspaces in $\mathbf{R}^2$ are lines through the origin, affine sets can be described as translations of subspaces (Fig. 31.2), a description that is equally useful in any linear space $\mathbf{L}$.

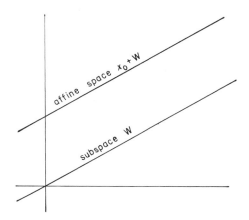

**Fig. 31.2**

**Theorem A.** A set $U \subseteq \mathbf{L}$ is affine if and only if it is a translate of a subspace of $\mathbf{L}$.

**Proof.** Suppose that $U$ is a **translate** of a subspace; that is, suppose

$$U = \mathbf{x}_0 + \mathbf{W} = \{\mathbf{x}_0 + \mathbf{w}: \mathbf{w} \in \mathbf{W}\}$$

where $\mathbf{x}_0$ is an arbitrary point of $\mathbf{L}$ and $\mathbf{W}$ is a subspace of $\mathbf{L}$. Then if $\mathbf{x}_1$, $\mathbf{x}_2 \in U$ so that $\mathbf{x}_1 = \mathbf{x}_0 + \mathbf{w}_1$ and $\mathbf{x}_2 = \mathbf{x}_0 + \mathbf{w}_2$, and if $\lambda \in \mathbf{R}$, then

$$\lambda \mathbf{x}_1 + (1 - \lambda) \mathbf{x}_2 = \lambda(\mathbf{x}_0 + \mathbf{w}_1) + (1 - \lambda)(\mathbf{x}_0 + \mathbf{w}_2)$$
$$= \mathbf{x}_0 + \lambda \mathbf{w}_1 + (1 - \lambda) \mathbf{w}_2$$

Since $\mathbf{W}$ is a subspace, the last expression is in $\mathbf{x}_0 + \mathbf{W}$.

Conversely, suppose that $U$ is affine. Let $\mathbf{x}_0$ be any element of $U$ and set $\mathbf{W} = -\mathbf{x}_0 + U$. If $\mathbf{w}_1$ and $\mathbf{w}_2$ are two elements of $\mathbf{W}$, say $\mathbf{w}_1 = -\mathbf{x}_0 + \mathbf{x}_1$, $\mathbf{w}_2 = -\mathbf{x}_0 + \mathbf{x}_2$, then

$$\mathbf{w}_1 + \lambda \mathbf{w}_2 = -\mathbf{x}_0 + \lambda[2(\tfrac{1}{2}\mathbf{x}_1 + \tfrac{1}{2}\mathbf{x}_2) - \mathbf{x}_0] + (1 - \lambda) \mathbf{x}_1$$
$$= -\mathbf{x}_0 + \mathbf{z}$$

## 31. Convex Sets and Affine Sets

Since $U$ is affine, $\mathbf{y} = \frac{1}{2}\mathbf{x}_1 + \frac{1}{2}\mathbf{x}_2$ is in $U$; then $[2\mathbf{y} + (-1)\mathbf{x}_0] \in U$, so $\mathbf{z} \in U$, and it follows that $\mathbf{w}_1 + \lambda \mathbf{w}_2$ is in $\mathbf{W}$. ●

If $\lambda_i \in \mathbf{R}$ and $\sum_1^n \lambda_i = 1$, then $\mathbf{x} = \sum_1^n \lambda_i \mathbf{x}_i$ is called an **affine combination** of $\mathbf{x}_1, \ldots, \mathbf{x}_n$, the latter being elements of $\mathbf{L}$. If in addition $\lambda_i \geqslant 0$ for all $i$, then $\mathbf{x}$ is called a **convex combination** of $\mathbf{x}_1, \ldots, \mathbf{x}_n$. Thus a set $U$ is affine or convex according as it is closed under affine or convex combinations of pairs of its elements. Actually, there is no need to restrict attention to pairs of elements as we now demonstrate.

**Theorem B.** A set $U \subseteq \mathbf{L}$ is convex [affine] if and only if every convex [affine] combination of points of $U$ lies in $U$.

**Proof.** Since a set that contains all convex combinations of its points is obviously convex, we only need to consider a convex set $U$ and show that it contains any convex combination of its points. Our proof is by induction on the number of points of $U$ occurring in a convex combination, the conclusion following from the definition for $n = 2$. Assuming the result true for any convex combination with $n$ or fewer points, we consider one with $n + 1$ points, $\mathbf{x} = \sum_1^{n+1} \lambda_i \mathbf{x}_i$. Not all the $\lambda_i$'s can be as great as one, so we relabel if necessary so that $\lambda_{n+1} < 1$. Then

$$\mathbf{x} = (1 - \lambda_{n+1}) \sum_1^n \frac{\lambda_i}{1 - \lambda_{n+1}} \mathbf{x}_i + \lambda_{n+1} \mathbf{x}_{n+1}$$
$$= (1 - \lambda_{n+1})\mathbf{y} + \lambda_{n+1} \mathbf{x}_{n+1}$$

Now $\mathbf{y} \in U$ by assumption, and thus so is $\mathbf{x}$, being a convex combination of two points of $U$. The proof in the affine case follows exactly the same pattern. ●

**Theorem C.** If $\{U_\alpha\}$, $\alpha \in A$, is any family of convex [affine] sets, then $M_1 = \bigcap_{\alpha \in A} U_\alpha$ is convex [affine]. If in addition $\{U_\alpha\}$ is a chain (meaning for $\alpha$, $\beta \in A$, either $U_\alpha \subseteq U_\beta$ or $U_\beta \subseteq U_\alpha$), then $M_2 = \bigcup_{\alpha \in A} U_\alpha$ is convex [affine].

**Proof.** One considers $\mathbf{x}$, $\mathbf{y} \in M_i$ and shows easily that the line segment [line] through $\mathbf{x}$ and $\mathbf{y}$ also lies in $M_i$. ●

We call the intersection of all convex sets containing a given set $U$ the **convex hull** of $U$, denoted by $H(U)$. Similarly, the intersection of all affine sets containing $U$ is called the **affine hull** of $U$. By Theorem C, the convex hull is convex; the affine hull is affine (Fig. 31.3).

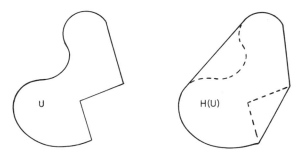

**Fig. 31.3**

**Theorem D.** For any $U \subseteq \mathbf{L}$, the convex [affine] hull of $U$ consists precisely of all convex [affine] combinations of elements of $U$.

**Proof.** We prove the statement for convex hulls, leaving the one for affine hulls to the reader. Let $H(U)$ denote the convex hull of $U$ and $K(U)$ the set of convex combinations of elements of $U$. Now $U \subseteq H(U)$, and since $H(U)$ is convex, Theorem B shows that $K(U) \subseteq H(U)$. Conversely, if $\mathbf{x} = \sum_1^n \alpha_i \mathbf{x}_i$ and $\mathbf{y} = \sum_1^m \beta_j \mathbf{y}_j$ are two elements of $K(U)$, then

$$\mathbf{z} = \lambda \mathbf{x} + (1 - \lambda)\mathbf{y} = \sum_1^n \lambda \alpha_i \mathbf{x}_i + \sum_1^m (1 - \lambda) \beta_j \mathbf{y}_j$$

is another element of $K(U)$ since

$$\sum_1^n \lambda \alpha_i + \sum_1^m (1 - \lambda) \beta_j = \lambda + (1 - \lambda) = 1$$

Thus $K(U)$ is a convex set containing $U$. Therefore $H(U) \subseteq K(U)$. •

This theorem can be improved if $\mathbf{L} = \mathbf{R}^n$. In this case, $H(U)$ consists of all convex combinations of $n + 1$ or fewer elements of $U$. Before proving a slightly more general version of this statement, let us introduce the concept of **dimension** for a convex set. First if $U$ is affine, we define its dimension to be that of the subspace of which it is a translate (Theorem A). More generally, if $U$ is convex, its dimension is the dimension of the affine hull of $U$. We are now ready to prove **Carathéodory's theorem** for convex sets.

**Theorem E.** If $U \subseteq \mathbf{L}$ and its convex hull $H(U)$ has dimension $m$, then for each $\mathbf{z} \in H(U)$, there exist $m + 1$

## 31. Convex Sets and Affine Sets

points $x_0, \ldots, x_m$ of $U$ such that $z$ is a convex combination of these points.

**Proof.** Let $z \in H(U)$. Then $z = \sum_0^n \alpha_i x_i$ where $x_i \in U$, $\alpha_i > 0$, and $\sum_0^n \alpha_i = 1$. Now suppose $n + 1$, the number of terms in the convex combination, is greater than $m + 1$ and let $B = \{x_0, \ldots, x_n\}$. Then

$$\dim(\text{affine hull } B) \leq \dim(\text{affine hull } U) = m \leq n - 1$$

and therefore $\{x_1 - x_0, x_2 - x_0, \ldots, x_n - x_0\}$ is a linearly dependent set (Problem B). Thus there are constants $\beta_1, \ldots, \beta_n$, not all 0, such that $\sum_1^n \beta_i(x_i - x_0) = \mathbf{O}$. Let $\beta_0 = -\sum_1^n \beta_i$. Then $\sum_0^n \beta_i x_i = \mathbf{O}$ and $\sum_0^n \beta_i = 0$. Since all the $\alpha_i$'s are positive, we may choose a positive number $t$ so that $\gamma_i = \alpha_i - t\beta_i \geq 0$ for $i = 0, 1, \ldots, n$ and so that $\gamma_k = 0$ for some $k$. Then

$$z = \sum_0^n \alpha_i x_i = \sum_0^n (\gamma_i + t\beta_i) x_i = \sum_{i \neq k} \gamma_i x_i$$

while

$$\sum_{i \neq k} \gamma_i = \sum_{i=0}^n \gamma_i = \sum_{i=0}^n (\alpha_i - t\beta_i) = \sum_{i=0}^n \alpha_i = 1$$

Thus $z$ is a convex combination of the $n$ points $x_0, \ldots, x_{k-1}, x_{k+1}, \ldots, x_n$. Now either $n = m + 1$ and we are done or the argument can be repeated. Eventually we are able to represent $z$ as a convex combination of $m + 1$ or fewer points of $U$. ●

The simplest convex sets are those which are convex hulls of a finite set of points, that is, sets of the form $U = H(\{x_0, \ldots, x_n\})$. Such a set is called a **polytope**. If $x_1 - x_0, \ldots, x_n - x_0$ are linearly independent (so dim $U = n$), then $U$ is called an $n$-**simplex** with vertices $x_0, \ldots, x_n$. A 0-simplex is a point, a 1-simplex is a line segment, a 2-simplex is a triangle, and so forth. A point $x$ in an $n$-simplex can be written in a unique way as a convex combination $x = \sum_0^n \lambda_i x_i$ of the vertices. The numbers $\lambda_0, \ldots, \lambda_n$ are called the **barycentric coordinates** of $x$.

So far we have not used any topology; only the linear properties of the space $\mathbf{L}$ have played a role. Suppose now that $\mathbf{L}$ is a normed linear space, so that notions like open and closed make sense.

**Theorem F.** If $U$ is a convex set, then its interior $U^0$ and its closure $\bar{U}$ are convex.

**Proof.** Let $\mathbf{x}, \mathbf{y} \in U^0$ and consider $\mathbf{z} = \lambda\mathbf{x} + (1 - \lambda)\mathbf{y}$ where $\lambda \in (0, 1)$. Then $\mathbf{z} + \mathbf{u} = \lambda(\mathbf{x} + \mathbf{u}/\lambda) + (1 - \lambda)\mathbf{y}$. Now if $\|\mathbf{u}\|$ is sufficiently small, $(\mathbf{x} + \mathbf{u}/\lambda) \in U$ and consequently $(\mathbf{z} + \mathbf{u}) \in U$; that is, $\mathbf{z} \in U^0$. It follows that $U^0$ is convex. Next let $\mathbf{x}, \mathbf{y} \in \bar{U}$. Then there exist sequences $\{\mathbf{x}_j\}$, $\{\mathbf{y}_j\}$ in $U$ converging to $\mathbf{x}, \mathbf{y}$, respectively. For any $\lambda \in (0, 1)$, $\lambda\mathbf{x}_j + (1 - \lambda)\mathbf{y}_j$ converges to $\lambda\mathbf{x} + (1 - \lambda)\mathbf{y}$ so $\bar{U}$ is also convex. ●

**Theorem G.** If $U$ is an open set in $\mathbf{L}$, then its convex hull $H(U)$ is open. If $U$ is compact and $\mathbf{L}$ is finite dimensional, then $H(U)$ is compact.

**Proof.** Let $U$ be open and $\mathbf{x} = \sum_1^n \lambda_i \mathbf{x}_i$, $\mathbf{x}_i \in U$, be an arbitrary element of $H(U)$. Then $\mathbf{x} + \mathbf{u} = \sum_1^n \lambda_i(\mathbf{x}_i + \mathbf{u})$, and since $U$ is open, it follows that $(\mathbf{x}_i + \mathbf{u}) \in U$ for $\|\mathbf{u}\|$ sufficiently small. We conclude that $(\mathbf{x} + \mathbf{u}) \in H(U)$ for $\|\mathbf{u}\|$ small, which shows that $H(U)$ is open.

Next suppose $U$ is compact, that is, closed and bounded. In particular, let $M$ be a bound for $\|\mathbf{x}\|$ for $\mathbf{x} \in U$. If $\mathbf{y} \in H(U)$, then by Theorem E, $\mathbf{y} = \sum_0^m \lambda_i \mathbf{x}_i$ where $m = \dim \mathbf{L}$. Then $\|\mathbf{y}\| \leqslant \sum_0^m \lambda_i \|\mathbf{x}\| \leqslant M$, so $H(U)$ is also bounded by $M$. To show that $H(U)$ is closed, let $\mathbf{x}$ be any limit point. There is a sequence $\{\mathbf{x}_j\}$ of points of $H(U)$ converging to $\mathbf{x}$, and again we appeal to Theorem E, this time to write $\mathbf{x}_j = \sum_0^m \lambda_{ij} \mathbf{x}_{ij}$ where $\mathbf{x}_{ij} \in U$. Now for each $i$, the sequences $\{\lambda_{ij}\}$ and $\{\mathbf{x}_{ij}\}$ are bounded sequences and therefore contain convergent subsequences. In fact since there are only finitely many $i$'s, we may select a subsequence $\{j_k\}$ of positive integers such that $\{\lambda_{ij_k}\}$ and $\{\mathbf{x}_{ij_k}\}$ converge for each $i$, say to $\lambda_i$ and $\mathbf{z}_i$, respectively. Thus

$$\mathbf{x} = \lim_{k \to \infty} \mathbf{x}_{j_k} = \lim_{k \to \infty} \sum_{i=0}^m \lambda_{ij_k} \mathbf{x}_{ij_k} = \sum_0^m \lambda_i \mathbf{z}_i$$

Now $\lambda_{ij_k} \geqslant 0$ and $\sum_{i=0}^m \lambda_{ij_k} = 1$ from which we conclude by taking limits as $k \to \infty$ that $\lambda_i \geqslant 0$ and $\sum_0^m \lambda_i = 1$. Finally since $U$ is closed, $\mathbf{z}_i \in U$ and $\mathbf{x} \in H(U)$ as desired. ●

We conclude this section with a few remarks about the relative topology. By the **relative interior** of a convex set $U$ we mean the interior of the set when viewed as a subset of its affine hull $A(U)$. It consists therefore of those points $\mathbf{x} \in A(U)$ for which there exists $\varepsilon > 0$ such that $\mathbf{y} \in U$ whenever $\mathbf{y} \in A(U)$ and $\|\mathbf{y} - \mathbf{x}\| < \varepsilon$. For example, the relative interior of the disk $\{(r, s, 1) \in \mathbf{R}^3: r^2 + s^2 \leqslant 1\}$ is the set $\{(r, s, 1) \in \mathbf{R}^3: r^2 + s^2 < 1\}$ while the interior in this case

# 31. Convex Sets and Affine Sets

is empty. A convex set is **relatively open** if it is equal to its relative interior. Many of our theorems in this and the next chapter deal with open sets. Usually they are valid if open is replaced by relatively open. One simply looks at $U$ (more precisely, a translate of $U$) in the subspace corresponding to the affine hull $A(U)$. One important fact, often needed, is that the relative interior of a nonempty finite-dimensional convex set is nonempty (Problem I).

## PROBLEMS AND REMARKS

A. Show that each of the following are convex.

  (1) Any interval in $\mathbf{R}$.
  (2) $\{(r, s) \in \mathbf{R}^2 : r^2 + s^2 < 4\}$.
  (3) $\{(r, s) \in \mathbf{R}^2 : |r| + |s| \leqslant 4\}$.
  (4) $\{\mathbf{x} \in \mathbf{L} : \|\mathbf{x}\| \leqslant \alpha\}$.
  (5) $\{(r, s) \in \mathbf{R}^2 : s \leqslant \log r, r > 0\}$.
  (6) $\{(r, s) \in \mathbf{R}^2 : s \geqslant f(r), f \text{ convex}\}$.
  (7) $\{r \in I : f(r) \leqslant \alpha, f : I \to \mathbf{R} \text{ convex}\}$.
  (8) The set of convex functions $f : I \to \mathbf{R}$.
  (9) The set of $n \times n$ matrices with nonnegative entries.

B. Let $M$ be a subset of a linear space $\mathbf{L}$.

  (1) If $M$ is affine, the subspace of Theorem A is unique.
  (2) $M$ is a maximal proper affine subset of $\mathbf{L}$ if and only if it is the translate of a maximal proper subspace.
  (3) If $M = A(\{\mathbf{x}_0, \ldots, \mathbf{x}_n\})$, the affine hull of a finite set, then it is the translate of the subspace $A(\{\mathbf{O}, \mathbf{x}_1 - \mathbf{x}_0, \ldots, \mathbf{x}_n - \mathbf{x}_0\})$. In this case $M$ has dimension $n$ if and only if $\{\mathbf{x}_1 - \mathbf{x}_0, \ldots, \mathbf{x}_n - \mathbf{x}_0\}$ is a linearly independent set.
  (4) If $\mathbf{L} = \mathbf{R}^n$ and $M$ is affine, then there is an $m \times n$ matrix $B$ and a vector $\mathbf{b} \in \mathbf{R}^n$ such that $M = \{\mathbf{x} \in \mathbf{R}^n : B\mathbf{x} = \mathbf{b}\}$. Moreover, any set of this form is affine.
  (5) If $\mathbf{L} = \mathbf{R}^n$ and $M$ is affine, then $M$ is closed.

C. For any sets $U$ and $V$ in $\mathbf{L}$, let $U + V = \{\mathbf{x} + \mathbf{y} : \mathbf{x} \in U, \mathbf{y} \in V\}$, and let $\alpha U = \{\alpha \mathbf{x} : \mathbf{x} \in U\}$. Show that if $U$ and $V$ are convex, then

  (1) $U + V$ is convex,
  (2) $\alpha U$ is convex for all $\alpha \in \mathbf{R}$,
  (3) $U = \lambda U + (1 - \lambda)U$ for $\lambda \in [0, 1]$,
  (4) $(\alpha_1 + \alpha_2)U = \alpha_1 U + \alpha_2 U$ for $\alpha_1 \geqslant 0, \alpha_2 \geqslant 0$.

D. Let $U_1, U_2, \ldots$ be convex.

  (1) If $U_j \subseteq U_{j+1}$, $j = 1, 2, \ldots$, then $\bigcup_{j=1}^{\infty} U_j$ is convex.

  (2) $\liminf_{j \to \infty} U_j = \bigcup_{k=1}^{\infty} \bigcap_{j=k}^{\infty} U_j$ is convex.

E. If $\{\mathbf{x}_0, \mathbf{x}_1, \mathbf{x}_2, \mathbf{y}, \mathbf{z}\}$ are, respectively, the three vertices, midpoint of the side opposite $x_0$, and intersection of the medians of a triangle, their barycentric coordinates are $(1, 0, 0)$, $(0, 1, 0)$, $(0, 0, 1)$, $(0, \frac{1}{2}, \frac{1}{2})$, $(\frac{1}{3}, \frac{1}{3}, \frac{1}{3})$.

**F.** The dimension of an $m$-simplex is $m$. The dimension of a convex set $U \subseteq \mathbf{R}^n$ is the maximum of the dimensions of the simplices contained in $U$.

**G.** The convex hull of a closed set need not be closed even in $\mathbf{R}^2$. Consider for example $U = \{(r, s): r^2 s^2 = 1, s \in (0, \infty)\}$. What about the convex hull of a closed set in $\mathbf{R}$?

**H.** For $\mathbf{x}, \mathbf{y} \in \mathbf{L}$, let seg$[\mathbf{x}, \mathbf{y}]$ denote the segment

$$\{\lambda \mathbf{x} + (1 - \lambda)\mathbf{y}: \lambda \in [0, 1]\}$$

and let seg$(\mathbf{x}, \mathbf{y})$ be the segment where $\lambda \in (0, 1)$. For any set $U \subseteq \mathbf{L}$, define four sets as follows.

$$\ker U = \{\mathbf{z} \in \mathbf{L}: \text{seg}[\mathbf{x}, \mathbf{z}] \subseteq U \text{ for each } \mathbf{x} \in U\}$$
$$\text{lina } U = \{\mathbf{z} \in \mathbf{L}: \text{ there is an } \mathbf{x} \in U, \mathbf{x} \neq \mathbf{z}, \text{ such that } \text{seg}(\mathbf{x}, \mathbf{z}) \subseteq U\}$$
$$\text{lin } U = U \cup \text{lina } U$$
$$\text{core } U = \{\mathbf{x} \in U: \text{ for each } \mathbf{y} \in \mathbf{L}, \mathbf{y} \neq \mathbf{x}, \text{ there is a } \mathbf{z} \in \text{seg}(\mathbf{x}, \mathbf{y}) \text{ such that seg}[\mathbf{x}, \mathbf{z}] \subseteq U\}$$

(1) Let $A, B, C$ be subsets of the plane represented in terms of polar coordinates $(\rho, \theta)$ by

$A =$ the closed set bounded by the cardioid $\rho = 1 - \cos \theta$
$B = $ seg$[(0, 0), (1, 0)]$
$C = $ the single point $(\pi, 3)$.

Let $U = A \cup B \cup C$ and describe each of the four sets defined above.

(2) Show that ker $U$ is always convex, and that if $U$ is convex, then lin $U$ and core $U$ are also convex.

(3) If $U$ is convex with a nonempty interior, then lina $U = $ lin $U = \bar{U}$ and core $U = U^0$.

These ideas, which make sense in any linear topological space (normed or not) are discussed by Valentine [1964, pp. 5–13].

**I.** The relative interior of a nonempty finite-dimensional convex set is nonempty [Eggleston, 1958, p. 16].

**J.** For $f: I \to \mathbf{R}$, let **epigraph** ($f$) be the set in $\mathbf{R}^2$ described by

$$\text{epi}(f) = \{(x, y): x \in I, y \geq f(x)\}$$

(1) $f$ is convex $\Leftrightarrow$ epi($f$) is a convex set in $\mathbf{R}^2$
(2) epi$(f \vee g) = $ epi$(f) \cap $ epi$(g)$
(3) $f_\alpha : I \to \mathbf{R}$ convex and pointwise bounded above $\Rightarrow$ epi(sup$_\alpha f_\alpha$) convex $\Rightarrow$ sup$_\alpha f_\alpha$ convex (cf. Theorem 13D).

**K.** [Danzer et al., 1963]. An alternate and elegant proof of the second half of Theorem G may be given. Let $\Lambda = \{\lambda = (\lambda_0, \ldots, \lambda_n) \in \mathbf{R}^{n+1}: \lambda_i \geq 0, \sum_0^n \lambda_i = 1\}$, $X = U \times \cdots \times U$ ($n + 1$ factors), and $\mathbf{x} = (\mathbf{x}_0, \ldots, \mathbf{x}_n)$, $n$ being the dimension of $\mathbf{L}$. For each $(\lambda, \mathbf{x}) \in \Lambda \times X$, let $f(\lambda, \mathbf{x}) = \sum_0^n \lambda_i \mathbf{x}_i$. Since $f$ is continuous and $\Lambda \times X$ is compact, the set $f(\Lambda \times X)$ is compact. But by Carathéodory's theorem, $f(\Lambda \times X) = H(U)$.

## 32. Hyperplanes and Extreme Points

The concept of a hyperplane in **L** results from a straightforward generalization of the notion of a line in $\mathbf{R}^2$ or a plane in $\mathbf{R}^3$. In terms of the geometric ideas of the previous section, we may define a **hyperplane** $H$ as a maximal proper affine subset of **L**, or equivalently (Problem 31B) as the translate of a maximal proper subspace of **L**. Just as a line in $\mathbf{R}^2$ is usefully described by an equation $ar + bs = c$, a plane in $\mathbf{R}^3$ by $ar + bs + ct = d$, so a hyperplane can be described by an equation. In Theorem 22D we saw that each maximal proper subspace of **L** is the null space **N** of a nonidentically zero linear functional $f: \mathbf{L} \to \mathbf{R}$. Since $H$ is a translate of **N**, every $\mathbf{z} \in H$ may be written in the form $\mathbf{z} = \mathbf{z}_0 + \mathbf{x}$ where $\mathbf{x} \in \mathbf{N}$, and since $f$ is zero on $\mathbf{N}, f(\mathbf{z}) = f(\mathbf{z}_0) + f(\mathbf{x}) = f(\mathbf{z}_0) = \alpha$. We see that every hyperplane can be described as a set in **L** of the form

$$H = \{\mathbf{z}: \ f(\mathbf{z}) = \alpha, \ f \text{ not identically zero}\} \tag{1}$$

The converse is also true. If $H \subseteq \mathbf{L}$ is a set described by (1) for some nonzero linear functional $f$ and real number $\alpha$, we may choose any $\mathbf{z}_0 \in H$ and form the set $\mathbf{N} = \{\mathbf{x}: \ \mathbf{x} = \mathbf{z} - \mathbf{z}_0, \mathbf{z} \in H\}$. It is easily seen that **N** is the null space of $f$, hence (Theorem 22D) that **N** is a maximal proper subspace. Since $H$ is the set of all vectors $\mathbf{z} = \mathbf{x} + \mathbf{z}_0$ where $\mathbf{x} \in \mathbf{N}$, it is a translate of **N**. Hence it is a hyperplane.

With the notion of hyperplane firmly established, we can introduce a number of related ideas. If $H$ is a hyperplane described by (1), we call the sets $\{\mathbf{z} \in \mathbf{L}: \ f(\mathbf{z}) \leq \alpha\}$ and $\{\mathbf{z} \in \mathbf{L}: \ f(\mathbf{z}) \geq \alpha\}$ **half-spaces** determined by $H$. Then $H$ **separates** $U$ and $V$ if $U$ and $V$ lie in opposite half-spaces and $H$ **strongly separates** $U$ and $V$ if $H$ lies strictly between two translates of $H$ that separate $U$ and $V$. Note that strong separation requires that $U$ and $V$ be disjoint while mere separation does not, since, for example, the definition tells us that a line tangent to a circle in $\mathbf{R}^2$ separates the closed disk from the point of tangency. On the other hand, disjointness even for closed convex sets does not guarantee strong separation as may be seen by considering

$$U = \{(r, s) \in \mathbf{R}^2: \ r > 0, s \geq 1/r\}, \qquad V = \{(r, s) \in \mathbf{R}^2: \ r > 0, s \leq -1/r\}$$

We do get strong separation if one of the sets is compact, at least in $\mathbf{R}^n$.

> **Theorem A.** Let $U$ and $V$ be disjoint nonempty convex subsets of $\mathbf{R}^n$ with $U$ compact and $V$ closed. Then there is a hyperplane which strongly separates $U$ and $V$.

**Proof.** First let $d(U, V) = \inf_{x \in U, y \in V} \| x - y \|$. By a standard compactness argument, it follows that there are points $\mathbf{x}_0 \in U$, $\mathbf{y}_0 \in V$ such that $d(U, V) = \| \mathbf{x}_0 - \mathbf{y}_0 \| > 0$ (Problem B). We are going to show that a hyperplane $H$ through any point $\mathbf{z}_0 \in \text{seg}(\mathbf{x}_0, \mathbf{y}_0)$ orthogonal to this segment strongly separates $U$ and $V$ (Fig. 32.1).

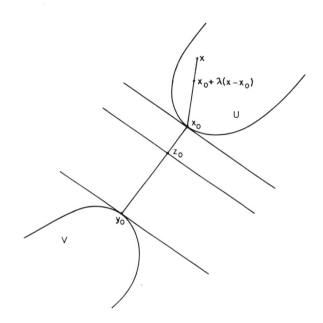

**Fig. 32.1**

Recalling that a linear functional on $\mathbf{R}^n$ may be written as an inner product [see Eq. (5), Section 22], we see that the hyperplane $H'$ through $\mathbf{x}_0$ orthogonal to $\text{seg}[\mathbf{x}_0, \mathbf{y}_0]$ has the equation $\langle \mathbf{y}_0 - \mathbf{x}_0, \mathbf{z} - \mathbf{x}_0 \rangle = 0$. For $\mathbf{x} \in U$, let

$$\phi(\lambda) = \| \mathbf{y}_0 - [\mathbf{x}_0 + \lambda(\mathbf{x} - \mathbf{x}_0)] \|^2$$
$$= \langle \mathbf{y}_0 - \mathbf{x}_0, \mathbf{y}_0 - \mathbf{x}_0 \rangle - 2\lambda \langle \mathbf{y}_0 - \mathbf{x}_0, \mathbf{x} - \mathbf{x}_0 \rangle + \lambda^2 \langle \mathbf{x} - \mathbf{x}_0, \mathbf{x} - \mathbf{x}_0 \rangle$$

Note that $\phi$ is differentiable and $\phi(\lambda) \geqslant \phi(0)$ for $\lambda \in [0, 1]$, the latter because $\mathbf{x}_0$ is a point of $U$ closest to $\mathbf{y}_0$. Thus, $\phi_+{'}(0) \geqslant 0$; that is, $\langle \mathbf{y}_0 - \mathbf{x}_0, \mathbf{x} - \mathbf{x}_0 \rangle \leqslant 0$ for all $\mathbf{x} \in U$. On the other hand, by an entirely similar argument applied to the plane $H''$ through $\mathbf{y}_0$ orthogonal to

## 32. Hyperplanes and Extreme Points

seg$[\mathbf{x}_0, \mathbf{y}_0]$, we find $\langle \mathbf{x}_0 - \mathbf{y}_0, \mathbf{y} - \mathbf{y}_0 \rangle \leq 0$ for all $\mathbf{y} \in V$. Thus, if $\mathbf{y} \in V$,

$$\langle \mathbf{y}_0 - \mathbf{x}_0, \mathbf{y} - \mathbf{x}_0 \rangle = \langle \mathbf{y}_0 - \mathbf{x}_0, \mathbf{y}_0 - \mathbf{x}_0 \rangle + \langle \mathbf{y}_0 - \mathbf{x}_0, \mathbf{y} - \mathbf{y}_0 \rangle \geq 0$$

Combined with the already noted fact that $\mathbf{x} \in U$ implies $\langle \mathbf{y}_0 - \mathbf{x}_0, \mathbf{x} - \mathbf{x}_0 \rangle \leq 0$, we have proved that $H'$ separates $U$ and $V$. A like result holds for $H''$, so $H$ strongly separates $U$ and $V$. ●

We now use this result on strong separation to prove the basic separation theorem for convex sets in $\mathbf{R}^n$.

**Theorem B.** Let $U$ and $V$ be convex sets in $\mathbf{R}^n$ with $U^0 \neq \varnothing$, $U^0 \cap V = \varnothing$. Then there is a hyperplane that separates $U$ and $V$.

**Proof.** The truth of the theorem for $\bar{U}$ and $\bar{V}$ would imply its truth for $U$ and $V$, so there is no loss of generality in supposing $U$ and $V$ to be closed. For $\mathbf{x}_0 \in U^0$, define

$$B_n = \{\mathbf{x} \in \mathbf{R}^n : \|\mathbf{x} - \mathbf{x}_0\| \leq n\}, \qquad D_n = \{\mathbf{x}_0 + (1 - 1/n)(\mathbf{x} - \mathbf{x}_0): \mathbf{x} \in U\}$$

Then set $U_n = B_n \cap D_n$ (Fig. 32.2). It is easy to check that $D_n$ is convex

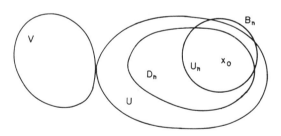

Fig. 32.2

and closed; thus $U_n$ is convex and compact. Moreover $U_n \cap V = \varnothing$, meaning that Theorem A may be invoked to guarantee a hyperplane $H_n$ with equation $\langle \mathbf{u}_n, \mathbf{x} \rangle = \alpha_n$ that strongly separates $U_n$ and $V$. Thus, for $\mathbf{x} \in U_n$, $\langle \mathbf{u}_n, \mathbf{x} \rangle \leq \alpha_n$, while for $\mathbf{y} \in V$, $\langle \mathbf{u}_n, \mathbf{y} \rangle \geq \alpha_n$. We may assume that in writing these relations, the $\mathbf{u}_n$'s have been normalized so that $\|\mathbf{u}_n\| = 1$. Now $\{\mathbf{u}_n\}$ and $\{\alpha_n\}$ are both bounded sequences, the latter because $\langle \mathbf{u}_n, \mathbf{x}_0 \rangle \leq \alpha_n \leq \langle \mathbf{u}_n, \mathbf{y}_0 \rangle$ where $\mathbf{x}_0$ and $\mathbf{y}_0$ are fixed points of $U$ and $V$, respectively. Thus we may select subsequences

$\{\mathbf{u}_{n_k}\}$ and $\{\alpha_{n_k}\}$ which converge, say to $\mathbf{u}$ and $\alpha$. But for any $\mathbf{x} \in U^0$, $\mathbf{x}$ will be in $U_{n_k}$ for $k$ sufficiently large. For these $k$, $\langle \mathbf{u}_{n_k}, \mathbf{x} \rangle \leqslant \alpha_{n_k}$ and consequently $\langle \mathbf{u}, \mathbf{x} \rangle \leqslant \alpha$ for all $\mathbf{x} \in U^0$. This in turn implies the same inequality for all $\mathbf{x} \in U$. On the other hand, $\alpha_{n_k} \leqslant \langle \mathbf{u}_{n_k}, \mathbf{y} \rangle$ for all $\mathbf{y} \in V$ which upon letting $k \to \infty$ gives $\alpha \leqslant \langle \mathbf{u}, \mathbf{y} \rangle$ for all $\mathbf{y} \in V$. The desired hyperplane is $H = \{\mathbf{z} \in \mathbf{L}: \langle \mathbf{u}, \mathbf{z} \rangle = \alpha\}$. •

Next we introduce the notion of a supporting hyperplane for a convex set. We say that the hyperplane $H$ **supports** $U$ at $\mathbf{x}_0 \in U$ if $\mathbf{x}_0 \in H$ and $U$ is a subset of one of the half-spaces determined by $H$. If we apply our last theorem to the situation where $V = \{\mathbf{x}_0\}$ is a boundary point of $U$, we obtain another fundamental result.

**Theorem C.** If $\mathbf{x}_0 \in U$ is a boundary point of a convex set $U \subseteq \mathbf{R}^n$ and $U^0 \neq \varnothing$, then there is a supporting hyperplane for $U$ at $\mathbf{x}_0$.

A point $\mathbf{x}_0$ of a convex set $U$ is called an **extreme point** if $\mathbf{x}_0$ is not an interior point of any line segment in $U$, that is, if there do not exist points $\mathbf{x}_1, \mathbf{x}_2 \in U$ and $\lambda \in (0, 1)$ such that $\mathbf{x}_0 = \lambda \mathbf{x}_1 + (1 - \lambda)\mathbf{x}_2$. The extreme points of a closed ball and a closed cube in $\mathbf{R}^3$ are its boundary points and the eight vertices, respectively. A half-space has no extreme points even if it is closed.

**Theorem D.** Let $U \subseteq \mathbf{R}^n$ be convex and compact. Then $U$ is the convex hull of its extreme points.

**Proof.** We use induction on the dimension $m$ of the set $U$. The cases $m = 0$ and $m = 1$ where $U$ is a point or a closed line segment are trivial. We suppose the result true for any compact set of dimension at most $m$ where $m \leqslant n - 1$. Let $U$ have dimension $m + 1$ and consider it to be embedded in $\mathbf{R}^{m+1}$. If $\mathbf{x}_0$ is a boundary point of $U$, then by Theorem C there is a supporting hyperplane $H$ (an $m$-dimensional affine set) for $U$ through $\mathbf{x}_0$. The set $U \cap H$ is compact and convex, and its dimension does not exceed $m$. By the induction hypothesis $\mathbf{x}_0$ is a convex combination of the extreme points of $U \cap H$ and hence also of $U$ (Problem J). We now turn to the case where $\mathbf{x}_0$ is an interior point of $U$, in which case any line through $\mathbf{x}_0$ intersects $U$ in a segment with endpoints $\mathbf{x}_1$ and $\mathbf{x}_2$ which are boundary points of $U$. Since $\mathbf{x}_1$ and $\mathbf{x}_2$ are convex combinations of extreme points, so is $\mathbf{x}_0$. •

The first theorem proved in this section differs from the others in that it specifically identifies a separating hyperplane in certain situations.

## 32. Hyperplanes and Extreme Points

The other three theorems assert what can be done relating to separation, support, and extreme points. These three topics are central to the study of convex sets. We have proved them in $\mathbf{R}^n$, and in light of Theorem 21F we may take them as proved for any finite-dimensional normed linear space. They can be used, along with notions related to the graph of a convex function, to develop most of the results of this text from a geometric point of view. Although this is not the program that we shall carry out, we do include here the infinite-dimensional generalizations of these theorems in a form that invites comparison with the results as we shall develop them for convex functions. For details and other generalizations we refer the reader to the books by Valentine [1964] and Kelly and Namioka [1963], and the survey article by Klee [1969a].

**Theorem E (Separation Theorem).** Let $U$ and $V$ be convex sets in a normed linear space $\mathbf{L}$ with $U^0$ nonempty. If $U^0 \cap V \neq \varnothing$, then there is a (closed) hyperplane that separates $U$ and $V$.

**Theorem F (Support Theorem).** Let $U$ be a convex set in a normed linear space $\mathbf{L}$ with $U^0 \neq \varnothing$. If $\mathbf{x}_0 \in U$ is a boundary point of $U$, then there is a (closed) supporting hyperplane for $U$ at $\mathbf{x}_0$.

**Theorem G (Krein–Millman Theorem).** Let $U$ be a compact convex set in a normed linear space $\mathbf{L}$. Then $U$ is the closure of the convex hull of its extreme points.

### PROBLEMS AND REMARKS

A. Let $H$ be a hyperplane in a normed linear space $\mathbf{L}$.
(1) If $\mathbf{L} = \mathbf{R}^n$, then $H$ is closed.
(2) If $\mathbf{L}$ is infinite dimensional, $H$ need not be closed.
(3) The closure of $H$ is always affine.
(4) $H$ is either closed or dense in $\mathbf{L}$.

B. If $U$ is closed and $V$ is compact in $\mathbf{R}^n$, then $d(U, V) = \|\mathbf{x}_0 - \mathbf{y}_0\|$ for some $\mathbf{x}_0 \in U$, $\mathbf{y}_0 \in V$.

C. $H$ separates $U$ and $V$ **properly** if it separates them and $U$ and $V$ are not both contained in $H$. If in addition one of the sets $U \cap H$ or $V \cap H$ is empty, it separates them **nicely**, and if both $U \cap H$ and $V \cap H$ are empty, it separates them **strictly**. Clearly, strong separation ⇒ strict separation ⇒ nice separation ⇒ proper separation ⇒ separation. Give examples in $\mathbf{R}^2$ showing that none of these implications can be reversed.

**D.** Give an alternate proof of Theorem C based on the following argument.

(1) Assume $\mathbf{x}_0 = \mathbf{O}$ and let $S = \{\mathbf{x} \in \mathbf{R}^n : \|\mathbf{x}\| = 1\}$. There exists a point $\mathbf{y}_0 \in S$ farthest from $U$.
(2) $\mathbf{x}_0$ is the point of $U$ closest to $\mathbf{y}_0$.
(3) The hyperplane $\{\mathbf{z} \in \mathbf{R}^n : \langle \mathbf{y}_0 - \mathbf{x}_0, \mathbf{z} - \mathbf{x}_0 \rangle\}$ supports $U$ at $\mathbf{x}_0$.

**\*E.** A supporting hyperplane for the convex set $U \subseteq \mathbf{R}^n$ at the boundary point $\mathbf{x}_0$ can contain no interior points of $U$. What if $\mathbf{R}^n$ is replaced with $\mathbf{L}$?

**F.** If $U$ and $V$ are convex, $U \cap V = \emptyset$, $U$ is closed, and $\mathbf{R}^n = U \cup V$, then $U$ is a half-space.

**G.** If the closed set $U \subseteq \mathbf{R}^n$ has a nonempty interior, and if through each point of its boundary there passes a supporting hyperplane, then $U$ is convex.

**H.** A closed convex set $U \subseteq \mathbf{R}^n$ is the intersection of all the half-spaces that contain $U$.

**I.** The point $\mathbf{x}_0 \in U$ is an extreme point of the convex set $U$ if and only if $U \setminus \{\mathbf{x}_0\}$ is convex.

**J.** Let $U$ be a closed convex subset of $\mathbf{R}^n$ and suppose $H$ is a supporting hyperplane at $\mathbf{x}_0$. If $\mathbf{x}$ is an extreme point of $H \cap U$, then $\mathbf{x}$ is an extreme point of $U$.

**K.** Let $\mathbf{x}_0 \in U$ be a boundary point of the convex set $U$. We call $\mathbf{x}_0$ an **exposed point** of $U$ if there is a supporting hyperplane $H$ for $U$ at $\mathbf{x}_0$ such that $H \cap U = \{\mathbf{x}_0\}$. Show that every exposed point is an extreme point, but not conversely even in $\mathbf{R}^2$.

**L.** If $n \geq 3$, the set of extreme points of a compact convex set in $\mathbf{R}^n$ need not be closed. What if $n = 2$?

**M.** Let $W$ be a convex subset of the convex set $U$. We call $W$ an **extreme subset** of $U$ if for $\mathbf{x}, \mathbf{y} \in U$ and $\lambda \in (0, 1)$, $(\lambda \mathbf{x} + (1 - \lambda)\mathbf{y}) \in W$ implies $\mathbf{x}, \mathbf{y} \in W$.

(1) Find the extreme subsets and the extreme points of $\{(r, s) \in \mathbf{R}^2 : |r| + |s| \leq 1\}$ and $\{(r, s) \in \mathbf{R}^2 : r^2 + s^2 \leq 1\}$.
(2) If $\{W_\alpha\}$ is a family of extreme subsets of $U$, then $\cap W_\alpha$ is an extreme subset of $U$.
(3) If $V$ is an extreme subset of $W$ and $W$ an extreme subset of $U$, then $V$ is an extreme subset of $U$.

**N.** Show that the compactness condition is needed in the Krein–Millman theorem by finding

(1) a closed convex set in $\mathbf{R}^2$ with no extreme points,
(2) a closed bounded convex set in

$$C_0(\mathbf{R}) = \{f : \mathbf{R} \to \mathbf{R} : f \text{ continuous}, f(x) \to 0 \text{ as } x \to \pm \infty\},$$

with $\|f\| = \max |f(x)|$, which has no extreme points.

**O.** Let $\Omega_2$ be the set of $2 \times 2$ **doubly stochastic matrices** $A = (a_{ij})$, that is, the set of matrices satisfying $a_{ij} \geq 0$ and $\sum_{i=1}^{2} a_{ij} = \sum_{j=1}^{2} a_{ij} = 1$.

(1) The set $\Omega_2$ is a convex subset of the linear space of all $2 \times 2$ matrices.

## 32. Hyperplanes and Extreme Points

(2) The matrices
$$I = \begin{bmatrix} 1 & 0 \\ 0 & 1 \end{bmatrix} \quad \text{and} \quad P = \begin{bmatrix} 0 & 1 \\ 1 & 0 \end{bmatrix}$$
are extreme points of $\Omega_2$.

(3) Any member of $\Omega_2$ can be written as a convex combination of the matrices $I$ and $P$. (Prove this as an application of Theorem D by showing that $I$ and $P$ are the only extreme points of $\Omega_2$).

Because of applications such as this of Theorem D, and because of other applications to appear in Chapter V, it is often useful to identify the extreme points of a given convex set. This is generally not easy, and there is extensive literature in mathematics addressed to this problem for specific convex sets. See, for example, Köthe [1969, pp. 333–337].

# IV

## *Convex Functions on a Normed Linear Space*

> The interplay between generality and individuality, deduction and construction, logic and imagination—this is the profound essence of live mathematics. Any one or another of these aspects of mathematics can be at the center of a given achievement. In a far reaching development all of them will be involved. Generally speaking, such a development will start from the "concrete" ground, then discard ballast by abstraction and rise to the lofty layers of thin air where navigation and observation are easy: after this flight comes the crucial test of landing and reaching specific goals in the newly surveyed low plains of individual "reality." In brief, the flight into abstract generality must start from and return again to the concrete and specific.
>
> RICHARD COURANT

## 40. Introduction

The definition of a convex function has a very natural generalization to real-valued functions defined on an arbitrary normed linear space **L**. We merely require that the domain $U$ of $f$ be convex. This assures us that for $\mathbf{x}_1, \mathbf{x}_2 \in U$, $\alpha \in (0, 1)$, $f$ will always be defined at $\alpha \mathbf{x}_1 + (1-\alpha)\mathbf{x}_2$. We then define $f$ to be **convex** on $U \subseteq \mathbf{L}$ if

$$f[\alpha \mathbf{x}_1 + (1-\alpha)\mathbf{x}_2] \leqslant \alpha f(\mathbf{x}_1) + (1-\alpha) f(\mathbf{x}_2)$$

Unless specifically stated to the contrary, we assume convex functions to be finite valued and defined on convex sets.

We note immediately that for three points $\mathbf{x}_1, \mathbf{x}_2, \mathbf{x}_3 \in U$ and three positive numbers $\alpha_1, \alpha_2, \alpha_3$ such that $\alpha_1 + \alpha_2 + \alpha_3 = 1$, a convex function satisfies

$$\begin{aligned} f(\alpha_1 \mathbf{x}_1 + \alpha_2 \mathbf{x}_2 + \alpha_3 \mathbf{x}_3) &= f\left[\alpha_1 \mathbf{x}_1 + (\alpha_2 + \alpha_3)\left(\frac{\alpha_2}{\alpha_2 + \alpha_3}\mathbf{x}_2 + \frac{\alpha_3}{\alpha_2 + \alpha_3}\mathbf{x}_3\right)\right] \\ &\leqslant \alpha_1 f(\mathbf{x}_1) + (\alpha_2 + \alpha_3) f\left(\frac{\alpha_2}{\alpha_2 + \alpha_3}\mathbf{x}_2 + \frac{\alpha_3}{\alpha_2 + \alpha_3}\mathbf{x}_3\right) \\ &\leqslant \alpha_1 f(\mathbf{x}_1) + \alpha_2 f(\mathbf{x}_2) + \alpha_3 f(\mathbf{x}_3) \end{aligned}$$

Following the same pattern, one easily establishes inductively that for $n$ points in $U$ and $n$ positive $\alpha_i$ with $\sum_1^n \alpha_i = 1$, a convex function satisfies

$$f\left[\sum_1^n \alpha_i \mathbf{x}_i\right] \leqslant \sum_1^n \alpha_i f(\mathbf{x}_i) \tag{1}$$

This relation, called **Jensen's inequality**, is sometimes taken as the definition of a convex function. The reader will note that this inequality, and indeed much of what we shall say in this chapter, makes no use of the norm and could be stated for more general spaces. But for reasons already cited in Section 20, we shall work in the context of normed linear spaces.

One of the first things we learned about functions of a real variable convex on an open interval was that they are continuous. This is not generally true on an infinite-dimensional space **L**, but a convex function defined on an open set $U \subseteq \mathbf{R}^n$ is continuous. In Section 41 we prove this fact and explore related ideas.

Convex functions that are differentiable can be characterized by properties of their derivatives. In Section 42 we explore this relationship. In choosing not to address ourselves in this section to questions of existence of the derivative, we have had in mind the development of a ready reference for those primarily interested in applications involving convex functions known to be differentiable.

This completed, we then turn to the question of what can be said about the existence of the derivative of a convex function. We have already seen that for functions of a real variable, convex functions are differentiable except at possibly a countable number of points. The answer to this question for functions defined on a normed linear space **L** demands some deep results from functional analysis, and it is for this reason that the material leading to an answer is grouped together in the final two sections. As already indicated, these sections can be omitted by readers more concerned with getting on to the applications of Chapters V and VI.

Many of the results of this chapter concern the behavior of $f$ at a particular point $\mathbf{x}_0$. Although we state our theorems in terms of an arbitrary point $\mathbf{x}_0 \in U$, we usually simplify the notation in the proof by assuming that $\mathbf{x}_0 = \mathbf{O}$ and that $f(\mathbf{O}) = 0$. We do this without loss of generality since the study of $f$ on a convex set $U$ containing $\mathbf{x}_0$ is equivalent

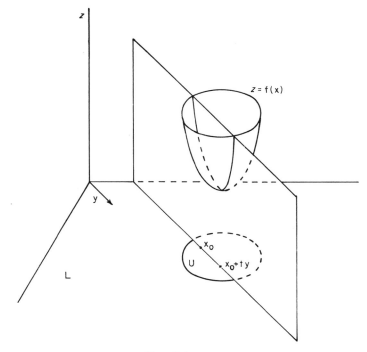

**Fig. 40.1**

## 41. Continuity of Convex Functions

to the study of $g$ on the set $V = \{\mathbf{x}: (\mathbf{x} + \mathbf{x}_0) \in U\}$ if we define $g(\mathbf{x}) = f(\mathbf{x} + \mathbf{x}_0) - f(\mathbf{x}_0)$. It is easily verified that $g$ is convex on $V$, and it is clear that $f$ is locally bounded (continuous, differentiable, etc.) at $\mathbf{x}_0$ if and only if $g$ is locally bounded (continuous, differentiable, etc.) at $\mathbf{O}$. Since $g(\mathbf{O}) = 0$, the simplifying assumptions we wish to make about $f$ really are satisfied by $g$.

Finally we note that if $f$ is convex on an open set $U \subseteq \mathbf{L}$, and if $\mathbf{x}_0 \in U$, we can often make use of what we know about convex functions of a real variable by noting that for arbitrary $\mathbf{y} \in \mathbf{L}$, $g(t) = f(\mathbf{x}_0 + t\mathbf{y})$ is convex for $t$ in some interval $(a, b)$ containing the origin (Fig. 40.1).

## 41. Continuity of Convex Functions

A real-valued linear function defined on $\mathbf{L}$ is convex on $\mathbf{L}$. Having already seen (Problem 22B) that such a function, defined on an infinite-dimensional space, need not be continuous, it follows that a convex function defined on $\mathbf{L}$ need not be continuous. There are two directions to go. One can ask what additional conditions need to be put on a convex function in order to guarantee its continuity. Or, one can ask what further restrictions must be placed on $\mathbf{L}$ in order to guarantee that a function convex on $U \subseteq \mathbf{L}$ will be continuous there. We shall take some steps in both directions.

The key to the proof of the continuity of a function convex on $(a, b) \subseteq \mathbf{R}$ was to establish the boundedness of $f$ on closed subintervals. We were then able to establish a Lipschitz condition and thus to conclude that $f$ was continuous. This suggests that we start with some boundedness requirement for $f$ in $U$. It turns out that it is enough to have $f$ bounded in a neighborhood of just one point of $U$. From this follows a Lipschitz condition and hence continuity.

> **Theorem A.** Let $f$ be convex on an open set $U$ in a normed linear space $\mathbf{L}$. If $f$ is bounded from above in a neighborhood of one point $\mathbf{x}_0 \in U$, then it is locally bounded; that is, each $\mathbf{x} \in U$ has a neighborhood on which $f$ is bounded.

**Proof.** We first show that if $f$ is bounded above in an $\varepsilon$-neighborhood of some point, it is bounded below in the same neighborhood. Taking the point to be $\mathbf{O}$ for convenience, suppose $f$ is bounded above by $B$ in a neighborhood $N_\varepsilon$ of the origin. Since $\mathbf{O} = \frac{1}{2}\mathbf{x} + \frac{1}{2}(-\mathbf{x})$, $f(\mathbf{O}) \leq \frac{1}{2}f(\mathbf{x}) + \frac{1}{2}f(-\mathbf{x})$, and therefore $f(\mathbf{x}) \geq 2f(\mathbf{O}) - f(-\mathbf{x})$. Now $\|\mathbf{x}\| < \varepsilon$ implies $\|-\mathbf{x}\| < \varepsilon$, so $-f(-\mathbf{x}) \geq -B$ and $f(\mathbf{x}) \geq 2f(\mathbf{O}) - B$, meaning $f$ is bounded from below.

We now return to our theorem, taking $f$ to be bounded from above by $B$ on an $\varepsilon$-neighborhood $N$ of the origin. We will show $f$ to be bounded in a neighborhood of $\mathbf{y} \in U$, $\mathbf{y} \neq \mathbf{O}$. Choose $\rho > 1$ so that $\mathbf{z} = \rho\mathbf{y} \in U$ and let $\lambda = 1/\rho$. Then

$$M = \{\mathbf{v} \in \mathbf{L}: \ \mathbf{v} = (1 - \lambda)\mathbf{x} + \lambda\mathbf{z}, \mathbf{x} \in N\}$$

is a neighborhood of $\lambda\mathbf{z} = \mathbf{y}$ with radius $(1 - \lambda)\varepsilon$ (Fig. 41.1). Moreover

$$f(\mathbf{v}) \leq (1 - \lambda)f(\mathbf{x}) + \lambda f(\mathbf{z}) \leq B + f(\mathbf{z})$$

That is, $f$ is bounded above on $M$; and by the first remark of this proof, $f$ is also bounded below on $M$. •

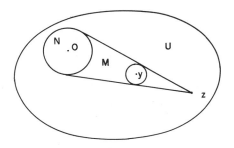

**Fig. 41.1**

We call attention to the fact that a slight modification can be made in the proof of Theorem A to prove directly that $f$ is continuous on $U$. This procedure is followed, in fact, in our proof of Theorem 72A.

A function defined on an open set $U$ is said to be **locally Lipschitz** if at each $\mathbf{x} \in U$, there is a neighborhood $N_\varepsilon(\mathbf{x})$ and a constant $K(\mathbf{x})$ such that if $\mathbf{y}, \mathbf{z} \in N$, then

$$|f(\mathbf{y}) - f(\mathbf{z})| \leq K\|\mathbf{y} - \mathbf{z}\|$$

If this inequality holds throughout a set $V \subseteq U$ with $K$ independent of $\mathbf{x}$, then we say that $f$ is **Lipschitz** on $V$.

## 41. Continuity of Convex Functions

**Theorem B.** Let $f$ be convex on an open set $U \subseteq \mathbf{L}$. If $f$ is bounded from above in a neighborhood of one point of $U$, then $f$ is locally Lipschitz in $U$, hence Lipschitz on any compact subset of $U$.

**Proof.** By Theorem A, $f$ is locally bounded, so given $\mathbf{x}_0$ we may find a neighborhood $N_{2\varepsilon}(\mathbf{x}_0) \subseteq U$ on which $f$ is bounded, say by $M$. Then $f$ satisfies the stated Lipschitz condition on $N_\varepsilon(\mathbf{x}_0)$, for if it does not, we may choose $\mathbf{x}_1, \mathbf{x}_2 \in N_\varepsilon(\mathbf{x}_0)$ such that

$$\frac{f(\mathbf{x}_2) - f(\mathbf{x}_1)}{\|\mathbf{x}_2 - \mathbf{x}_1\|} > \frac{2M}{\varepsilon}$$

Then we may choose $\alpha > 0$ so that $\mathbf{x}_3 = \mathbf{x}_2 + \alpha(\mathbf{x}_2 - \mathbf{x}_1)$ is in $N_{2\varepsilon}(\mathbf{x}_0)$ and such that $\|\mathbf{x}_3 - \mathbf{x}_2\| = \varepsilon$. Because $f$ is convex on the line through $\mathbf{x}_1, \mathbf{x}_2$, and $\mathbf{x}_3$, we may use what we know about functions convex on a line (10.2) to write

$$\frac{f(\mathbf{x}_3) - f(\mathbf{x}_2)}{\|\mathbf{x}_3 - \mathbf{x}_2\|} \geqslant \frac{f(\mathbf{x}_2) - f(\mathbf{x}_1)}{\|\mathbf{x}_2 - \mathbf{x}_1\|} > \frac{2M}{\varepsilon}$$

This says $f(\mathbf{x}_3) - f(\mathbf{x}_2) > 2M$, contradicting the fact that $|f| \leqslant M$.
For a compact subset of $U$, argue as outlined in Problem A. ●

**Theorem C.** Let $f$ be convex on an open set $U \subseteq \mathbf{L}$. If $f$ is bounded from above in a neighborhood of one point of $U$, then $f$ is continuous on $U$.

**Proof.** Theorem B implies that $f$ is locally Lipschitz, from which continuity follows immediately. ●

**Theorem D.** Let $f$ be convex on the open set $U \subseteq \mathbf{R}^n$. Then $f$ is Lipschitz on every compact subset of $U$ and continuous on $U$.

**Proof.** We may and do suppose that $\mathbf{0} \in U$. Choose $\alpha > 0$ small enough so that the convex hull

$$V = H(\mathbf{0}, \alpha \mathbf{e}_1, \ldots, \alpha \mathbf{e}_n) \subseteq U$$

and note that $V$ has a nonempty interior $V^0$ (Problem 31I). Any $\mathbf{x} \in V$

has the representation $\mathbf{x} = \lambda_0 \mathbf{O} + \lambda_1(\alpha \mathbf{e}_1) + \cdots + \lambda_n(\alpha \mathbf{e}_n)$ where $\lambda_i \geq 0$, $\sum_0^n \lambda_i = 1$. Hence,

$$f(\mathbf{x}) \leq \lambda_0 f(\mathbf{O}) + \sum_1^n \lambda_i f(\alpha \mathbf{e}_i)$$
$$\leq \max\{f(\mathbf{O}), f(\alpha \mathbf{e}_1),\ldots, f(\alpha \mathbf{e}_n)\}$$

Thus $f$ is bounded above on the nonempty open set $V^0$. The result now follows from Theorems B and C. ●

## PROBLEMS AND REMARKS

**A.** Prove the second assertion of Theorem B.

*Hint.* Use the first assertion and compactness to get a finite open cover $N_1, N_2, \ldots, N_r$ so that for $\mathbf{u}, \mathbf{v} \in N_i$, $|f(\mathbf{u}) - f(\mathbf{v})| \leq K \|\mathbf{u} - \mathbf{v}\|$. Now for $\mathbf{x}, \mathbf{y}$ in the compact set, $\mathbf{x} \in N_i$, $\mathbf{y} \in N_j$ and we may choose $\mathbf{w} \in N_i$, $\mathbf{z} \in N_j$ so that $\mathbf{x}, \mathbf{y} \in \text{Seg}(\mathbf{w}, \mathbf{z})$. Finally, apply Eq. (11.2) to $g(t) = f[\mathbf{w} + t(\mathbf{z} - \mathbf{w})]$.

**B.** Section 13 was devoted to showing that the class of functions convex on an interval remains closed under certain functional operations. Similar results hold for functions convex on a set $U \subseteq L$. For more discussion, see Rockafellar [1970a, pp. 32–40].

(1) If $f$ and $g$ are convex on $U$ and $\alpha \geq 0$, then $f + g$, $\alpha f$, and $f \vee g = \max(f, g)$ are convex.

(2) If $f_\alpha : U_\alpha \to \mathbf{R}$ is convex and $U = \cap U_\alpha \neq \phi$, then the subset of $U$ on which $f(\mathbf{x}) = \sup_\alpha f_\alpha(\mathbf{x}) < \infty$ is convex and $f$ is convex on it.

(3) Let $f_n : U \to \mathbf{R}$ be a sequence of convex functions converging to a finite limit function $f$ on $U$. Then $f$ is convex. (The convergence need not be uniform, even if $U$ is a compact set in $\mathbf{R}^n$, but conditions can be given to assure uniform convergence [Guberman, 1970]. See also Rockafellar [1970a, p. 90].)

(4) Let $f: U \to \mathbf{R}$ and $g: V \to \mathbf{R}$ be convex where range $(f) \subseteq V \subseteq \mathbf{R}$ and $g$ is increasing. Then $g \circ f$ is convex on $U$.

**C.** The following facts help one recognize convex functions.

(1) If $f$ is convex, then so is $g(\mathbf{x}) = f(\mathbf{x}) + L(\mathbf{x}) + a$ where $L$ is linear and $a$ is a constant.

(2) If $f(\mathbf{x}) = L(\mathbf{x}) + a$ where $L$ is linear and $a$ is constant, then $|f|^p$ is convex for $p \geq 1$.

(3) If $f(r, s)$ is convex on $U \subseteq \mathbf{R}^2$, then $f(r, s_0)$ is convex on $V = \{r : (r, s_0) \in U\}$. The converse is false; $f(r, s)$ may be convex in $r$ for each fixed $s$ and convex in $s$ for each fixed $r$, but not be convex.

(4) If $f$ is convex on $U \subseteq \mathbf{R}^m$ and $g: U \times \mathbf{R}^k \to \mathbf{R}$ is defined by $g(x_1, \ldots, x_m, x_{m+1}, \ldots, x_{m+k}) = f(x_1, \ldots, x_m)$, then $g$ is convex.

**D.** Show that each of the following is convex on $\mathbf{R}^n$.

(1) $f(x_1, \ldots, x_n) = x_j$.
(2) $g(x_1, \ldots, x_n) = \sum_1^k |x_i + b_i|, k \leq n$.

## 41. Continuity of Convex Functions

(3) $h(x_1, \ldots, x_n) = \sum_1^k a_i \,|\, x_i + b_i \,|^p$, $a_i \geqslant 0$, $p \geqslant 1$, $k \leqslant n$.
(4) $\theta(x_1, \ldots, x_n) = \sum_1^k \{[a_i \,|\, x_i + b_i \,|^p + c_i] \vee 0\}$, $a_i \geqslant 0$, $p \geqslant 1$, $k \leqslant n$.
(5) $\phi(x_1, \ldots, x_n) = \exp[\theta(k_1, x_2, \ldots, x_n)] + k_2 h(x_1, \ldots, x_{n-1}, k_3)$ where $k_i$ are constants, $k_2 \geqslant 0$, and $\theta$ and $h$ are defined as above.

**E.** Let $U$ be a convex set in $\mathbf{R}^n$. The following functions associated with $U$ are all convex:

(1) the **support function** $\delta(\mathbf{x}) = \sup\{\langle \mathbf{x}, \mathbf{y} \rangle : \mathbf{y} \in U\}$,
(2) the **gauge function** $\gamma(\mathbf{x}) = \inf\{\lambda \geqslant 0 : \mathbf{x} \in \lambda U\}$,
(3) the **distance function** $d(\mathbf{x}) = \inf\{\|\mathbf{x} - \mathbf{y}\| : \mathbf{y} \in U\}$.

**\*F.** Let $f$ be convex on a set $U \subseteq \mathbf{R}^n$. The behavior of $f$ at the boundary of $U$ is summarized as follows.

(1) If $\mathbf{x}_0$ is a boundary point of $U$,
$$\liminf_{\mathbf{x} \to \mathbf{x}_0} f(\mathbf{x}) > -\infty$$

(2) If the boundary point $\mathbf{x}_0 \in U$,
$$\liminf_{\mathbf{x} \to \mathbf{x}_0} f(\mathbf{x}) \leqslant f(\mathbf{x}_0)$$

(3) We cannot replace lim inf by lim in parts (1) and (2) since the latter may not exist. Consider $\lim_{\mathbf{x} \to 0} f(\mathbf{x})$ for
$$f(r, s) = \frac{r^2 + s^2}{2s}, \quad s > 0, \quad f(0, 0) = 1$$

(4) If $U$ is open, we may extend $f$ to the set $V$ obtained by adding to $U$ all boundary points $\mathbf{y}$ for which
$$\liminf_{\mathbf{x} \to \mathbf{y}} f(\mathbf{x}) < \infty$$

At such boundary points, we define
$$f(\mathbf{y}) = \liminf_{\mathbf{x} \to \mathbf{y}} f(\mathbf{x})$$

Having done this, $V$ is a convex set and $f$ is convex on $V$.

(5) A convex function defined on a convex set $U$ is **closed** if at each boundary point $\mathbf{y}$ of $U$,
$$\liminf_{\mathbf{x} \to \mathbf{y}} f(\mathbf{x}) = \begin{cases} \infty & \text{if } \mathbf{y} \notin U \\ f(\mathbf{y}) & \text{if } \mathbf{y} \in U \end{cases}$$

It is clear that the convex function defined in (4) above is closed, and that by a possible redefining on part or all of its boundary, any convex function can be altered so as to be closed. If $f$ is a closed convex function on $U$, then for any point $\mathbf{y} \in U$, $\lim_{\mathbf{x} \to \mathbf{y}} f(\mathbf{x}) = f(\mathbf{y})$ as $\mathbf{x} \to \mathbf{y}$ along a line segment contained in $U$.

(6) $f: U \to \mathbf{R}$ is a closed convex function if and only if the **epigraph** = epi($f$) = $\{(\mathbf{x}, y): \mathbf{x} \in U, y \geqslant f(\mathbf{x})\}$ is a closed convex subset of $\mathbf{R}^{n+1}$.

(7) In Section 15 a convex function $g: I \to \mathbf{R}$ was called closed if $\{x \in I: g(x) \leqslant \alpha\}$ was a closed subset of $\mathbf{R}$ for each real $\alpha$. This suggests, for $f: U \to \mathbf{R}$, that $f$ is closed $\Leftrightarrow$ $\{\mathbf{x} \in U: f(\mathbf{x}) \leqslant \alpha\}$ is a closed subset of $\mathbf{R}^n$ for each $\alpha$. This is the case.

These observations about the behavior of a convex function on its boundary follow

Fenchel, [1953, pp. 75–79]. A fuller discussion may be found in Rockafellar [1970a, pp. 51–59] and Fenchel [1949].

*G. A convex function $f: (a, b) \to \mathbf{R}$ is absolutely continuous on any closed subinterval of $(a, b)$. Several suggestions have been made for extending the notion of absolute continuity to functions of several variables. Explore, for various definitions, the absolute continuity of a convex function defined on $U \subseteq \mathbf{R}^2$ [Friedman, 1940].

*H. For $\mathbf{x} = (x_1, \ldots, x_n) \in \mathbf{R}^n$, we write $\mathbf{x} \geqslant 0$ to mean that $x_i \geqslant 0$ for $i = 1, \ldots, n$. Let $I^n = \{\mathbf{x}: x_i \in [0, 1]\}$. Let $f: I^n \to \mathbf{R}$ be such that $f(\mathbf{x}) \geqslant 0$, $f(\mathbf{x} + \mathbf{h}) - f(\mathbf{x}) \geqslant 0$ for $\mathbf{h} \geqslant \mathbf{O}$ and $(\mathbf{x} + \mathbf{h}) \in I^n$. Then there exist two convex functions $g_1$ and $g_2$ such that $0 \leqslant g_1 \leqslant f \leqslant g_2$ and

$$(n+1)! \int g_1(\mathbf{x})\, d\mathbf{x} \geqslant \int f(\mathbf{x})\, d\mathbf{x} \geqslant \frac{n!}{(n+1)^n} \int g_2(\mathbf{x})\, d\mathbf{x}$$

These constants are the best possible [Nishiura and Schnitzer, 1965]. (Note that this generalizes Problem 13M).

*I. Suppose $f$ is a continuous convex function defined on an open set $U \subseteq \mathbf{L}$. Corresponding to $\mathbf{x}_0 \in U$, there is a $\delta > 0$ and a function $F$ that is convex and Lipschitz on $U$ such that $F(\mathbf{y}) = f(\mathbf{y})$ for all $\mathbf{y} \in N_\delta(\mathbf{x}_0)$ [Asplund, 1968]. (Rockafellar [1970a, pp. 87–88] discusses a number of conditions under which a convex function defined on $\mathbf{R}^n$ will be Lipschitz).

*J. If $f$ is convex on a bounded open set $C$ in $\mathbf{R}^n$, then it is Lipschitz on any closed subset $B$ of $C$ (Theorem D). One might hope for an extension to infinite-dimensional spaces, at least if $f$ is continuous and $B$ is kept at a positive distance from the boundary of $C$. The following example, due to Victor Klee, ends speculation along these lines.

Let $A$, $B$, and $C$ be the closed balls in $l^2$ that are centered at the origin and have radii of $1 - 2\varepsilon$, $1 - \varepsilon$, and $1$, respectively. For each $i$, let $\mathbf{z}_i$ be the point at which the $i$th coordinate axis emerges from the ball $A$, and let $H_i$ be the hyperplane supporting $A$ at that point. Let $X_i$ be the set of all points in the ball $C$ that lie on the opposite side of $H_i$ from $A$. We choose $\varepsilon$ small enough so that the sets $X_i$ are pairwise disjoint. Finally, let $Y$ consist of all points of $C$ not lying in any of the sets $X_i$. We now define a function $f$ on $C$ as follows. On $Y$, $f$ is zero. At $\mathbf{x} \in X_i$, $f(\mathbf{x})$ is $i$ times the distance from $\mathbf{x}$ to the set $Y$. Then $f$ is a continuous convex function on the bounded convex set $C$, and $B$ is a closed bounded convex set whose $\varepsilon$-neighborhood is contained in $C$. But $f$ is not Lipschitz on $B$.

*K. Suppose $f$ is a convex function defined on a polytope $D \subseteq \mathbf{R}^n$.

(1) Let $M$ be the maximum of $f$ on the (finite) set of all extreme points of $D$. Then for all $\mathbf{x} \in D$, $f(\mathbf{x}) \leqslant M$.

(2) $f$ is **upper-Lipschitzian** at each point of $D$; that is, corresponding to each $\mathbf{x}_0 \in D$, there is an $L < \infty$ such that $f(\mathbf{x}) - f(\mathbf{x}_0) \leqslant L \|\mathbf{x} - \mathbf{x}_0\|$ for all $\mathbf{x} \in D$.

(3) $f$ is **upper semicontinuous** on $D$; that is, corresponding to each $\mathbf{x}_0 \in D$ and each $\varepsilon > 0$, there is a $\delta > 0$ such that $\|\mathbf{x} - \mathbf{x}_0\| < \delta$ implies $f(\mathbf{x}) - f(\mathbf{x}_0) < \varepsilon$.

(4) The condition that $D$ be a polytope is essential to the proving of parts (2) and (3). Consider in a neighborhood of $\mathbf{O}$,

$$f(r, s) = r^2/s \quad \text{if} \quad s \geqslant r^2, \quad (r, s) \neq \mathbf{O}; \quad f(\mathbf{O}) = 0$$

A subset is called **boundedly polyhedral** provided that its intersection with any polytope is a polytope. After discussing the properties above, Gale *et al.* [1968] prove that if $D$ is boundedly polyhedral and $f$ is a convex function defined on $D^0$ so that $f$ is bounded on bounded sets in $D^0$, then $f$ can be extended in a unique way to a continuous convex function on $D$.

*L. We say an affine function $A$ is **majorized** on $U$ by $f: U \to \mathbf{R}$ if $A(\mathbf{x}) \leq f(\mathbf{x})$ for all $\mathbf{x} \in U$. Let $f$ be a function (not necessarily convex) defined on a closed convex set $U \subseteq \mathbf{R}^n$ so that $f$ majorizes at least one affine function, and let $\mathcal{A}(f)$ be the set of all affine functions majorized by $f$. Define the **envelope function** of $f$ by

$$\mathrm{env}\, f(\mathbf{x}) = \sup_{A \in \mathcal{A}(f)} \{A(\mathbf{x})\}$$

(1) Env $f$ is convex on $U$, hence continuous on $U^0$.

(2) Env $f$ is lower semicontinuous everywhere on $U$. The function env $f$ appears in control theory where it is of interest to know when env $f$ is continuous on the boundary as well as the interior of $U$.

(3) [Witsenhausen, 1968]. If $U$ is a convex polytope and $f$ is continuous on $U$, then env $f$ is continuous on $U$.

(4) [Kruskal, 1969]. If it is only known that $U$ is closed and convex, continuity of $f$ on $U$ is not enough to guarantee the continuity of env $f$. Consider in $\mathbf{R}^3$ the function $f(r, s, t) = -t^2$ defined on $U$ the convex hull of the two-dimensional circle $r^2 + s^2 = 1$ and the points $(0, 1, 1)$, $(0, 1, -1)$. Then env $f$ is discontinuous at $(0, 1, 0)$.

Note that this problem generalizes Problem 13J. Affine minorants are used by Aggeri [1966] and Brøndsted [1966b] to extend the Krein–Milman theorem and its converse to convex functions. Klee and Martin [1971] give further results about the continuity of the envelope function.

## 42. Differentiable Convex Functions

To give necessary and sufficient conditions for the derivative of a convex function to exist at $\mathbf{x}_0 \in U \subseteq \mathbf{L}$, we need to draw on some deep results from functional analysis. Keeping in mind the needs of those primarily interested in applications that involve convex functions, we have chosen to leave the existence questions for Sections 43 and 44, devoting this section to results about convex functions known to be differentiable at a point or in a region. We begin with three theorems that hold in the setting of a normed linear space $\mathbf{L}$. Then we turn to results that can be proved when $\mathbf{L} = \mathbf{R}^n$. Again in an effort to provide a useful reference here, we include a restatement of the first three theorems in the terminology of $\mathbf{R}^n$.

For functions of a real variable, we saw that a differentiable function was convex if and only if $f'$ was increasing (Theorem 12B). For differentiable functions on $\mathbf{L}$, we again are able to characterize convexity in terms of the first derivative.

**Theorem A.** Suppose $f$ is defined on the open convex set $U \subseteq \mathbf{L}$. If $f$ is convex on $U$ and differentiable at $\mathbf{x}_0$, then for $\mathbf{x} \in U$,

$$f(\mathbf{x}) - f(\mathbf{x}_0) \geqslant f'(\mathbf{x}_0)(\mathbf{x} - \mathbf{x}_0) \tag{1}$$

If $f$ is differentiable throughout $U$, then $f$ is convex if and only if (1) holds for all $\mathbf{x}, \mathbf{x}_0 \in U$. Moreover, $f$ is strictly convex if and only if the inequality is strict.

**Proof.** If $f$ is convex, then for $t \in (0, 1)$,

$$f[\mathbf{x}_0 + t(\mathbf{x} - \mathbf{x}_0)] = f[(1-t)\mathbf{x}_0 + t\mathbf{x}] \leqslant (1-t)f(\mathbf{x}_0) + tf(\mathbf{x})$$

Setting $\mathbf{h} = \mathbf{x} - \mathbf{x}_0$, we get

$$f(\mathbf{x}_0 + t\mathbf{h}) - f(\mathbf{x}_0) \leqslant t[f(\mathbf{x}_0 + \mathbf{h}) - f(\mathbf{x}_0)] \tag{2}$$

Subtracting $f'(\mathbf{x}_0)(t\mathbf{h})$ from both sides and dividing by $t$ gives

$$\frac{f(\mathbf{x}_0 + t\mathbf{h}) - f(\mathbf{x}_0) - f'(\mathbf{x}_0)(t\mathbf{h})}{t} \leqslant f(\mathbf{x}_0 + \mathbf{h}) - f(\mathbf{x}_0) - f'(\mathbf{x}_0)(\mathbf{h})$$

Now as $t \to 0$, the left side goes to zero while the right side, being independent of $t$, remains constant. This establishes (1). If $f$ is strictly convex, (2) is a strict inequality which when used along with (1) where $\mathbf{x} = \mathbf{x}_0 + t\mathbf{h}$ gives

$$t[f(\mathbf{x}_0 + \mathbf{h}) - f(\mathbf{x}_0)] > f(\mathbf{x}_0 + t\mathbf{h}) - f(\mathbf{x}_0) \geqslant f'(\mathbf{x}_0)(t\mathbf{h})$$

Division by $t$ gives the desired strict inequality.

Now suppose we know that $f$ is differentiable and satisfies (1) throughout $U$. Given $\mathbf{x}_1, \mathbf{x}_2 \in U$, $t \in (0, 1)$, we set $\mathbf{x}_0 = t\mathbf{x}_1 + (1-t)\mathbf{x}_2$. Then

$$f(\mathbf{x}_0) = f(\mathbf{x}_0) + f'(\mathbf{x}_0)[t(\mathbf{x}_1 - \mathbf{x}_0) + (1-t)(\mathbf{x}_2 - \mathbf{x}_0)]$$

and using the linearity of $f'(\mathbf{x}_0)$ we can write this as

$$f(\mathbf{x}_0) = t[f(\mathbf{x}_0) + f'(\mathbf{x}_0)(\mathbf{x}_1 - \mathbf{x}_0)] + (1-t)[f(\mathbf{x}_0) + f'(\mathbf{x}_0)(\mathbf{x}_2 - \mathbf{x}_0)]$$

Inequality (1) holds for $\mathbf{x} = \mathbf{x}_1$ and $\mathbf{x} = \mathbf{x}_2$, so

$$f(\mathbf{x}_0) \leqslant tf(\mathbf{x}_1) + (1-t)f(\mathbf{x}_2)$$

proving that $f$ is convex. Moreover, if (1) is a strict inequality, the last inequality is also strict. ●

## 42. Differentiable Convex Functions

If $f$ is convex on $U \subseteq \mathbf{L}$, we shall see that a one-sided directional derivative always exists, enabling us to obtain an inequality similar to (1) without making any assumptions about the differentiability of $f$ (Problem 44A).

Theorem A does not exactly parallel the real variable theorem which asserts that $f$ is convex if and only if $f'$ is increasing, but we can use this theorem to prove such a result if we first get a suitable definition for an increasing function on $U \subseteq \mathbf{L}$. The key is to observe that for a function of a real variable, $f'$ is monotone increasing if and only if for any $x$, $y \in (a, b)$, $[f'(x) - f'(y)][x - y] \geqslant 0$. This expression remains meaningful for $f$ defined on $U \subseteq \mathbf{L}$, so we say that $f'$ is **monotone increasing** if for $\mathbf{x}, \mathbf{y} \in U$,

$$[f'(\mathbf{x}) - f'(\mathbf{y})][\mathbf{x} - \mathbf{y}] \geqslant 0 \qquad (3)$$

$f'$ is **strictly monotone increasing** if the inequality (3) is strict for $\mathbf{x} \neq \mathbf{y}$.

**Theorem B.** *Let $f: U \to \mathbf{R}$ be continuous and differentiable on the open convex set $U \subseteq \mathbf{L}$. Then $f$ is convex [strictly convex] if and only if $f'$ is monotone [strictly monotone] increasing on $U$.*

**Proof.** For a convex function differentiable on $U$, Theorem A gives us

$$f(\mathbf{x}) - f(\mathbf{y}) \geqslant f'(\mathbf{y})(\mathbf{x} - \mathbf{y})$$
$$f(\mathbf{y}) - f(\mathbf{x}) \geqslant f'(\mathbf{x})(\mathbf{y} - \mathbf{x})$$

Addition and simplification gives the desired result. Strict inequalities may be substituted when $f$ is strictly convex.

Now suppose $f'$ is monotone increasing. Let $\phi: [0, 1] \to \mathbf{R}$ be defined by $\phi(\lambda) = f[\lambda \mathbf{x} + (1 - \lambda)\mathbf{y}]$. For $0 \leqslant \lambda_1 < \lambda_2 \leqslant 1$, let $\mathbf{u}_1 = \lambda_1 \mathbf{x} + (1 - \lambda_1)\mathbf{y}$ and $\mathbf{u}_2 = \lambda_2 \mathbf{x} + (1 - \lambda_2)\mathbf{y}$. Then $\mathbf{u}_2 - \mathbf{u}_1 = (\lambda_2 - \lambda_1)(\mathbf{x} - \mathbf{y})$, so

$$0 \leqslant [f'(\mathbf{u}_2) - f'(\mathbf{u}_1)][\mathbf{u}_2 - \mathbf{u}_1] = (\lambda_2 - \lambda_1)[f'(\mathbf{u}_2) - f'(\mathbf{u}_1)](\mathbf{x} - \mathbf{y})$$

which ensures that $f'(\mathbf{u}_1)(\mathbf{x} - \mathbf{y}) \leqslant f'(\mathbf{u}_2)(\mathbf{x} - \mathbf{y})$. From the chain rule,

$$\phi'(\lambda_1) = f'(\mathbf{u}_1)(\mathbf{x} - \mathbf{y}) \leqslant f'(\mathbf{u}_2)(\mathbf{x} - \mathbf{y}) = \phi'(\lambda_2)$$

It now follows that $\phi$ is convex (strictly convex if the inequalities are strict) so we can write

$$f[\lambda \mathbf{x} + (1 - \lambda)\mathbf{y}] = \phi(\lambda) = \phi[\lambda(1) + (1 - \lambda)0]$$
$$\leqslant \lambda \phi(1) + (1 - \lambda)\phi(0) = \lambda f(\mathbf{x}) + (1 - \lambda)f(\mathbf{y}) \quad \bullet$$

Having now characterized convex functions in terms of the first derivative, we turn to the second derivative. Twice differentiable functions of a real variable are convex if and only if the second derivative is nonnegative. Our review of nonnegative matrices in Section 23 puts us in position to anticipate the correct generalization for functions with domain in $\mathbf{R}^n$. They are convex if and only if the matrix $[f''(\mathbf{x})]$ is nonnegative definite. Even more generally, we might anticipate that a function twice differentiable on a convex set $U \subseteq \mathbf{L}$ is convex if and only if the bilinear symmetric transformation $f''(\mathbf{x})$ is nonnegative definite.

Before proving that these conjectures are good, we refer once again to the illustrative function $f(r, s) = r^2 + 3rs + 5s^2$ introduced in Section 23. We have already seen that $f''(r, s)$ is positive definite. We therefore anticipate that the graph of $w = r^2 + 3rs + 5s^2$ is convex. This is the case; in fact by a simple rotation of the $rs$ plane through $\theta = \arctan 3$, the equation becomes $w = \frac{11}{2}r'^2 + \frac{1}{2}s'^2$, obviously describing a convex surface.

**Theorem C.** Let $f$ be continuously differentiable and suppose that the second derivative exists throughout an open convex set $U \subseteq \mathbf{L}$. Then $f$ is convex on $U$ if and only if $f''(\mathbf{x})$ is nonnegative definite for each $\mathbf{x} \in U$. And if $f''(\mathbf{x})$ is positive definite on $U$, then $f$ is strictly convex.

**Proof.** The proof depends on Theorem 23D which says that for any $\mathbf{x}$, $\mathbf{x}_0 \in U$,

$$f(\mathbf{x}) = f(\mathbf{x}_0) + f'(\mathbf{x}_0)(\mathbf{h}) + \tfrac{1}{2}f''(\mathbf{x}_0 + s\mathbf{h})(\mathbf{h}, \mathbf{h})$$

where $s \in (0, 1)$ and $\mathbf{h} = \mathbf{x} - \mathbf{x}_0$. Suppose that $f''(\mathbf{x})$ is nonnegative definite. Then it is immediate that

$$f(\mathbf{x}) \geq f(\mathbf{x}_0) + f'(\mathbf{x}_0)(\mathbf{x} - \mathbf{x}_0)$$

from which we conclude (Theorem A) that $f$ is convex. It also follows from Theorem A that if $f''(\mathbf{x}_0)$ is positive definite so that the inequality above is strict, then $f$ is strictly convex.

Conversely, if $f$ is known to be convex, then we use the technique mentioned in the introduction to this chapter, defining for $\mathbf{x} \in U$ and $\mathbf{h} \in \mathbf{L}$, $g(t) = f(\mathbf{x} + t\mathbf{h})$. Then $g$ is convex in a neighborhood of the origin and

$$g'(t) = f'(\mathbf{x} + t\mathbf{h})(\mathbf{h})$$
$$g''(t) = f''(\mathbf{x} + t\mathbf{h})(\mathbf{h}, \mathbf{h})$$

## 42. Differentiable Convex Functions

The convexity of $g$ means (Theorem 12C) that for each $t$ in its domain, $g''(t) \geq 0$. In particular, $g''(0) \geq 0$, so $f''(\mathbf{x})(\mathbf{h}, \mathbf{h}) \geq 0$. Since $\mathbf{h}$ was arbitrary, $f''(\mathbf{x})$ is nonnegative definite. ∎

We turn now to the situation in $\mathbf{R}^n$. For a function of several variables in which all the partials exist, we can always define the linear transformation with matrix

$$\nabla f(\mathbf{x}_0) = \left[ \frac{\partial f}{\partial x_1}(\mathbf{x}_0) \cdots \frac{\partial f}{\partial x_n}(\mathbf{x}_0) \right] = [f_1(\mathbf{x}_0) \cdots f_n(\mathbf{x}_0)]$$

commonly called the **gradient** of $f$. When $f$ is differentiable at $\mathbf{x}_0$, then this linear transformation is the Fréchet derivative $f'(\mathbf{x}_0)$. We have seen that existence of the gradient $\nabla f(\mathbf{x}_0)$ does not imply the existence of $f'(\mathbf{x}_0)$, but it turns out that when $f$ is convex, then existence of $\nabla f(\mathbf{x}_0)$ does imply that $f'(\mathbf{x}_0)$ exists.

**Theorem D.** If $f$ is convex on an open set $U \subseteq \mathbf{R}^n$ and all partial derivatives exist at $\mathbf{x}_0 \in U$, then $f'(\mathbf{x}_0)$ exists.

**Proof.** It is natural to think that the linear transformation $T$ determined by the partial derivatives should be the derivative. To establish that this is the case, we must show that

$$\varepsilon(\mathbf{h}) = \frac{1}{\|\mathbf{h}\|}[f(\mathbf{x}_0 + \mathbf{h}) - f(\mathbf{x}_0) - T(\mathbf{h})]$$

goes to zero as $\|\mathbf{h}\| \to 0$. We find it convenient to work with $\phi(\mathbf{h}) = \|\mathbf{h}\|\varepsilon(\mathbf{h})$ on $N_\delta(\mathbf{O})$ chosen so that $\mathbf{h} \in N$ implies $n\mathbf{h} \in U$. The function $\phi$ is convex, being the difference of a convex and a linear function, so for $\mathbf{h} = h_1 \mathbf{e}_1 + \cdots + h_n \mathbf{e}_n$ expressed in terms of the standard basis of $\mathbf{R}^n$,

$$\phi(\mathbf{h}) = \phi\left(\sum_1^n \frac{1}{n} h_i n \mathbf{e}_i\right) \leq \frac{1}{n} \sum_1^n \phi(h_i n \mathbf{e}_i)$$

Now

$$\phi(h_i n \mathbf{e}_i) = f(\mathbf{x}_0 + h_i n \mathbf{e}_i) - f(\mathbf{x}_0) - f_i(\mathbf{x}_0) h_i n$$

so from the definition of the partial derivative, we have

$$\lim_{h_i \to 0} \frac{\phi(h_i n \mathbf{e}_i)}{h_i n} = 0$$

From the CBS inequality, we conclude that for two vectors $\mathbf{u}, \mathbf{v} \in \mathbf{R}^n$,

$\sum_1^n u_i v_i \leq \|\mathbf{u}\| \|\mathbf{v}\| \leq \|\mathbf{u}\| \sum_1^n |v_i|$. Thus, summing over $i$ for which $h_i \neq 0$, we get

$$\phi(\mathbf{h}) \leq \sum h_i \frac{\phi(h_i n \mathbf{e}_i)}{h_i n} \leq \|\mathbf{h}\| \sum \left| \frac{\phi(h_i n \mathbf{e}_i)}{h_i n} \right|$$

Similarly,

$$\phi(-\mathbf{h}) \leq \|\mathbf{h}\| \sum \left| \frac{\phi(-h_i n \mathbf{e}_i)}{h_i n} \right|$$

From the definition of $\phi$ and its convexity,

$$0 = \phi\left[\frac{\mathbf{h} + (-\mathbf{h})}{2}\right] \leq \tfrac{1}{2}[\phi(\mathbf{h}) + \phi(-\mathbf{h})]$$

or $\phi(\mathbf{h}) \geq -\phi(-\mathbf{h})$. Thus,

$$-\|\mathbf{h}\| \sum \left|\frac{\phi(-h_i n \mathbf{e}_i)}{h_i n}\right| \leq -\phi(-\mathbf{h}) \leq \phi(\mathbf{h}) \leq \|\mathbf{h}\| \sum \left|\frac{\phi(h_i n \mathbf{e}_i)}{h_i n}\right|$$

It follows that

$$\lim_{\|\mathbf{h}\| \to 0} \varepsilon(\mathbf{h}) = \lim_{\|\mathbf{h}\| \to 0} \frac{\phi(\mathbf{h})}{\|\mathbf{h}\|} = 0 \quad \bullet$$

Taking advantage of Theorem D, we may now restate Theorems A and B which we combine for the case $\mathbf{L} = \mathbf{R}^n$.

**Theorem E.** Suppose $f$ is defined on the open convex set $U \subseteq \mathbf{R}^n$. If $f$ is convex on $U$ and the gradient $\nabla f(\mathbf{x}_0)$ exists, then for $\mathbf{x} \in U$,

$$f(\mathbf{x}) - f(\mathbf{x}_0) \geq \nabla f(\mathbf{x}_0)(\mathbf{x} - \mathbf{x}_0)$$

If $f$ is convex [strictly convex] and $\nabla f(\mathbf{x})$ exists throughout $U$, then $\nabla f$ is monotone [strictly monotone] increasing on $U$. Conversely, if the partial derivatives of $f$ exist and are continuous throughout $U$ and if $\nabla f$ is monotone [strictly monotone] increasing, then $f$ is convex [strictly convex].

The careful reader will note that while the last statement certainly follows from Theorem B, it is weaker in the sense that we only need require that $f'$ exist, a condition not as strong as requiring continuity of the partials. The same observations can be made about our next theorem. It follows from Theorem C, but it seems to be somewhat weaker in the sense that we here require continuity of all the second partial derivatives.

## 42. Differentiable Convex Functions

Continuity of all the second partial derivatives certainly is enough to guarantee that the derivative of

$$f'(\mathbf{x}): \begin{aligned} y_1 &= \frac{\partial f}{\partial x_1}(\mathbf{x}) \\ &\vdots \\ y_n &= \frac{\partial f}{\partial x_n}(\mathbf{x}) \end{aligned}$$

exists (Theorem 23B); that is, the second derivative $f''(\mathbf{x})$ exists throughout $U$. This enables us to appeal to Theorem C, but we know of functions where the derivative exists even when the partials involved are not continuous. We therefore conjecture that Theorem F can be proved with a weakened hypothesis (Problem 44I).

**Theorem F.** Let $f$ have continuous second partial derivatives $\partial^2 f / \partial x_i \, \partial x_j = f_{ji}$ throughout an open convex set $U \subseteq \mathbf{R}^n$. Then $f$ is convex on $U$ if and only if the **Hessian** matrix

$$\begin{bmatrix} f_{11}(\mathbf{x}) & \cdots & f_{1n}(\mathbf{x}) \\ \vdots & & \vdots \\ f_{n1}(\mathbf{x}) & \cdots & f_{nn}(\mathbf{x}) \end{bmatrix}$$

is nonnegative definite for each $\mathbf{x} \in U$. Moreover, if the Hessian matrix is positive definite on $U$, then $f$ is strictly convex.

We pointed out after stating the single real variable form of this theorem (Theorem 12C) that the last statement of this theorem is not reversible; that is, strict convexity of $f$ on $U$ does not mean the Hessian matrix will be positive definite on $U$. Bernstein and Toupin [1962] have proved, however, that if $f$ is strictly convex and twice continuously differentiable, then $[f''(\mathbf{x})]$ will be positive definite with possible exceptions on a nowhere dense subset of $U$.

### PROBLEMS AND REMARKS

**A.** We saw in this section that $f(r, s) = r^2 + 3rs + 5s^2$ is convex. Verify the inequality of Theorem A for $\mathbf{x}_0 = (1, 2)$ and $\mathbf{x} = (\tfrac{5}{4}, \tfrac{9}{4})$.

**B.** According to Theorem $D$, the function of Problem 23B cannot be convex. Show that this is the case without appeal to Theorem $D$.

**C.** Use Theorem F to verify that $f(r, s) = \exp(r^2 + s^2)$ is convex for all $(r, s)$. Verify the same thing with reasoning that involves less computation.

**D.** Let $f$ be defined and continuously differentiable on the open convex set $U \subseteq \mathbf{R}^n$. Then $f$ is **pseudoconvex** on $U$ if for $\mathbf{x}_1, \mathbf{x}_2 \in U$,
$$f(\mathbf{x}_2) < f(\mathbf{x}_1) \qquad \text{implies} \qquad \langle \nabla f(\mathbf{x}_1), \mathbf{x}_2 - \mathbf{x}_1 \rangle < 0$$
and $f$ is **quasiconvex** on $U$ if for $\mathbf{x}_1, \mathbf{x}_2 \in U$,
$$f(\mathbf{x}_2) \leq f(\mathbf{x}_1) \qquad \text{implies} \qquad \langle \nabla f(\mathbf{x}_1), \mathbf{x}_2 - \mathbf{x}_1 \rangle \leq 0$$

(1) If $f$ is pseudoconvex, then it is quasiconvex.
(2) If $f$ is convex, then it is quasiconvex.

We discuss quasiconvex functions more fully in Section 81. The notions of quasiconvexity, pseudoconvexity, and others are discussed by Ponstein [1967] in a paper titled *Seven kinds of convexity*.

**E.** Let $f(x_1, \ldots, x_n) = -\prod_{i=1}^{n}(1 - e^{x_i})^{\lambda_i}$ where $\lambda_i > 0$ on the set $K = \{\mathbf{x} \in \mathbf{R}^n : x_i < 0, i = 1, \ldots, n, \sum_{i=1}^{n} \lambda_i e^{x_i} \leq 1\}$.

(1) $K$ is convex.
(2) The Hessian matrix for $f$ has entries
$$f_{ij}(\mathbf{x}) = -f(\mathbf{x}) \frac{\lambda_i \lambda_j \exp(x_i + x_j)}{(1 - \exp(x_i))(1 - \exp(x_j))} \left[ -1 + \delta_{ij} \frac{\exp(-x_i)}{\lambda_i} \right]$$
where $\delta_{ij} = 1$ if $i = j$, $\delta_{ij} = 0$ if $i \neq j$.
(3) The Hessian matrix is nonnegative definite. This is most easily established by noting that it is sufficient to show the matrix $[-1 + \delta_{ij} (e^{-x_i}/\lambda_i)]$ to be nonnegative definite. This function enters into a paper by Chaundy and Evelyn [1967].

**\*F.** Let $F: \mathbf{R}^m \times \mathbf{R}^n \to \mathbf{R}^n$ be described by
$$y_1 = f_1(\mathbf{x}, \mathbf{y})$$
$$\vdots \qquad \vdots$$
$$y_n = f_n(\mathbf{x}, \mathbf{y})$$

Standard theorems in advanced calculus [Buck, 1965, p. 285] state conditions on the matrix

$$\begin{bmatrix} \dfrac{\partial f_1}{\partial y_1} & \cdots & \dfrac{\partial f_1}{\partial y_n} \\ \vdots & & \vdots \\ \dfrac{\partial f_n}{\partial y_1} & \cdots & \dfrac{\partial f_n}{\partial y_n} \end{bmatrix}$$

under which the equation $F(\mathbf{x}, \mathbf{y}) = \mathbf{0}$ defines implicitly a function $G: \mathbf{R}^m \to \mathbf{R}^n$ such that $G$ is differentiable and satisfies $F(\mathbf{x}, G(\mathbf{x})) = 0$. Similar conditions can be used to guarantee the convexity of the coordinate functions defining $G$ [Brock and Thompson, 1966].

## 43. The Support of Convex Functions

We saw in Chapter I (Theorem 12D) that a convex function $f: (a, b) \to \mathbf{R}$ is characterized by having a line of support at each point

## 43. The Support of Convex Functions

of $(a, b)$. It is this theorem that we now wish to generalize. Toward this end, we recall that straight lines in $\mathbf{R}^2$ are graphs of affine functions. The natural extension to $\mathbf{L}$ then anticipates that support functions will be affine. An affine function $A$ known to pass through $(\mathbf{x}_0, f(\mathbf{x}_0))$ may be represented in the form $A(\mathbf{x}) = f(\mathbf{x}_0) + T(\mathbf{x} - \mathbf{x}_0)$ where $T$ is linear. In order to prove the fundamental theorem on the support of convex functions, we need a version of the **Hahn–Banach** theorem which we now prove.

**Theorem A.** Let $f$ be convex on an open set $U$ of a normed linear space $\mathbf{L}$ and let $V_0$ be a nontrivial subspace such that $V_0 \cap U \neq \varnothing$. If $A_0 \colon V_0 \to \mathbf{R}$ is affine and $A_0(\mathbf{x}) \leqslant f(\mathbf{x})$ on $V_0 \cap U$, then there is an affine extension $A \colon \mathbf{L} \to \mathbf{R}$ of $A_0$ such that $A(\mathbf{x}) \leqslant f(\mathbf{x})$ on $U$.

**Proof.** Define $f$ to be $+\infty$ if $x \notin U$ and note that $f$ now satisfies the defining inequality for convex functions on all of $\mathbf{L}$. Choose a fixed $\mathbf{w} \in \mathbf{L}$, $\mathbf{w} \notin V_0$. Then for $\mathbf{x}, \mathbf{y} \in V_0$, $r > 0$, $s > 0$ (Fig. 43.1),

$$\frac{r}{r+s} A_0(\mathbf{x}) + \frac{s}{r+s} A_0(\mathbf{y}) = A_0 \left( \frac{r}{r+s} \mathbf{x} + \frac{s}{r+s} \mathbf{y} \right)$$

$$\leqslant f \left( \frac{r}{r+s} \mathbf{x} + \frac{s}{r+s} \mathbf{y} \right)$$

$$= f \left[ \frac{r}{r+s} (\mathbf{x} - s\mathbf{w}) + \frac{s}{r+s} (\mathbf{y} + r\mathbf{w}) \right]$$

$$\leqslant \frac{r}{r+s} f(\mathbf{x} - s\mathbf{w}) + \frac{s}{r+s} f(\mathbf{y} + r\mathbf{w})$$

Multiplying by $r + s$ gives

$$r A_0(\mathbf{x}) + s A_0(\mathbf{y}) \leqslant r f(\mathbf{x} - s\mathbf{w}) + s f(\mathbf{y} + r\mathbf{w})$$

or

$$g(\mathbf{x}, s) = \frac{A_0(\mathbf{x}) - f(\mathbf{x} - s\mathbf{w})}{s} \leqslant \frac{f(\mathbf{y} + r\mathbf{w}) - A_0(\mathbf{y})}{r} = h(\mathbf{y}, r)$$

It follows that $\sup g \leqslant \inf h$ on $V_0 \times P$ where $P$ denotes the positive real line. Moreover, if $\mathbf{x}_0 \in V_0 \cap U$ and $s_0$ is small enough so that both $\mathbf{x}_0 - s_0 \mathbf{w}$ and $\mathbf{x}_0 + s_0 \mathbf{w}$ are in $U$, then both $g(\mathbf{x}_0, s_0)$ and $h(\mathbf{x}_0, s_0)$ are finite; hence $\sup g$ and $\inf h$ are also. We may therefore select a finite real number $\alpha$ between $\sup g$ and $\inf h$. In particular

$$\frac{A_0(\mathbf{x}) - f(\mathbf{x} - s\mathbf{w})}{s} \leqslant \alpha \leqslant \frac{f(\mathbf{x} + r\mathbf{w}) - A_0(\mathbf{x})}{r}$$

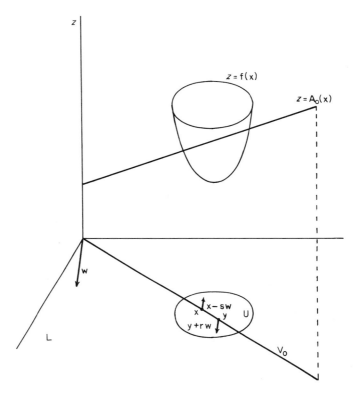

**Fig. 43.1**

for all $\mathbf{x} \in V_0$, $r > 0$, $s > 0$. Substituting $t = -s$ when $t < 0$ and $t = r$ for $t > 0$ leads immediately to

$$A_0(\mathbf{x}) + \alpha t \leqslant f(\mathbf{x} + t\mathbf{w})$$

for all $\mathbf{x} \in V_0$, $t \in \mathbf{R}$.

Now let $V_1 = \{\mathbf{x} + t\mathbf{w} : \mathbf{x} \in V_0, t \in \mathbf{R}\}$; $V_1$ is a subspace which properly contains $V_0$. On $V_1$ define $A_1$ by $A_1(\mathbf{x} + t\mathbf{w}) = A_0(\mathbf{x}) + t\alpha$. Then $A_1$ is affine on $V_1$, $A_1 = A_0$ on $V_0$, and we have just established that $A_1(\mathbf{x} + t\mathbf{w}) \leqslant f(\mathbf{x} + t\mathbf{w})$. If $V_1 = \mathbf{L}$, our theorem is proved. If not and the dimension of $\mathbf{L}$ is finite, we may proceed by mathematical induction to extend $A_0$ to all of $\mathbf{L}$ since $\dim V_n = \dim V_{n-1} + 1$.

If the dimension of $\mathbf{L}$ is infinite, we need a form of transfinite induction to complete the proof. Suppose we consider pairs $\{V_\alpha, A_\alpha\}$ where $V_\alpha$ is a subspace of $\mathbf{L}$, $V_\alpha \supseteq V_0$, and $A_\alpha$ is an affine function defined on $V_\alpha$ in

### 43. The Support of Convex Functions

such a way as to be an extension of $A_0$ that satisfies $A_\alpha(\mathbf{x}) \leq f(\mathbf{x})$ for any $\mathbf{x} \in V_\alpha \cap U$. For two such pairs, we shall write $\{V_\alpha, A_\alpha\} \leq \{V_\beta, A_\beta\}$ if $V_\alpha \subseteq V_\beta$ and $A_\beta$ is an extension of $A_\alpha$ to $V_\beta$. The collection $\mathscr{P}$ of all such pairs is **partially ordered** by $\leq$; that is,

(O1) $\{V_\alpha, A_\alpha\} \leq \{V_\alpha, A_\alpha\}$ for any pair in $\mathscr{P}$.
(O2) If $\{V_\alpha, A_\alpha\} \leq \{V_\beta, A_\beta\}$ and $\{V_\beta, A_\beta\} \leq \{V_\gamma, A_\gamma\}$, then $\{V_\alpha, A_\alpha\} \leq \{V_\gamma, A_\gamma\}$.

Let us now consider a subcollection $\mathscr{C} \subseteq \mathscr{P}$ that is totally ordered so that for pairs in $\mathscr{C}$ we have

(O3) $\{V_\alpha, A_\alpha\} \leq \{V_\beta, A_\beta\}$ and $\{V_\beta, A_\beta\} \leq \{V_\alpha, A_\alpha\}$ imply $V_\alpha = V_\beta$ and $A_\alpha = A_\beta$.
(O4) Given two members of $\mathscr{C}$, either $\{V_\alpha, A_\alpha\} \leq \{V_\beta, A_\beta\}$ or $\{V_\beta, A_\beta\} \leq \{V_\alpha, A_\alpha\}$.

Such a collection is called a **chain**. The original pair $\{V_0, A_0\}$ together with the pair $\{V_1, A_1\}$ obtained above form a chain of two elements. For any chain $\mathscr{C} \subseteq \mathscr{P}$, even one having an infinite number of elements, we can form

$$V_C = \bigcup V, \qquad A_C = \bigcup A.$$

where the unions are over all the members of $\mathscr{C}$ and the union of $A_\alpha$ is understood as follows. For $\mathbf{x} \in V_C$, there exists an $\alpha$ such that $\mathbf{x} \in V_\alpha$. Then we understand $A_C(\mathbf{x}) = A_\alpha(\mathbf{x})$, and the fact that we are in a chain assures us that $A_C(\mathbf{x})$ is uniquely defined. Moreover, for $\mathbf{x} \in V_C \cap U$, $A_C(\mathbf{x}) = A_\alpha(\mathbf{x}) \leq f(\mathbf{x})$, so $\{V_C, A_C\} \in \mathscr{C}$. Now note that for any $\{V_\alpha, A_\alpha\} \in \mathscr{C}$, $\{V_\alpha, A_\alpha\} \leq \{V_C, A_C\}$. That is, $\mathscr{C}$ has $\{V_C, A_C\}$ as a **maximal element**. Since $\mathscr{C}$ was an arbitrary chain in $\mathscr{P}$, we are now in a position to invoke our axiom of transfinite induction [Taylor, 1958, p. 39].

> **Zorn's Lemma.** If each chain $\mathscr{C}$ in a partially ordered set $\mathscr{P}$ has a maximal element, then there is a maximal element in the set $\mathscr{P}$.

We conclude that $\mathscr{P}$ has a maximal element, say $\{V_M, A_M\}$. Now $V_M = \mathbf{L}$, for if not, a repeat of the first part of our argument would extend $A_M$ to a subset of $\mathbf{L}$ properly containing $V_M$, thus violating the maximal character of $\{V_M, A_M\}$. Since $A_M$ is thus defined on $\mathbf{L}$ with the property (common to all members of $\mathscr{P}$) that $A_M = A_0$ on $V_0$ and $A_M(\mathbf{x}) \leq f(\mathbf{x})$ for $x \in \mathbf{L} \cap U = U$, our proof is completed by taking $A = A_M$. ●

We are now ready to prove the fundamental theorem on the support of a convex function. We say a function $f : \mathbf{U} \to \mathbf{R}$ has **support** at $\mathbf{x}_0 \in \mathbf{U}$ if there is an affine function $A : \mathbf{L} \to \mathbf{R}$ such that $A(\mathbf{x}_0) = f(\mathbf{x}_0)$ and $A(\mathbf{x}) \leq f(\mathbf{x})$ for every $\mathbf{x} \in \mathbf{U}$.

**Theorem B.** The function $f$ is convex on an open convex set $U$ of a normed linear space $\mathbf{L}$ if and only if $f$ has support at each point of $U$.

**Proof.** Suppose $f$ has support at each point of $U$. Let $\mathbf{x}_1, \mathbf{x}_2 \in U$, $\mathbf{x}_0 = t\mathbf{x}_1 + (1-t)\mathbf{x}_2$ where $t \in [0, 1]$, and let $A(\mathbf{x}) = f(\mathbf{x}_0) + T(\mathbf{x} - \mathbf{x}_0)$ be an affine function that supports $f$ at $\mathbf{x}_0$. Then

$$f(\mathbf{x}_0) = A(\mathbf{x}_0) = tA(\mathbf{x}_1) + (1-t)A(\mathbf{x}_2)$$
$$\leq tf(\mathbf{x}_1) + (1-t)f(\mathbf{x}_2)$$

which establishes the convexity of $f$.

Now suppose $f$ is convex on $U$, and choose $\mathbf{x}_0 \in U$. Take $\mathbf{u}$ to be a unit vector with the same direction as $\mathbf{x}_0$ (or an arbitrary unit vector if $\mathbf{x}_0 = \mathbf{O}$), and define the subspace $V_0 = \{\mathbf{x}_0 + t\mathbf{u} : t \in \mathbf{R}\}$. The open interval $I = \{t : (\mathbf{x}_0 + t\mathbf{u}) \in U\}$ contains 0 and we may define a convex function $g(t) = f(\mathbf{x}_0 + t\mathbf{u})$ in the now customary way. According to Theorem 12D there is a support functional $G : \mathbf{R} \to \mathbf{R}$ such that $G(0) = g(0)$ and $G(t) \leq g(t)$ on $I$. We use this $G$ to define $A_0 : V_0 \to \mathbf{R}$ by $A_0(\mathbf{x}_0 + t\mathbf{u}) = G(t)$. Now $A_0$ is affine, $A_0(\mathbf{x}_0) = f(\mathbf{x}_0)$, and

$$A_0(\mathbf{x}_0 + t\mathbf{u}) = G(t) \leq g(t) = f(\mathbf{x}_0 + t\mathbf{u})$$

for $(\mathbf{x}_0 + t\mathbf{u}) \in V_0 \cap U$. It follows from Theorem A that $A_0$ can be extended to a function $A : \mathbf{L} \to \mathbf{R}$ where $A(\mathbf{x}) \leq f(\mathbf{x})$ on $U$. That is, $A$ is a support function for $f$ at $\mathbf{x}_0$. ●

We now know that if $f$ is convex on an open set $U \subseteq \mathbf{L}$, then $f$ has support at each $\mathbf{x} \in U$. Although we have said nothing about the continuity of either $f$ or its support functions, it is easy (Problem A) to show that if $f$ is continuous on $U$, then all the support functions are continuous. Consideration of a discontinuous linear functional $f$ (which is its own support at all $\mathbf{x} \in \mathbf{L}$) shows that we may have $f$ and all of its support functions discontinuous. It is natural to ask whether continuity of the support functions implies continuity of $f$. This is not an easy question to answer. Theorem C answers it affirmatively for the case where $\mathbf{L}$ is a Banach space.

## 43. The Support of Convex Functions

**Theorem C.** Suppose $f: U \to \mathbf{R}$ has continuous support at each point of a convex open set $U$ in a Banach space $\mathbf{L}$. Then $f$ is a continuous convex function.

**Proof.** Since $f$ is convex (Theorem B), its continuity is established if we can show that it is bounded above in some open subset of $U$ (Theorem 41C). Let us choose an open ball $S$ of radius no more than 1 such that $\bar{S} \subseteq U$, and then define

$$F_m = \{\mathbf{x} \in \bar{S}: f(\mathbf{x}) \leqslant m\}, \quad m = 1, 2, \ldots$$

Thus, $\bar{S} = \bigcup_{m=1}^{\infty} F_m$. We will prove our theorem by showing that at least one $F_m$ has a nonempty interior.

First we show that $F_m$ is closed. Let $\{\mathbf{x}_n\}$ be a sequence in $F_m$ with $\mathbf{x}_n \to \bar{\mathbf{x}}$. Now $\bar{\mathbf{x}} \in \bar{S} \subseteq U$, so there is a continuous linear functional $T$ such that $f(\mathbf{x}) \geqslant f(\bar{\mathbf{x}}) + T(\mathbf{x} - \bar{\mathbf{x}})$. And because $T$ is continuous, we can for arbitrary $\varepsilon > 0$ find a neighborhood $N$ of $\bar{\mathbf{x}}$ on which $f(\mathbf{x}) \geqslant f(\bar{\mathbf{x}}) - \varepsilon$. But for sufficiently large $n$, $\mathbf{x}_n \in N$; so for such $n$, $f(\bar{\mathbf{x}}) \leqslant f(\mathbf{x}_n) + \varepsilon \leqslant m + \varepsilon$. We conclude that $f(\bar{\mathbf{x}}) \leqslant m$, hence that $\bar{\mathbf{x}} \in F_m$ and $F_m$ is closed.

We shall now assume that each $F_m$ has an empty interior $F_m^0$, showing that this leads to a contradiction. Note that $S \setminus F_1$ is open (Problem 21C) and it is nonempty (for $S \setminus F_1 = \varnothing$ implies $F_1 \supseteq S$ implies $F_1^0 \neq \varnothing$). Choose an open ball $S_1$ of radius at most $\frac{1}{2}$ such that $\bar{S}_1 \subseteq S \setminus F_1$. Repeating the arguments above, $S_1 \setminus F_2$ is open and nonempty. We may therefore choose an open ball $S_2$ of radius at most $\frac{1}{4}$ such that $\bar{S}_2 \subseteq S_1 \setminus F_2$. Proceeding in this way, we obtain a sequence of open balls $S = S_0, S_1, S_2, \ldots$ in which $S_m$ has a radius of at most $1/2^m$ and $\bar{S}_m \subseteq S_{m-1} \setminus F_m$. Form $V = \bigcap_{m=1}^{\infty} \bar{S}_m$. A sequence $\{\mathbf{y}_m\}$ formed by choosing $\mathbf{y}_i \in S_i$ will be Cauchy and must, owing to the completeness of a Banach space, converge to a point that in turn must be in $V$; thus, $V \neq \varnothing$. Let $\mathbf{z} \in V$. Then of course $\mathbf{z} \in \bar{S}_m \subseteq S_{m-1} \setminus F_m$ for $m = 1, 2, \ldots$, so $\mathbf{z} \in \bar{S}$. But $\mathbf{z} \notin \bigcup_{m=1}^{\infty} F_m$, violating the fact that $\bar{S} = \bigcup F_m$. We have our contradiction. ●

We pause now to consider the situation in $\mathbf{R}^n$ where things are more simple than they are in a general space $\mathbf{L}$. Our aim, of course, is to extend as many of the results of Chapter I as we can. How far have we come? If $f$ is convex on $U \subseteq \mathbf{R}^n$, then Section 41 assures us that $f$ is continuous on $U^0$ and Lipschitz on any compact subset of $U^0$. In Section 42 we learned how to characterize the convexity of $f$ in terms of first and second derivatives much as we did in Chapter I. And now we know that $f$ has support at each of the points of $U^0$ just as before. Of course we should not expect this support to be unique; it was not in the one-dimensional case except at points where $f: I \to \mathbf{R}$ was differentiable. And as a matter of fact, uniqueness of support for functions convex on $U \subseteq \mathbf{R}^n$ is intimately related to Fréchet differentiability as we shall see in the next section. Even without uniqueness, however, we can introduce the notion of derivative that played such an important role in Section 15 in connection with conjugate convex functions.

Let us recall that any linear functional $T: \mathbf{R}^n \to \mathbf{R}$ is uniquely determined by a vector $\mathbf{y} \in \mathbf{R}^n$ according to the relation $T(\mathbf{u}) = \langle \mathbf{y}, \mathbf{u} \rangle$, and conversely each $\mathbf{y} \in \mathbf{R}^n$ determines such a functional. For convex $f: U \to \mathbf{R}$ we define the **subdifferential** $\partial f$ of $f$ by

$$\partial f(\mathbf{x}_0) = \{\mathbf{y} \in \mathbf{R}^n : f(\mathbf{x}) \geq f(\mathbf{x}_0) + \langle \mathbf{y}, \mathbf{x} - \mathbf{x}_0 \rangle, \mathbf{x} \in U\}$$

Abusing the language just slightly, we may say that $\partial f(\mathbf{x}_0)$ is the set of supports for $f$ at $\mathbf{x}_0$. In this context, we can generalize practically everything we did in Section 15. For example, if $f: U \to \mathbf{R}$ is convex and closed, then $\partial f$ is a maximal monotone increasing relation. If we define the **conjugate** $f^*: U^* \to \mathbf{R}$ by

$$f^*(\mathbf{y}) = \sup_{\mathbf{x} \in U} [\langle \mathbf{y}, \mathbf{x} \rangle - f(\mathbf{x})]$$

on the set $U^*$ where $f^*$ is finite, then the facts we learned about $f^*$ are still valid, as was pointed out by Fenchel [1949]. This material is treated by Rockafellar [1970a] in much greater detail; we limit ourselves to the precise statement of a few results in Problems C and D.

If $\partial f(\mathbf{x}_0)$ has just one member (that is, if the support of $f$ at $\mathbf{x}_0$ is unique), and if $\mathbf{x}_n \to \mathbf{x}_0$, then $\partial f(\mathbf{x}_n)$ must converge to $\partial f(\mathbf{x}_0)$. This is the content of our last theorem.

**Theorem D.** Suppose $f: U \to \mathbf{R}$ is convex on an open set $U \subseteq \mathbf{R}^n$, and let $f$ have unique support

$$A_0(\mathbf{x}) = f(\mathbf{x}_0) + T_0(\mathbf{x} - \mathbf{x}_0)$$

at $\mathbf{x}_0 \in U$. If $\mathbf{x}_n \to \mathbf{x}_0$ and $A_n(\mathbf{x}) = f(\mathbf{x}_n) + T_n(\mathbf{x} - \mathbf{x}_n)$ is a support at $\mathbf{x}_n$, then $\| T_n - T_0 \| \to 0$.

**Proof.** Choose $\rho$ so that $\overline{N_\rho(\mathbf{x}_0)} \subseteq U$ and such that $f$ satisfies a Lipschitz condition on $\overline{N_\rho(\mathbf{x}_0)}$ with constant $B$ (Theorem 41B). Choose $r \in (0, \rho)$ and let $n$ be large enough so that $\mathbf{x}_n + r\mathbf{u} \in \overline{N_\rho(\mathbf{x}_0)}$ for all unit vectors $\mathbf{u}$. Then since $A_n(\mathbf{x}_n + r\mathbf{u}) \leq f(\mathbf{x}_n + r\mathbf{u})$,

$$rT_n(\mathbf{u}) = T_n(r\mathbf{u}) \leq f(\mathbf{x}_n + r\mathbf{u}) - f(\mathbf{x}_n) \leq Br$$

### 43. The Support of Convex Functions

We have $T_n(\mathbf{u}) \leqslant B$ for any unit vector, and in particular then $T_n(-\mathbf{u}) \leqslant B$ so $T_n(\mathbf{u}) \geqslant -B$, and we conclude that $\| T_n \| \leqslant B$. This means that $\{T_n\}$ is a bounded sequence in the dual space of $\mathbf{R}^n$ (see Section 22); hence some subsequence $\{T_{n_k}\}$ must converge, say to $S$. Let $A(\mathbf{x}) = f(\mathbf{x}_0) + S(\mathbf{x} - \mathbf{x}_0)$. Since

$$A(\mathbf{x}) = \lim_{k \to \infty} [f(\mathbf{x}_{n_k}) + T_{n_k}(\mathbf{x} - \mathbf{x}_{n_k})] \leqslant f(\mathbf{x})$$

$A$ supports $f$ at $\mathbf{x}_0$. Uniqueness of the support at $\mathbf{x}_0$ then implies that $S = T_0$. The same argument shows that any other convergent subsequence of $\{T_n\}$ must converge to $T_0$, so we conclude that $\{T_n\}$ itself converges to $T_0$. ●

### PROBLEMS AND REMARKS

**A.** Suppose $f$ is convex and continuous on $U \subseteq \mathbf{L}$. Then the support to $f$ at any point $\mathbf{x}_0 \in U$ is continuous.

\***B.** Consider the convex function $f$ defined on a subset of $\mathbf{R}^2$ by $f(r, s) = r^2/s$, $s > 0$; $f(\mathbf{O}) = 0$. It is discontinuous at $\mathbf{O}$ but has (continuous) support at this boundary point. Can you find an example (necessarily on an incomplete infinite-dimensional space $\mathbf{L}$) of a function that has continuous support at each point of a convex set $U$, and yet is discontinuous at an interior point of $U$?

\***C.** Consider the subdifferential $\partial f$ of a convex function $f \colon U \to \mathbf{R}$, $U \subseteq \mathbf{R}^n$, as introduced in this section.

(1) If $U = \mathbf{R}^n$ and $f(\mathbf{x}) = \| \mathbf{x} \|$, then $\partial f(\mathbf{O}) = \{\mathbf{x} \in \mathbf{R}^n \colon \| \mathbf{x} \| \leqslant 1\}$.
(2) If $\mathbf{x} \in U^0$, then $\partial f(\mathbf{x}) \neq \varnothing$. What if $\mathbf{x}$ is a boundary point?
(3) If $U$ is open and $\mathbf{x} \in U$, then $\mathbf{y} \in \partial f(\mathbf{x}) \Leftrightarrow f_+'(\mathbf{x}; \mathbf{v}) \geqslant \langle \mathbf{y}, \mathbf{v} \rangle$ for all $\mathbf{v} \in \mathbf{R}^n$.
(4) We say $f$ is **closed** if $L_\alpha = \{\mathbf{x} \in U \colon f(\mathbf{x}) \leqslant \alpha\}$ is closed for all $\alpha \in \mathbf{R}$. If $f$ is closed, then $\partial f$ is a maximal monotone increasing relation; that is, $\partial f$ is monotone increasing ($\mathbf{x}_1$, $\mathbf{x}_2 \in \mathrm{dom}\ \partial f$ implies $\langle \mathbf{x}_2 - \mathbf{x}_1,\ \partial f(\mathbf{x}_2) - \partial f(\mathbf{x}_1) \rangle \geqslant 0$) and cannot be properly embedded in any monotone increasing relation in $\mathbf{R}^n \times \mathbf{R}^n$.

See Rockafellar [1970a] for the finite-dimensional case stated here. There is also an extension to the infinite-dimensional case [Minty, 1964; Rockafellar, 1970b].

\***D.** Let $f \colon U \to \mathbf{R}$ be a closed convex function defined on $U \subseteq \mathbf{R}^n$, and let $f^* \colon U^* \to \mathbf{R}$ be the conjugate. Then

(1) $\langle \mathbf{x}, \mathbf{y} \rangle \leqslant f(\mathbf{x}) + f^*(\mathbf{y})$ for all $\mathbf{x} \in U$, $\mathbf{y} \in U^*$,
(2) $\langle \mathbf{x}, \mathbf{y} \rangle = f(\mathbf{x}) + f^*(\mathbf{y}) \Leftrightarrow \mathbf{y} \in \partial f(\mathbf{x})$,
(3) $\partial(f^*) = (\partial f)^{-1}$,
(4) $f^{**} = f$.

See Fenchel [1949] and Rockafellar [1970a] for the finite-dimensional case stated here. Again there is an extension to the infinite-dimensional case [Moreau, 1962; Brøndsted, 1964; Ioffe and Tikhomirov, 1968].

E. Theorem B is of course related to the fundamental support theorem for convex sets. (Theorem 32F). Based on this theorem, we outline a proof of Theorem B for the case in which the convex function is continuous on the open set $U \subseteq \mathbf{L}$.

(1) The set $\operatorname{epi}(f) = \{(\mathbf{x}, y): \mathbf{x} \in U, y \in \mathbf{R}, y \geqslant f(\mathbf{x})\}$ is convex, it has interior points, and the points $(\mathbf{x}, f(\mathbf{x}))$ are boundary points.

(2) Corresponding to $\mathbf{x}_0 \in U$, there is according to Theorem 32F a closed hyperplane of support that passes through $(\mathbf{x}_0, f(\mathbf{x}_0))$. This means there is a linear functional $g$: $\mathbf{L} \times \mathbf{R} \to \mathbf{R}$ such that $(\mathbf{x}, y) \in \operatorname{epi}(f)$ implies that $g(\mathbf{x}, y) \geqslant g(\mathbf{x}_0, f(\mathbf{x}_0)) = \lambda$.

(3) Argue that $g(\mathbf{O}, 1) \neq 0$.

(4) Define $h: \mathbf{L} \to \mathbf{R}$ by

$$h(\mathbf{x}) = \frac{1}{g(\mathbf{O}, 1)} [\lambda - g(\mathbf{x}, 0)]$$

Show that $h$ supports $f$ at $\mathbf{x}_0$.

Though this proof seems to avaid using Theorem A and its dependence on Zorn's lemma, the simplicity is only apparent since the support theorem for convex sets (Theorem 32F) itself rests on an application of Zorn's lemma. We have also avoided difficulties by restricting ourselves here to a continuous function, thereby assuring the existence of interior points for $\operatorname{epi}(f)$. In this case, the resulting support function $h$, corresponding to the closed hyperplane of support, is also continuous.

F. A function $f: U \to \mathbf{R}$ is said to be **lower semicontinuous** at $\mathbf{x}_0 \in U \subseteq \mathbf{L}$ if, given any $\varepsilon > 0$, there exists a neighborhood of $\mathbf{x}_0$ in which $f(\mathbf{x}) > f(\mathbf{x}_0) - \varepsilon$. Using arguments similar to those of Theorem C, show that a lower semicontinuous convex function $f$ defined on an open convex set $U$ of a Banach space $\mathbf{L}$ must be continuous on $U$.

G. Here is an alternate proof for Theorem C. At each $\mathbf{z} \in U$, let $A_\mathbf{z}$ be a continuous support function for $f$. Define $\phi(\mathbf{x}) = \sup_{\mathbf{z} \in U} A_\mathbf{z}(\mathbf{x})$. (How does $\phi$ compare with $\operatorname{env} f$ defined in Problem 41L?)

(1) Show that $\phi$ is convex and lower semicontinuous on $U$, hence that it is continuous on $U$ by Problem F. Then note that the restriction of $\phi$ to $U$ is $f$.

*(2) This proof, valid in any space in which a lower semicontinuous convex function is continuous, can be used to extend Theorem C to more general spaces. Rockafellar [1966] describes such spaces.

*H. [Dines, 1938]. Suppose $f$ is convex on an open set $U \subseteq \mathbf{R}^n$, and that $\{\mathbf{y}_n\}$ is a sequence of points in $U$ converging to $\mathbf{y}_0 \in U$. If for each $i$, we let

$$a_{i1}x_1 + \cdots + a_{in}x_n + a_{i,n+1}x_{n+1} = d_i$$

be the normal form of the equation of a supporting plane $S_i$ at $(\mathbf{y}_i, f(\mathbf{y}_i))$, then there is a subsequence of $\{S_i\}$ converging to a plane $S_0$,

$$a_{01}x_1 + \cdots + a_{0n}x_n + a_{0,n+1}x_{n+1} = d_0$$

that supports $f$ at $\mathbf{y}_0$. Convergence is in the sense that vector $(a_{i1}, \ldots, a_{in}, a_{i,n+1}, d_i)$ converges to $(a_{01}, \ldots, a_{0n}, a_{0,n+1}, d_0)$ in the space $\mathbf{R}^{n+2}$.

## 44. Differentiability of Convex Functions

We begin our discussion of the existence of the derivative of a function convex on $U \subseteq \mathbf{L}$ with some facts about one- and two-sided directional derivatives that follow directly from Section 11. Consideration of two-sided derivatives at a point $\mathbf{x}_0$ leads us to the concept of the Gateaux differential which we define and about which we prove our first theorem. After discussing the relation of the Gateaux differential to the Fréchet derivative $f'(\mathbf{x}_0)$, we see that a necessary condition for the existence of $f'(\mathbf{x}_0)$ is uniqueness of support at $\mathbf{x}_0$. We state this fact as Theorem B. As is usually the case, our results can be sharpened when $\mathbf{L} = \mathbf{R}^n$, and Theorem C points out that uniqueness of support at $\mathbf{x}_0$ is both necessary and sufficient for the existence of $f'(\mathbf{x}_0)$ in $\mathbf{R}^n$. We finally address ourselves to the question of whether or not a convex function must have a Fréchet derivative at any point of its domain.

Let $f$ be convex on an open set $U \subseteq \mathbf{L}$. Corresponding to any two points $\mathbf{x}_0, \mathbf{x}_1 \in U$, $\text{seg}[\mathbf{x}_0, \mathbf{x}_1] \subseteq U$ and we may define $g: [0, 1] \to \mathbf{R}$ by $g(t) = f[\mathbf{x}_0 + t(\mathbf{x}_1 - \mathbf{x}_0)]$. Then $g$ is convex on $[0, 1]$. It follows from what we know about $g$ that $f$ has one-sided derivatives in the direction $\mathbf{v} = \mathbf{x}_1 - \mathbf{x}_0$ at every point of $\text{seg}[\mathbf{x}_0, \mathbf{x}_1]$ and the directional derivative $f'(\mathbf{x}; \mathbf{v})$ exists at all but possibly a countable number of points of the segment.

We say that $f$ is **Gateaux differentiable** at $\mathbf{x}_0$ if the two-sided directional derivative $f'(\mathbf{x}_0; \mathbf{v})$ exists for each $\mathbf{v} \in \mathbf{L}$; that is, if

$$\lim_{t \to 0} \frac{f(\mathbf{x}_0 + t\mathbf{v}) - f(\mathbf{x}_0)}{t} = f'(\mathbf{x}_0; \mathbf{v})$$

exists for each $\mathbf{v} \in \mathbf{L}$.

> **Theorem A.** Let $f$ be convex on an open set $U \subseteq \mathbf{L}$. Then $f$ has a Gateaux differential at $\mathbf{x}_0$ if and only if $f$ has unique support at $\mathbf{x}_0$. Moreover, if the Gateaux differential $f'(\mathbf{x}_0; \mathbf{v})$ exists, $A(\mathbf{v}) = f'(\mathbf{x}_0; \mathbf{v})$ is linear in $\mathbf{v}$.

**Proof.** We shall assume $\mathbf{x}_0 = \mathbf{O}$, $f(\mathbf{O}) = 0$, and we shall define, for fixed but arbitrary $\mathbf{v}$, $g(t) = f(t\mathbf{v})$. Then $g$ is convex on an interval containing 0 and

$$f_+'(\mathbf{O}; \mathbf{v}) = g_+'(0); \quad f_-'(\mathbf{O}; \mathbf{v}) = g_-'(0)$$

Choose $m$ so that $g_-'(0) \leqslant m \leqslant g_+'(0)$. Then $a(t) = mt$ is a support to $g$

at 0 (Theorem 12D). The linear function $A_0(t\mathbf{v}) = mt$ defined on the subspace $V_0$ spanned by $\mathbf{v}$ satisfies

$$A_0(t\mathbf{v}) = mt = a(t) \leqslant g(t) = f(t\mathbf{v})$$

and (by Theorem 43A) can be extended to an affine function $A$ (in this case linear) defined on $\mathbf{L}$ and supporting $f$ at $\mathbf{O}$. Now if the support to $f$ at $\mathbf{O}$ is unique, it follows that only one $m$ can be found satisfying $g_-'(0) \leqslant m \leqslant g_+'(0)$. Hence $f_+'(\mathbf{O}; \mathbf{v}) = f_-'(\mathbf{O}; \mathbf{v})$. Since $\mathbf{v}$ was arbitrary, $f$ has a Gateaux differential at $\mathbf{x}_0$.

Now let us suppose that $f$ has a Gateaux differential at $\mathbf{O}$. Let $A$ be a support for $f$ at $\mathbf{O}$, necessarily linear since $f(0) = 0$. Then for any $\mathbf{v}$, $t > 0$

$$A(\mathbf{v}) = \frac{1}{t} A(t\mathbf{v}) \leqslant \frac{f(t\mathbf{v})}{t}$$

Taking limits as $t \downarrow 0$ gives $A(\mathbf{v}) \leqslant f'(\mathbf{O}; \mathbf{v})$. Replacing $\mathbf{v}$ by $-\mathbf{v}$ enables us to write $A(-\mathbf{v}) \leqslant f'(\mathbf{O}; -\mathbf{v})$ and so

$$-f'(\mathbf{O}, -\mathbf{v}) \leqslant A(\mathbf{v}) \leqslant f'(\mathbf{O}; \mathbf{v})$$

But when $f'(\mathbf{O}; \mathbf{v})$ exists as it does by hypothesis here, we recall (23.1) that $-f'(\mathbf{O}; -\mathbf{v}) = f'(\mathbf{O}; \mathbf{v})$ so

$$A(\mathbf{v}) = f'(\mathbf{O}; \mathbf{v}) \qquad (1)$$

is completely determined. This being true for all $\mathbf{v}$, only one $A$ can exist. ●

If $f$ has a Fréchet derivative at $\mathbf{x}_0$, then $f$ is Gateaux differentiable at $\mathbf{x}_0$ with $f'(\mathbf{x}_0; \mathbf{v}) = f'(\mathbf{x}_0)(\mathbf{v})$ because

$$f'(\mathbf{x}_0; \mathbf{v}) = \lim_{t \to 0} \frac{f(\mathbf{x}_0 + t\mathbf{v}) - f(\mathbf{x}_0)}{t}$$

$$= \lim_{t \to 0} \frac{f'(\mathbf{x}_0)(t\mathbf{v}) + \|t\mathbf{v}\| \varepsilon(\mathbf{x}_0, t\mathbf{v})}{t} = f'(\mathbf{x}_0)(\mathbf{v})$$

Gateaux differentiability of $f$ at $\mathbf{x}_0$ does not, however, guarantee the existence of $f'(\mathbf{x}_0)$ without some further conditions on $f$. To see what they are, define for $f$ Gateaux differentiable at $\mathbf{x}_0$, $\mathbf{h} = t\mathbf{v}$,

$$\varepsilon(\mathbf{x}_0, \mathbf{h}) = \varepsilon(\mathbf{x}_0, t\mathbf{v}) = \frac{t}{\|t\mathbf{v}\|} \left[ \frac{f(\mathbf{x}_0 + t\mathbf{v}) - f(\mathbf{x}_0)}{t} - f'(\mathbf{x}_0; \mathbf{v}) \right]$$

## 44. Differentiability of Convex Functions

Then since $f'(\mathbf{x}_0; t\mathbf{v}) = tf'(\mathbf{x}_0; \mathbf{v})$, we may rewrite this as

$$f(\mathbf{x}_0 + \mathbf{h}) = f(\mathbf{x}_0) + f'(\mathbf{x}_0; \mathbf{h}) + \|\mathbf{h}\| \varepsilon(\mathbf{x}_0, \mathbf{h}) \qquad \text{where}$$

$\varepsilon(\mathbf{x}_0, \mathbf{h})$ goes to zero as $\mathbf{h}$ approaches $\mathbf{O}$ along any fixed line through $\mathbf{O}$ (2)

Thus, to assure Fréchet differentiability at $\mathbf{O}$, we must impose the further conditions that

$$\varepsilon(\mathbf{x}_0, \mathbf{h}) \text{ must go to zero for any mode of approach of } \mathbf{h} \text{ to } \mathbf{O} \qquad (3)$$

and

$$f'(\mathbf{x}_0; \mathbf{h}) \text{ must be linear in } \mathbf{h} \qquad (4)$$

Examples exist (Problem B) showing that for a function Gateaux differentiable at $\mathbf{x}_0$, $f'(\mathbf{x}_0; \mathbf{v})$ may not be linear in $\mathbf{v}$. But since existence of the (Fréchet) derivative $f'(\mathbf{x}_0)$ always implies Gateaux differentiability, it is clear from Theorem A that existence of $f'(\mathbf{x}_0)$ implies uniqueness of the support of $f$ at $\mathbf{x}_0$. Moreover, from (1) and the fact that $f'(\mathbf{O}; \mathbf{v}) = f'(\mathbf{O})(\mathbf{v})$, we can specify the supporting function.

**Theorem B.** Let $f$ be convex on an open set $U \subseteq \mathbf{L}$. Then if $f'(\mathbf{x}_0)$ exists, $f$ has $A(\mathbf{x}) = f(\mathbf{x}_0) + f'(\mathbf{x}_0)(\mathbf{x} - \mathbf{x}_0)$ as its unique support at $\mathbf{x}_0$.

As indicated in the introduction, Theorem B is a two-way street if $\mathbf{L} = \mathbf{R}^n$.

**Theorem C.** Suppose $f : U \to \mathbf{R}$ is convex on an open set $U \subseteq \mathbf{R}^n$, and let $\mathbf{x}_0 \in U$. Then $f$ has unique support at $\mathbf{x}_0$ if and only if $f$ is (Fréchet) differentiable at $\mathbf{x}_0$.

**Proof.** Suppose $f$ has unique support at $\mathbf{x}_0$. Then it is Gateaux differentiable, meaning certainly that the partial derivatives all exist. The existence of $f'(\mathbf{x}_0)$ then follows from Theorem 42D. The reverse implication is a special case of Theorem B. ●

A convex function of a real variable was differentiable except perhaps on a countable set. There is no hope of carrying this theorem over to functions on $\mathbf{R}^n$ since the obviously convex function $f(r, s) = |r|$ (Fig. 44.1), failing to have unique support at any point on the $s$ axis, is not differentiable anywhere on this uncoutable set. We can generalize, however, in the following way, using the notion of almost everywhere as it is used in Lebesgue measure theory.

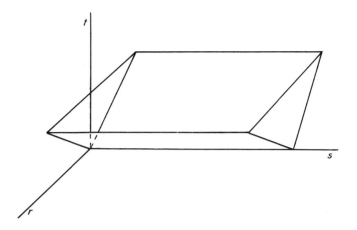

**Fig. 44.1**

**Theorem D.** Let $f: U \to \mathbf{R}$ be convex on the open set $U \subseteq \mathbf{R}^n$. Then $f$ is (Fréchet) differentiable almost everywhere in $U$.

**Proof.** On the basis of Theorem 42D, it will suffice to show that for each $j = 1, 2,..., n$, the set

$$E_j = \left\{ \mathbf{x} \in U: \; \frac{\partial f}{\partial x_j}(\mathbf{x}) \text{ does not exist} \right\}$$

has $m(E_j) = 0$. We first observe that $E_j$ is measurable (Problem F), thus assuring ourselves that in the case where $U$ is bounded, the characteristic function $\chi$ of $E_j$ will be integrable. Thus, for $\mathbf{x} = (x_1,..., x_n)$, Fubini's theorem [Natanson II, 1961, p. 86] assures us that

$$m(E_j) = \int \chi(\mathbf{x}) \, dm = \int \cdots \int [\chi(\mathbf{x}) \, dx_j] \prod_{i \neq j} dx_i$$

But if all the variables except $x_j$ are held fixed, $f(x_1,..., x_n)$ is convex as a function of $x_j$ and is therefore differentiable except at a countable number of points (Theorem 11C). Thus $\int \chi(\mathbf{x}) \, dx_j = 0$, and in turn, $m(E_j) = 0$. Finally, in the case where $U$ is not bounded, we arrive at the same conclusion by considering $U \cap B_r$, where $B_r$ is the ball of radius $r$ centered at $\mathbf{O}$, and then letting $r \to \infty$. ●

## 44. Differentiability of Convex Functions

Just as in the case of a function of a real variable, $f'$ is continuous on the set where it exists in $\mathbf{R}^n$.

**Theorem E.** Let $f$ be a convex function defined on an open convex set $U \subseteq \mathbf{R}^n$, and let $E$ be the subset of $U$ where $f$ is (Fréchet) differentiable. Then $f'$ is continuous on $E$.

**Proof.** Suppose $\mathbf{x}_0 \in E$ and $\{\mathbf{x}_n\}$ is a sequence of points in $E$ converging to $\mathbf{x}_0$. By Theorem B, $A_n(\mathbf{x}) = f(\mathbf{x}_n) + f'(\mathbf{x}_n)(\mathbf{x} - \mathbf{x}_n)$ is the unique support for $f$ at $\mathbf{x}_n$. By Theorem 43D, $\|f'(\mathbf{x}_n) - f'(\mathbf{x}_0)\| \to 0$ as $n \to \infty$. ●

### PROBLEMS AND REMARKS

**A.** Let $f$ be convex on an open set $U \subseteq \mathbf{L}$. For any $\mathbf{x}, \mathbf{x}_0 \in U$,

$$f(\mathbf{x}) - f(\mathbf{x}_0) \geqslant f_+'(\mathbf{x}_0 ; \mathbf{x} - \mathbf{x}_0)$$

and if $f$ is strictly convex, the inequality is strict. This provides a slight strengthening of Theorem 42A.

**B.** Suppose $f$ is Gateaux differentiable at $\mathbf{x}_0$ so that it satisfies condition (2) of the previous section.

(1) $f$ may fail to satisfy condition (4). Consider Problem 23B where for $\mathbf{v} = (a, b)$, $f'(\mathbf{O}; \mathbf{v}) = a^2 b/(a^2 + b^2)$.

(2) $f$ may satisfy condition (4) but not condition (3). Consider $f(r, s) = (r/s)(r^2 + s^2)$.

*(3) Can you find an example of a function satisfying conditions (2) and (3) but not (4)?

**C.** Some writers [Nashed, 1966] require that $f'(\mathbf{x}_0 ; \mathbf{v})$, in addition to existing, must be linear and continuous in $\mathbf{v}$ before they will say $f$ is Gateaux differentiable at $\mathbf{x}_0$. Let us call this **strong Gateaux differentiability**.

(1) The existence of $f'(\mathbf{x}_0)$ does not imply that $f$ is strongly Gateaux differentiable at $\mathbf{x}_0$ (Problem 22B).

(2) A function may be strongly Gateaux differentiable at a point $\mathbf{x}_0$ where $f'(\mathbf{x}_0)$ fails to exist.

(3) Let $f$ be convex in an open set $U \subseteq \mathbf{R}^n$. If $f$ is Gateaux differentiable at $\mathbf{x}_0$, then it is strongly Gateaux differentiable at $\mathbf{x}_0$.

*(4) If $f'(\mathbf{x}; \mathbf{v})$ is continuous in $\mathbf{x}$ at $\mathbf{x}_0$ and if $f'(\mathbf{x}_0 ; \mathbf{v})$ is continuous in $\mathbf{v}$ at $\mathbf{O}$, then $f$ is strongly Gateaux differentiable at $\mathbf{x}_0$ [Nashed, 1966, Theorem 4].

*(5) In any normed linear space $\mathbf{L}$, the norm function $N(\mathbf{x}) = \|\mathbf{x}\|$, if Gateaux differentiable, is strongly Gateaux differentiable [Ascoli, 1932; Mazur, 1933].

*D. Examples of convex functions that are Gateaux differentiable at some points Fréchet differentiable nowhere can be constructed.

(1) In the space $C[0, 1]$, we have defined the norm function

$$N(f) = \sup_{x \in [0,1]} |f(x)|$$

This function is not Fréchet differentiable at any point [Dieudonné, 1960, p. 147], but it is Gateaux differentiable at those points $f$ of the unit sphere for which only one $x_0$ exists such that $|f(x_0)| = 1$ [Köthe, 1969, p. 350].

(2) In the space $l^1$ we have defined the norm function

$$N(\{x_n\}) = \sum_1^\infty |x_n|$$

This function is not Fréchet differentiable at any point [Dieudonné, 1960, p. 147], but it is Gateaux differentiable at those points of the unit sphere for which $\{x_n\}$ has all nonzero coordinates [Köthe, 1969, p. 351–352].

(3) Phelps [1960] has, with a renorming of $l^1$, constructed an example of a function that is Gateaux differentiable at all points of the unit sphere $S$ but Fréchet differentiable at no points of $S$.

Asplund [1968] has suggested the classification of all Banach spaces according to whether continuous convex functions are necessarily differentiable (on a dense $G_\delta$ subset of their domain of continuity) in the sense of Gateaux [a weak differentiability space (WDS)] or Fréchet [a strong differentiability space (SDS)]. He remarks that $l^1$ is a WDS but not a SDS. See also Asplund and Rockafellar [1969].

*E. Let **L** be a normed linear space with norm $N(x) = \|x\|$. Various measures of smoothness of the unit sphere $S = \{x: \|x\| = 1\}$ are related to the differentiability properties of the norm function $N: \mathbf{L} \to \mathbf{R}$. Since norm functions are convex functions, some very nice applications, examples, and counterexamples relevant to our work are to be found in the literature dealing with the differentiability of norms.

(1) A normed linear space and its norm are called **smooth** if at each point of the unit sphere there is a unique hyperplane of support for the unit ball $B = \{x: \|x\| \leq 1\}$. The space **L** is smooth if and only if the norm function $N$ is Gateaux differentiable at every $x \neq \mathbf{O}$.

(2) A normed linear space and its norm are called **uniformly smooth** if for each $\varepsilon > 0$ there is a $\delta > 0$ so that $\|x\| = 1$, $\|y\| \leq \delta$ implies $\|x + y\| + \|x - y\| \leq 2 + \varepsilon \|y\|$. A normed linear space is uniformly smooth if and only if the norm function $N$ is (Fréchet) differentiable at every $x \neq \mathbf{O}$, and in the expression

$$N(x + h) = N(x) + N'(x)(h) + \|h\| \varepsilon(x, h)$$

we have $\lim_{h \to \mathbf{O}} \varepsilon(x, h) = 0$ uniformly for $x \in S$ [Smulian, 1940].

(3) An inner product space with norm $N(x) = \langle x, x \rangle^{1/2}$ is uniformly smooth.

$$N'(x_0)(h) = \frac{\langle x_0, h \rangle}{\|x_0\|} \quad \text{for all} \quad x_0 \neq \mathbf{O}$$

It is natural to wonder, in light of part (2), what measure of smoothness of $S$ corresponds exactly to Fréchet differentiability of the norm function. The question is answered [Cudia, 1964] using the notion of **weak uniform rotundity** of the unit sphere.

F. The following argument establishes that the sets $E_j$ used in the proof of Theorem D are measurable.

(1)
$$f_+'(\mathbf{x}; e_j) = \lim_{k \to \infty} k[f(x_1, ..., x_j + 1/k, ..., x_n) - f(x_1, ..., x_n)]$$

## 44. Differentiability of Convex Functions

Therefore, $f_+'(\mathbf{x}; \mathbf{e}_j)$ is measurable on $U$ [Natanson I, 1961, p. 94].

(2) $f_-'(\mathbf{x}; \mathbf{e}_j)$ is measurable on $U$.

(3) $E_j = \{\mathbf{x} \in U: f_+'(\mathbf{x}; \mathbf{e}_j) - f_-'(\mathbf{x}; \mathbf{e}_j) > 0\}$.

*G. An alternate proof of Theorem D can be given using a theorem of Rademacher [1919] saying that if $f$ satisfies a Lipschitz condition, then it is differentiable almost everywhere. The nature of the set of points where a convex function on $\mathbf{R}^n$ is not differentiable has been studied by Anderson and Klee [1952].

*H. This problem explores the possibility of defining the second derivative as a bilinear transformation exhibiting approximation properties required by the Taylor theorem. The situation for functions of a single real variable suggests what to expect and what not to expect.

(1) Let $f: \mathbf{R} \to \mathbf{R}$ be differentiable in a neighborhood of $\mathbf{x}_0$. Then if $f''(\mathbf{x}_0)$ exists, there is a real number $b$ such that

$$f(x_0 + h) = f(x_0) + f'(x_0)h + (b/2)h^2 + h^2\,\varepsilon(h)$$

where $\varepsilon(h) \to 0$ as $h \to 0$.

(2) The converse of the assertion in part (1) is false, as is illustrated by $f(x) = x^3 \sin(1/x)$ for $x \neq 0$, $f(0) = 0$.

(3) Let $f: U \to \mathbf{R}$ be continuously differentiable on the open convex set $U$ in the Banach space $\mathbf{L}$. Then if $f''(\mathbf{x}_0)$ exists, there is a continuous bilinear transformation $B$ such that

$$f(\mathbf{x}_0 + \mathbf{h}) = f(\mathbf{x}_0) + f'(\mathbf{x}_0)(\mathbf{h}) + \tfrac{1}{2}B(\mathbf{h}, \mathbf{h}) + \|\mathbf{h}\|^2\,\varepsilon(\mathbf{h})$$

where $\varepsilon(\mathbf{h}) \to 0$ as $\mathbf{h} \to 0$.

(4) The converse of the assertion in part (3) is of course false, but if we add the hypothesis that $f: U \to \mathbf{R}$ is convex on $U \subseteq \mathbf{L}$, then the existence of a bilinear transformation $B$ with the stated approximation property does imply the existence of $f''(\mathbf{x}_0)$; indeed $B = f''(\mathbf{x}_0)$.

The interplay of these ideas is used by Sundaresan [1967] in his study of twice differentiable norm functions on a Banach space. He has supplied the example cited in (2) above.

*I. Little has been written about the existence of the second derivative $f''(\mathbf{x}_0)$ of a convex function, even when $\mathbf{L} = \mathbf{R}^n$. The following questions may be investigated.

(1) Suppose that for $f$ defined on an open set $U \subseteq \mathbf{R}^n$, all the second partial derivatives exist so that we can write down the Hessian matrix

$$H(\mathbf{x}) = \begin{bmatrix} f_{11}(\mathbf{x}) & \cdots & f_{1n}(\mathbf{x}) \\ \vdots & & \vdots \\ f_{n1}(\mathbf{x}) & \cdots & f_{nn}(\mathbf{x}) \end{bmatrix}$$

If this matrix is positive definite in $U$, must $f$ be convex there?

(2) Suppose $f(\mathbf{x})$ is convex and the Hessian matrix $H(\mathbf{x}_0)$ exists. Must $f''(\mathbf{x}_0)$ exist?

(3) For the function

$$f(r, s) = \begin{cases} \dfrac{rs(r^2 - s^2)}{r^2 + s^2}, & (r, s) \neq (0, 0) \\ 0, & (r, s) = (0, 0) \end{cases}$$

the matrix $H(\mathbf{x})$ exists, but $f_{12}(\mathbf{O}) \neq f_{21}(\mathbf{O})$ [Buck, 1965, p. 249]. For a convex function $f$ where $H(\mathbf{x})$ exists, will it always be true that $f_{ij}(\mathbf{x}) = f_{ji}(\mathbf{x})$? Stoer and Witzgall [1970] claim the answer to this question is no, using $g(r, s) = f(r, s) + 13r^2 + 13s^2$ as their counterexample.

Discussions of the second derivative of a function defined on a subset of **L** or even $\mathbf{R}^n$ often address themselves to the existence of certain kinds of generalized derivatives (a circumstance not unlike the real variable case; see Problem 12N). Papers to be mentioned in this regard are by Busemann and Feller [1936] and Alexandorff [1939]. Another paper addressing a problem related to second derivatives is by Lorch [1951].

# V

# *Optimization*

When a branch of mathematics ceases to interest any but the specialists, it is very near its death, or at any rate dangerously close to a paralysis, from which it can be rescued only by being plunged back into the vivifying source of the science.

A. WEIL

True optimization is the revolutionary contribution of modern research to decision processes. In the entire history of mankind, a great gulf has always existed between a man's aspirations and his actions ... but planning staffs freed from the drudgery of computing are beginning to express themselves in terms of overall objectives and to ask the computers to find them the "best."

G. B. DANTZIG

## 50. Introduction

Convexity theory can hardly be called a youthful subject. Most of the results we have encountered so far are, at least in their essence, more than 40 years old. Perhaps our subject is more aptly described as middle-aged. If so, it shows little of the lethargy and dulled vision often associated with that period of life. Rather, it reaches out in all directions with youthful vigor.

Why is this so? Surely any answer must take note of the tremendous impetus the subject has received from outside of mathematics, from such diverse fields as economics, agriculture, military planning, and flows in networks. With the invention of high-speed computers, large-scale problems from these fields became at least potentially solvable. Whole new areas of mathematics (game theory, linear and nonlinear programming, control theory) aimed at solving these problems appeared almost overnight. And in each of them, convexity theory turned out to be at the core. The result has been a tremendous spurt in interest in convexity theory and a host of new results. We discuss some of them in this chapter.

For general functions, the study of maxima and minima is quite complicated. Convex functions, however, exhibit a particularly simple extremal structure, a property we explore in Section 51. The highs of concave functions and the lows of convex functions meet in our discussion of saddle points in Section 52 where we use several simple games to introduce the von Neumann minimax theorem. This theorem proves the existence of an optimal strategy for certain games, but actually to find this strategy leads to what is known as a linear programming problem. Our discussion of linear programming in Section 53 includes much of the theoretical development of the subject and the duality theorem, and in Section 54 we describe a powerful computational tool known as the simplex method. We treat convex programming in Section 55, keeping the development parallel to that of linear programming insofar as possible. Finally, in Section 56, we introduce the important problem of approximating in the best way a given point by another from a given set. Convexity enters this topic because the given set is usually convex and the measure of the distance between points is a convex function.

## 51. Maxima and Minima

Any course in advanced calculus treats the subject of maxima and minima for continuous and differentiable functions. It is proved that a

## 51. Maxima and Minima

function continuous on a compact set must attain both a maximum and a minimum value on that set. A differentiable function is shown to have a vanishing derivative at interior local maximum and local minimum points. For convex functions we can say considerably more; this is the subject of the present section. Our functions will be defined on a subset $U$ of a normed linear space $\mathbf{L}$. To clarify terms that sometimes are used in different ways, let us be specific as to our use of the terms local and global. We say that $f: U \to \mathbf{R}$ has a **local maximum** at $\bar{\mathbf{x}} \in U$ if there is a neighborhood $N_\varepsilon(\bar{\mathbf{x}})$ such that $f(\mathbf{x}) \leqslant f(\bar{\mathbf{x}})$ for all $\mathbf{x} \in N_\varepsilon(\bar{\mathbf{x}}) \cap U$. It is a **global maximum** if $f(\mathbf{x}) \leqslant f(\bar{\mathbf{x}})$ for all $\mathbf{x} \in U$. Obvious analogous definitions hold for minima.

**Theorem A.** Let $f: U \to \mathbf{R}$ be convex on a convex set $U \subseteq \mathbf{L}$. If $f$ has a local minimum at $\bar{\mathbf{x}}$, then $f(\bar{\mathbf{x}})$ is also a global minimum. The set $V$ (conceivably empty) on which $f$ attains its minimum is convex. And if $f$ is strictly convex in a neighborhood of a minimum point $\bar{\mathbf{x}}$, then $V = \{\bar{\mathbf{x}}\}$; that is, the minimum point is unique.

**Proof.** Suppose $f$ has a local minimum at $\bar{\mathbf{x}} \in U$. Then for $\mathbf{x} \in U$ and $\alpha > 0$ sufficiently small,

$$f(\bar{\mathbf{x}}) \leqslant f[(1-\alpha)\bar{\mathbf{x}} + \alpha\mathbf{x}] \leqslant (1-\alpha)f(\bar{\mathbf{x}}) + \alpha f(\mathbf{x}) \tag{1}$$

This says

$$0 \leqslant \alpha[f(\mathbf{x}) - f(\bar{\mathbf{x}})] \quad \text{or} \quad f(\mathbf{x}) \geqslant f(\bar{\mathbf{x}}) \tag{2}$$

so $f(\bar{\mathbf{x}})$ is a global minimum.

If $f$ attains its minimum $m$ at $\mathbf{x}_1$ and $\mathbf{x}_2$, then for $\alpha \in (0, 1)$,

$$m \leqslant f[(1-\alpha)\mathbf{x}_1 + \alpha\mathbf{x}_2] \leqslant (1-\alpha)m + \alpha m = m$$

Thus $f$ also attains its minimum at $(1-\alpha)\mathbf{x}_1 + \alpha\mathbf{x}_2$, and $V$ is convex.

If $f$ is strictly convex in a neighborhood of a minimum point $\bar{\mathbf{x}}$, then the second inequality in (1) is strict and (2) becomes $f(\mathbf{x}) > f(\bar{\mathbf{x}})$ for all $\mathbf{x} \in U$, $\mathbf{x} \neq \bar{\mathbf{x}}$. ●

Note that Theorem A does not assert that there is a minimum. A convex function may fail to have a minimum for several reasons. The set $U$ may not be closed [$f(x) = 1/x$ on $(1, 2)$] or it may be closed but not bounded [$f(x) = 1/x$ on $[1, \infty)$]. Even if $U$ is compact, there may be no minimum because $f$ is discontinuous [$f(x) = 1/x$ on $[1, 2)$, $f(2) = 1$]. In spite of these difficulties, however, Rockafellar [1970a, pp. 263–272] discusses a number of situations in which a convex function attains a minimum value, and we have the following result.

> **Theorem B.** Let $f$ be convex on $U \subseteq \mathbf{L}$. If $f'(\bar{\mathbf{x}}) = \mathbf{O}$ at $\bar{\mathbf{x}} \in U^0$, then $f(\bar{\mathbf{x}})$ is a global minimum for $f$. If in addition $f$ is continuously differentiable in a neighborhood of $\bar{\mathbf{x}}$ and $f''(\mathbf{x})$ exists and is positive definite there, then $\bar{\mathbf{x}}$ is the unique minimum point for $f$ on $U$.

**Proof.** By Theorem 42A,

$$f(\mathbf{x}) - f(\bar{\mathbf{x}}) \geq f'(\bar{\mathbf{x}})(\mathbf{x} - \bar{\mathbf{x}}) = 0$$

which establishes the first statement. The conditions in the second ensure, according to Theorem 42C, that $f$ is strictly convex in a neighborhood of $\bar{\mathbf{x}}$. Hence, by Theorem A, $\bar{\mathbf{x}}$ is the unique minimum point. ●

We turn now to the other extreme, studying the points at which a convex function assumes maximum values.

> **Theorem C.** If $f$ is convex on $U \subseteq \mathbf{L}$ and attains a global maximum at $\bar{\mathbf{x}} \in U^0$, then $f$ is constant on $U$.

**Proof.** Suppose $f$ is not constant on $U$. Choose $\mathbf{y} \in U$ such that $f(\mathbf{y}) < f(\bar{\mathbf{x}})$ and then $\alpha > 1$ so that $\mathbf{z} = \mathbf{y} + \alpha(\bar{\mathbf{x}} - \mathbf{y}) \in U$. Solving for $\bar{\mathbf{x}}$, we get

$$\bar{\mathbf{x}} = \frac{1}{\alpha}\mathbf{z} + \frac{\alpha - 1}{\alpha}\mathbf{y}$$

from which by convexity,

$$f(\bar{\mathbf{x}}) \leq \frac{1}{\alpha}f(\mathbf{z}) + \frac{\alpha - 1}{\alpha}f(\mathbf{y}) < \frac{1}{\alpha}f(\bar{\mathbf{x}}) + \frac{\alpha - 1}{\alpha}f(\bar{\mathbf{x}}) = f(\bar{\mathbf{x}})$$

We have our contradiction. ●

> **Theorem D.** If $f$ is continuous and convex on a compact set $U$ in a finite-dimensional normed linear space $\mathbf{L}^n$, then $f$ attains a global maximum at an extreme point of $U$.

**Proof.** It is clear from the general considerations mentioned in the initial paragraph of this section that $f$ attains a maximum at some point $\bar{\mathbf{x}}$. Since $\mathbf{L}^n$ is topologically isomorphic to $\mathbf{R}^n$ (Theorem 21F), we conclude

## 51. Maxima and Minima

from Theorem 32D that the compact convex set $U$ is the convex hull of its extreme points. Write $\bar{\mathbf{x}} = \sum_1^m \alpha_i \mathbf{v}_i$ where $\mathbf{v}_1, \ldots, \mathbf{v}_m$ are extreme points. Then

$$f(\bar{\mathbf{x}}) \leq \sum_i^m \alpha_i f(\mathbf{v}_i) \leq \max_{1 \leq i \leq m} f(\mathbf{v}_i)$$

But $f(\bar{\mathbf{x}}) \geq \max_{1 \leq i \leq m} f(\mathbf{v}_i)$, so $f$ must attain the value $f(\bar{\mathbf{x}})$ at some point $\mathbf{v}_i$. ●

While Theorem D includes a statement about the existence of a global maximum, the existence really has nothing to do with convexity. Our next result will exploit convexity in an essential way. First we need a new notion. A convex set is called **polyhedral** if it is the intersection of a finite number of closed half-spaces.

**Theorem E.** Let $f: U \to \mathbf{R}$ be convex on a closed finite-dimensional convex set $U \subseteq \mathbf{L}$ which contains no lines. If $f$ attains a global maximum somewhere on $U$, it is also attained at an extreme point of $U$. And if $U$ is polyhedral with $f$ bounded above on $U$, the attainment of a global maximum is assured.

**Proof.** We prove the first assertion by induction on $n$, the dimension of $U$. If $n = 1$, $U$ is either a closed interval or a closed half-line and the conclusion follows easily (Problem A). Suppose the result has been established for sets of dimension less than $n$, and consider a set $U$ of dimension $n$. The affine hull $A(U)$ is a translate of a subspace $V$ of dimension $n$, so we may assume without loss of generality that $A(U) = V$. Furthermore, since the concepts involved here are preserved under topological isomorphism, we may (Theorem 21F) suppose $V = \mathbf{R}^n$. In short, we take $U$ to be embedded in $\mathbf{R}^n$. Finally, we note that when this is done, $U^0 \neq \emptyset$ (Problem 32I).

Now let $\bar{\mathbf{x}}$ be a point where $f$ attains a maximum. By Theorem C, $\bar{\mathbf{x}}$ is on the boundary. (If $f$ is constant, we simply choose an $\bar{\mathbf{x}}$ on the boundary.) From Theorem 32C, we know there is a hyperplane $H$ through $\bar{\mathbf{x}}$ that supports $U$. Consider $H \cap U$, a closed convex set of dimension less than $n$ which certainly contains no lines. By our induction hypothesis, $f$ attains its maximum value $f(\bar{\mathbf{x}})$ at an extreme point $\mathbf{v} \in H \cap U$. But as we know (Problem 32J), $\mathbf{v}$ is also an extreme point of $U$, thus proving the first assertion of our theorem.

The proof of the existence of a maximum point under the stated conditions is also by induction, following a similar pattern, but with a slight complication. Proceed as before to embed $U$ in $\mathbf{R}^n$. Since $U$ is closed, it is easy to show (Problem B) that

$$M = \sup_{\mathbf{x} \in U} f(\mathbf{x}) = \sup_{\mathbf{x} \in \partial U} f(\mathbf{x})$$

where $\partial U$ is the boundary of $U$. But $U$ is polyhedral; its boundary is $(U \cap H_1) \cup \cdots \cup (U \cap H_m)$ where $H_1, \ldots, H_m$ are the hyperplanes (boundaries of half-spaces) which determine $U$. Thus

$$M = \sup_{\mathbf{x} \in U \cap H_i} f(\mathbf{x})$$

for some $i$. But $U \cap H_i$ is polyhedral and $\dim(U \cap H_i) < n$; by the induction hypothesis, $M = f(\mathbf{x})$ for some $\mathbf{x} \in U \cap H_i$. ●

Though our theory has been developed for convex functions, it has wider application. It obviously applies to concave functions $g(\mathbf{x})$; one merely studies $f(\mathbf{x}) = -g(\mathbf{x})$. More significantly, the theory developed here may be used to study the local behavior of rather general functions. Many functions met in practice are either convex or concave in neighborhoods of all but a few points. In particular, the material we have studied may be used to develop the theory of local maxima and minima of differentiable functions, a topic we explore further in Problem I.

Finally we note that it is sometimes possible by a change of variable(s) to transform a nonconvex function into a convex function, a technique that might be considered whenever the function to be minimized arises from an application in which physical considerations suggest the existence of a unique minimum point. The procedure is illustrated by consideration of the generalized polynomial

$$f(x_1, \ldots, x_n) = a_1 x_1^{r_{11}} x_2^{r_{12}} \cdots x_n^{r_{1n}} + \cdots + a_k x_1^{r_{k1}} x_2^{r_{k2}} \cdots x_n^{r_{kn}}$$

where $a_i > 0$, $x_i > 0$, and $r_{ij}$ is real. Though $f$ is generally (very) nonconvex,

$$g(y_1, \ldots, y_n) = \log f(e^{y_1}, \ldots, e^{y_n})$$

is a convex function to which our theorems apply. More is said about this example in Problem G.

## PROBLEMS AND REMARKS

**A.** The maximum of a convex function of a real variable was discussed briefly in Chapter I (Problem 12I). In this case, one can demonstrate directly that

## 51. Maxima and Minima

(1)  if $f: [a, b] \to \mathbf{R}$ is convex, $f$ attains a global maximum at $a$ or $b$,
(2)  if $f: [a, \infty) \to \mathbf{R}$ is convex and bounded above, then $f$ attains a global maximum at $a$.

**B.**  Let $f : U \to \mathbf{R}$ be convex and bounded above in a closed set $U \subseteq \mathbf{L}$ with a non-empty boundary $\partial U$. Then
$$\sup_{\mathbf{x} \in U} f(\mathbf{x}) = \sup_{\mathbf{x} \in \partial U} f(\mathbf{x})$$

**C.**  Let $f: U \to \mathbf{R}$ be convex on a set $U = H(S)$ where $H(S)$ is the convex hull of an arbitrary set $S$. Then
$$\sup_{\mathbf{x} \in U} f(\mathbf{x}) = \sup_{\mathbf{x} \in S} f(\mathbf{x})$$
and the first supremum is attained only if the second is attained.

*__D.__  Examples can be constructed showing the necessity of the continuity requirement in Theorem D,
 (1)  where $f$ is bounded above but does not attain its supremum,
 (2)  where the supremum is $+\infty$.

Now consider the case in which $f$ is convex and continuous on a compact set $U$ in an infinite-dimensional space $\mathbf{L}$. It follows from the Krein–Millman theorem that
$$\max_{\mathbf{x} \in U} f(\mathbf{x}) \leqslant \sup_{\mathbf{x} \in E} f(\mathbf{x})$$
where $E$ is the set of extreme points of $U$. Is Theorem D still true?

*__E.__  A polytope in $\mathbf{R}^n$ is polyhedral, but not conversely.

**F.**  If $f(\mathbf{x}) = \langle A\mathbf{x}, \mathbf{x} \rangle$ where $\mathbf{x}$ is a column vector $\mathbf{R}^n$ and $A$ is an $n \times n$ nonnegative definite matrix, then $f$ attains its minimum on any convex polyhedral set $P$. The same conclusion holds for any convex polynomial function which is bounded below on $P$ [Rockafellar, 1970a, p. 168].

**G.**  As indicated in the previous section, a change of variable transforms any generalized polynomial into a convex function.

 (1)  If $f(x) = 3x^2 + 2x^{1/2} + 4x^{-2}$ on $(0, \infty)$, then $g(y) = \log f(e^y)$ is convex on $\mathbf{R}$.
 (2)  If $f(x) = \sum_1^k a_i x^{r_i}$ on $(0, \infty)$ where $a_i > 0$, $r_i \in \mathbf{R}$, then Theorem 13F can be used to argue that $g(y) = \log f(e^y)$ is convex on $\mathbf{R}$.
 (3)  Let
$$f(\mathbf{x}) = f(x_1, \ldots, x_n) = \sum_{i=1}^{k} a_i x_1^{r_{1i}} x_2^{r_{2i}} \cdots x_n^{r_{ni}}$$
where $a_i > 0$, $r_{ij} \in \mathbf{R}$, and $\mathbf{x} = (x_1, \ldots, x_n)$ ranges over the positive orthant of $\mathbf{R}^n$. Then $g(\mathbf{y}) = \log f(e^{y_1}, \ldots, e^{y_n})$ is convex on $\mathbf{R}^n$.
 (4)  The change of variable used in part (3) is described by $\mathbf{y} = T(\mathbf{x})$ where $y_i = \log x_i$, $i = 1, \ldots, n$. In general $T(U)$ need not be a convex set even if $U$ is convex. However if $U = I_1 \times I_2 \times \cdots \times I_n$ where $I_i$ is an interval contained in $(0, \infty)$, then $V = T(U)$ is convex. What if $U$ is a ball?
 (5)  If $f : U \to \mathbf{R}$ is a generalized polynomial as in part (3) on a set $U$ as in part (4), then every local minimum is a global minimum and $f$ cannot attain an interior maximum unless it is constant.

Practical methods for optimizing a generalized polynomial have been developed under the topical heading of **geometric programming**. See Duffin et al. [1967].

**H.** A function $f: \mathbf{R} \to \mathbf{R}$ may have a local minimum at each point of $R$, but in this case $f(R)$ is at most countable. Construct an unbounded function of this type.

**I.** Theorem B can be used to obtain classical theorems on local minima (or maxima) of functions of several variables.

(1) Let $f: U \to \mathbf{R}$ have continuous second partial derivatives in a neighborhood of $\bar{x} \in U \subseteq \mathbf{R}^n$, and suppose $f'(\bar{x}) = \mathbf{O}$. If $f''(\bar{x})$ is positive definite [recall that $f''(\bar{x})$ is just the Hessian matrix of second partial derivatives], then $f(\bar{x})$ is a local minimum for $f$.

(2) Let $f(\mathbf{x}) = f(r, s)$ have continuous second partial derivatives in a neighborhood of $\bar{\mathbf{x}} = (r, s) \in \mathbf{R}^2$ and suppose $f_1(\bar{\mathbf{x}}) = f_2(\bar{\mathbf{x}}) = 0$. Let

$$\Delta = [f_{12}(\bar{\mathbf{x}})]^2 - f_{11}(\bar{\mathbf{x}}) f_{22}(\bar{\mathbf{x}})$$

If $\Delta < 0$ and $f_{11}(\bar{\mathbf{x}}) > 0$, then $f(\bar{\mathbf{x}})$ is a local minimum for $f$ (cf. Buck [1965, p. 353]).

★ ★ ★ ★ ★

A survey of more sophisticated results concerning the minimization of convex functionals on a domain in an arbitrary Hilbert space is given by Lyubich and Maistrovskii [1970]. Theorem E apparently first appeared in a paper by Hirsch and Hoffman [1961] where related material may be found.

## 52. Minimax Theorems and the Theory of Games

The graph of $z = K(x, y) = y^2 - x^2$ is a saddle-shaped surface called in analytic geometry a hyperbolic paraboloid (Fig. 52.1). The origin is called a saddle point and is of interest to us here because in the cross section $x = 0$, it appears to be a minimum of a convex function, while in the cross section $y = 0$ it appears to be a maximum of a concave function. Such points are of importance in many practical problems since it often happens that while we seek to maximize one thing, we wish to minimize something else. A manufacturer naturally wants to maximize his profits, but he may also wish (or be forced) to minimize the ecological damage caused by his production process. By way of introducing concepts and methods that have wide application in economics and the social sciences, we consider here an example of two gamblers engaged in a simple game.

Players A and B have each secured a large supply of pennies, nickels, and dimes. At a given signal, both will display a coin. If the sum of the displayed coins is odd, A wins B's coin; if even, B wins A's coin. The game is played repeatedly with each player aiming to maximize his winnings (or minimize his losses). How shall each player proceed?

## 52. Minimax Theorems and the Theory of Games

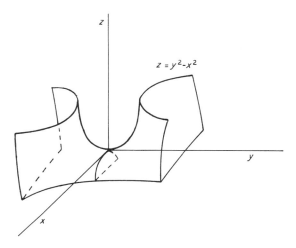

**Fig. 52.1**

Introduce a matrix, the **payoff matrix** $P$, which we take to represent A's winnings under the $3 \times 3 = 9$ alternatives that may occur.

$$P: \quad \begin{array}{c|ccc} {}_A\diagdown{}^B & p & n & d \\ \hline p & -1 & -1 & 10 \\ n & -5 & -5 & 10 \\ d & 1 & 5 & -10 \end{array}$$

Player A, noting the payoff of 10 in the first row for which he need only risk a penny, decides to play his penny and does so consistently. Player B initially plays his dime (the $-10$ in the third column appeals to him), but after observing A's behavior for awhile (meantime losing a dime each time), he shifts to his penny. Player A soon sees through this strategy and counters with his dime. And so it goes. Can either player develop a winning strategy?

Before trying to answer, let's change the rules by introducing a different payoff matrix $Q$, again representing A's winnings.

$$Q: \quad \begin{array}{c|ccc} {}_A\diagdown{}^B & p & n & d \\ \hline p & 10 & 0 & -10 \\ n & 0 & -5 & 5 \\ d & 5 & 1 & 5 \end{array}$$

Let the two players play the new game. Player A, being conservative, decides to play his dime. He reasons: No matter what B does, I shall always win at least a penny, whereas I could lose if I play anything else. Player B, being of a similar nature, decides to play his nickel. He realizes that he cannot guarantee winning, but that playing his nickel at least minimizes his losses. What has each player done? Player A has computed the minimum of each row (the worst that can happen to him) and chosen the row for which this minimum is largest. Similarly, B has taken the maximum of each column (the worst that can happen to him) and selected the column for which this maximum is smallest. And now the situation is stable. A will continue to play his dime, B his nickel. B will continue to lose, but nothing he can do will improve things. His fate is sealed unless A gets greedy and shifts to another strategy.

The situation in the first game is quite different. Let's see why. Let $Q(i,j)$ be the number in the $i$th row and $j$th column of the matrix $Q$. Note first that the stable position for $Q$, namely the $(3, 2)$ position, has the property that the entry there is the smallest in its row and the largest in its column. It is a so-called saddle point for $Q$. Note secondly that

$$\max_i \min_j Q(i,j) = \min_j \max_i Q(i,j) = 1$$

On the other hand, $P$ has no saddle point and moreover

$$-1 = \max_i \min_j P(i,j) < \min_j \max_i P(i,j) = 1$$

What we have observed here is true in an exceedingly more general context. The existence of a saddle point is intimately related to the equality of the maximin and the minimax.

**Theorem A.** Let $U$ and $V$ be arbitrary sets and suppose $K: U \times V \to \mathbf{R}$. Then

$$\sup_{\mathbf{x} \in U} \inf_{\mathbf{y} \in V} K(\mathbf{x}, \mathbf{y}) \leq \inf_{\mathbf{y} \in V} \sup_{\mathbf{x} \in U} K(\mathbf{x}, \mathbf{y}) \tag{1}$$

If there is a point $(\bar{\mathbf{x}}, \bar{\mathbf{y}}) \in U \times V$ such that

$$K(\mathbf{x}, \bar{\mathbf{y}}) \leq K(\bar{\mathbf{x}}, \bar{\mathbf{y}}) \leq K(\bar{\mathbf{x}}, \mathbf{y}) \tag{2}$$

for all $(\mathbf{x}, \mathbf{y}) \in U \times V$, then (1) is an equality and both terms are equal to $K(\bar{\mathbf{x}}, \bar{\mathbf{y}})$.

## 52. Minimax Theorems and the Theory of Games

**Proof.** Set $f(\mathbf{x}) = \inf_{\mathbf{y} \in V} K(\mathbf{x}, \mathbf{y})$ and $g(\mathbf{y}) = \sup_{\mathbf{x} \in U} K(\mathbf{x}, \mathbf{y})$. Clearly $f(\mathbf{x}) \leq K(\mathbf{x}, \mathbf{y}) \leq g(\mathbf{y})$ for all $(\mathbf{x}, \mathbf{y}) \in U \times V$, and consequently

$$\sup_{\mathbf{x} \in U} f(\mathbf{x}) \leq \inf_{\mathbf{y} \in V} g(\mathbf{y})$$

which is (1). Now suppose $(\bar{\mathbf{x}}, \bar{\mathbf{y}})$ satisfies (2). Then

$$K(\bar{\mathbf{x}}, \bar{\mathbf{y}}) = \inf_{\mathbf{y} \in V} K(\bar{\mathbf{x}}, \mathbf{y}) = f(\bar{\mathbf{x}}) \leq \sup_{\mathbf{x} \in U} f(\mathbf{x}) \qquad (3)$$

and

$$K(\bar{\mathbf{x}}, \bar{\mathbf{y}}) = \sup_{\mathbf{x} \in U} K(\mathbf{x}, \bar{\mathbf{y}}) = g(\bar{\mathbf{y}}) \geq \inf_{\mathbf{y} \in V} g(\mathbf{y}) \qquad (4)$$

Putting (3) and (4) together yields

$$\inf_{\mathbf{y} \in V} \sup_{\mathbf{x} \in U} K(\mathbf{x}, \mathbf{y}) \leq \sup_{\mathbf{x} \in U} \inf_{\mathbf{y} \in V} K(\mathbf{x}, \mathbf{y})$$

This together with (1) gives the desired equality. ●

A point $(\bar{\mathbf{x}}, \bar{\mathbf{y}})$ satisfying (2) is called a **saddle point** for $K$ on $U \times V$. If one wants to know when a saddle point exists, Theorem A is of no help whatever. And in fact, the existence of saddle points is hard to establish unless one puts severe restrictions on the function $K$ and the sets $U$ and $V$. We state one such result, the **von Neumann minimax theorem** which, as we shall see, is just what we need to carry our study of games a step further.

**Theorem B.** Let $U \subseteq \mathbf{R}^m$ and $V \subseteq \mathbf{R}^n$ be nonempty compact convex sets, and let $K: U \times V \to \mathbf{R}$ be continuous. If for each fixed $\mathbf{y}$, $K(\mathbf{x}, \mathbf{y})$ is concave on $U$ and for each fixed $\mathbf{x}$, $K(\mathbf{x}, \mathbf{y})$ is convex on $V$, then there is a saddle point $(\bar{\mathbf{x}}, \bar{\mathbf{y}})$ for $K$ on $U \times V$ and hence

$$\max_{\mathbf{x} \in U} \min_{\mathbf{y} \in V} K(\mathbf{x}, \mathbf{y}) = \min_{\mathbf{y} \in V} \max_{\mathbf{x} \in U} K(\mathbf{x}, \mathbf{y}) = K(\bar{\mathbf{x}}, \bar{\mathbf{y}})$$

**Proof.** We first demonstrate the existence of a saddle point under the added assumption of strict concavity and convexity for $K(\mathbf{x}, \mathbf{y})$ (Fig. 52.2). Set

$$g(\mathbf{y}) = \max_{\mathbf{x} \in U} K(\mathbf{x}, \mathbf{y})$$

and note that our assumption, for fixed $\mathbf{y}$, that $K(\mathbf{x}, \mathbf{y})$ is strictly concave in $\mathbf{x}$ means that this maximum is attained at a unique point in $U$ (Theorem 51A). Denote this point by $h(\mathbf{y})$.

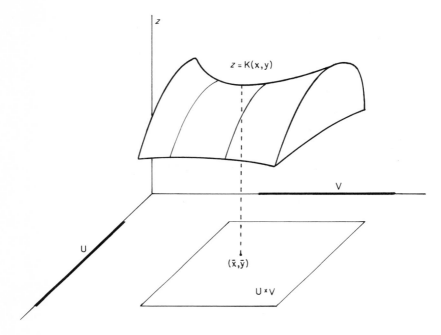

**Fig. 52.2**

Now $h: V \to U$ must be continuous on $V$. For if not, there is a $\mathbf{v}_0 \in V$, an $\varepsilon > 0$, and a sequence $\{\mathbf{v}_i\}_1^\infty$ in $V$ converging to $\mathbf{v}_0$ for which $\|h(\mathbf{v}_i) - h(\mathbf{v}_0)\| > \varepsilon$. If $\mathbf{u}_i = h(\mathbf{v}_i)$, $i = 0, 1, 2,...$, then from the strict concavity of $K(\mathbf{x}, \mathbf{v}_0)$ we infer the existence of a $\delta > 0$ for which $K(\mathbf{u}_0, \mathbf{v}_0) - K(\mathbf{u}_i, \mathbf{v}_0) > \delta$, $i = 1, 2,...$ . On the other hand,

$$K(\mathbf{u}_0, \mathbf{v}_0) - K(\mathbf{u}_i, \mathbf{v}_0) = [K(\mathbf{u}_0, \mathbf{v}_0) - K(\mathbf{u}_0, \mathbf{v}_i)] + [K(\mathbf{u}_0, \mathbf{v}_i) - K(\mathbf{u}_i, \mathbf{v}_0)]$$

and since $K(\mathbf{u}_i, \mathbf{v}_i)$ is the maximum value of $K(\mathbf{x}, \mathbf{v}_i)$ on $U$, the latter is less than

$$[K(\mathbf{u}_0, \mathbf{v}_0) - K(\mathbf{u}_0, \mathbf{v}_i)] + [K(\mathbf{u}_i, \mathbf{v}_i) - K(\mathbf{u}_i, \mathbf{v}_0)]$$

But the uniform continuity of $K$ (it is continuous on a compact set) ensures that both of these terms can be made less than $\delta/2$ by taking $i$ sufficiently large. We have our contradiction.

The function $g$ is also continuous. While this may be argued directly from its definition (Problem I), it is most easily seen from the fact that $g(\mathbf{y}) = K(h(\mathbf{y}), \mathbf{y})$ is the composite of two continuous functions.

## 52. Minimax Theorems and the Theory of Games

Let $\bar{\mathbf{y}}$ be the point where $g$ assumes its minimum on $V$; that is,

$$g(\bar{\mathbf{y}}) = \min_{\mathbf{y} \in V} g(\mathbf{y}) = \min_{\mathbf{y} \in V} \max_{\mathbf{x} \in U} K(\mathbf{x}, \mathbf{y})$$

For any point $\mathbf{y} \in V$ and for $\lambda \in (0, 1)$, let $\mathbf{y}_1 = (1 - \lambda)\bar{\mathbf{y}} + \lambda\mathbf{y}$ and set $\bar{\mathbf{x}} = h(\bar{\mathbf{y}})$, $\mathbf{x}_1 = h(\mathbf{y}_1)$. By definition of $\bar{\mathbf{y}}$, $g(\bar{\mathbf{y}}) \leqslant g(\mathbf{y}_1)$, or equivalently, $K(\bar{\mathbf{x}}, \bar{\mathbf{y}}) \leqslant K(\mathbf{x}_1, \mathbf{y}_1)$. Combining this with the convexity of $K(\mathbf{x}, \mathbf{y})$ in $\mathbf{y}$, we get

$$K(\bar{\mathbf{x}}, \bar{\mathbf{y}}) \leqslant K(\mathbf{x}_1, \mathbf{y}_1) \leqslant (1 - \lambda) K(\mathbf{x}_1, \bar{\mathbf{y}}) + \lambda K(\mathbf{x}_1, \mathbf{y})$$
$$\leqslant (1 - \lambda) K(\bar{\mathbf{x}}, \bar{\mathbf{y}}) + \lambda K(\mathbf{x}_1, \mathbf{y})$$

A bit of algebra leads to $K(\bar{\mathbf{x}}, \bar{\mathbf{y}}) \leqslant K(\mathbf{x}_1, \mathbf{y})$. Now let $\lambda \downarrow 0$ which makes $\mathbf{y}_1 \to \bar{\mathbf{y}}$ and, because of the continuity of $h$, also makes $\mathbf{x}_1 \to \bar{\mathbf{x}}$. Our last inequality becomes $K(\bar{\mathbf{x}}, \bar{\mathbf{y}}) \leqslant K(\bar{\mathbf{x}}, \mathbf{y})$, and with the obvious $K(\mathbf{x}, \bar{\mathbf{y}}) \leqslant K(\bar{\mathbf{x}}, \bar{\mathbf{y}})$, we have

$$K(\mathbf{x}, \bar{\mathbf{y}}) \leqslant K(\bar{\mathbf{x}}, \bar{\mathbf{y}}) \leqslant K(\bar{\mathbf{x}}, \mathbf{y}) \tag{5}$$

We have found a saddle point.

Now let us return to the original statement of the theorem in which no assumption was made about the concavity and convexity of $K(\mathbf{x}, \mathbf{y})$ being strict. Define

$$K_\varepsilon(\mathbf{x}, \mathbf{y}) = K(\mathbf{x}, \mathbf{y}) - \varepsilon \|\mathbf{x}\| + \varepsilon \|\mathbf{y}\|$$

where $\|\ \|$ denotes the Euclidean norm. This modified function meets the strictness requirement and so there is a point $(\mathbf{x}_\varepsilon, \mathbf{y}_\varepsilon) \in U \times V$ for which

$$K_\varepsilon(\mathbf{x}, \mathbf{y}_\varepsilon) \leqslant K_\varepsilon(\mathbf{x}_\varepsilon, \mathbf{y}_\varepsilon) \leqslant K_\varepsilon(\mathbf{x}_\varepsilon, \mathbf{y})$$

Take a sequence $\{\varepsilon_j\}$ converging to 0, and then a subsequence for which $\{(\mathbf{x}_{\varepsilon_j}, \mathbf{y}_{\varepsilon_j})\}$ converges (in the compact set $U \times V$), say, to $(\bar{\mathbf{x}}, \bar{\mathbf{y}})$, which then satisfies (5); it is a saddle point for $K$.

The final statement of equality in our theorem follows from Theorem A if we observe that in the present context, sup and inf can be replaced with max and min (Problem I). ●

We return to the penny, nickel, dime game with payoff matrix $P$ which, we recall, had no saddle point. Call the alternatives available to each player (play a penny, nickel, or dime) the pure strategies. Neither player has a pure strategy which will guarantee winning at each play. But perhaps A can think of a sequence of pure strategies that will so confuse B that at least A will win in the long run. Clearly this sequence should not follow a pattern; if it does, B will catch on and counter with an appropriate sequence of strategies of his own. The one possibility open to A is to choose his pure strategies at random, deciding only with what probability he will use each pure strategy. Is there an optimal way of doing this?

We may as well look at the problem more generally. Suppose that players A and B have, respectively, $m$ and $n$ **pure strategies** (possible actions at each play) and that a payoff matrix ($p_{ij}$), $i = 1,..., m$; $j = 1,..., n$, has been assigned. A **mixed strategy** for A is a probability vector $\mathbf{x} = (x_1,..., x_m)$, that is, a vector such that $x_i \geqslant 0$, $\sum_1^m x_i = 1$. The understanding is that A will use the $i$th pure strategy with probability $x_i$. Similarly $\mathbf{y} = (y_1,..., y_n)$ is a mixed strategy for B. From simple probability theory, we have the definition of the **expected payoff function**

$$E(\mathbf{x}, \mathbf{y}) = \sum_{i=1}^{m} \sum_{j=1}^{n} x_i p_{ij} y_j$$

Let

$$f(\mathbf{x}) = \min_{\mathbf{y} \in V} E(\mathbf{x}, \mathbf{y}), \qquad g(\mathbf{y}) = \max_{\mathbf{x} \in U} E(\mathbf{x}, \mathbf{y})$$

and call $\bar{\mathbf{x}}$ and $\bar{\mathbf{y}}$ **optimal mixed strategies** for players A and B if

$$f(\bar{\mathbf{x}}) = \max_{\mathbf{x} \in U} f(\mathbf{x}) = \max_{\mathbf{x} \in U} \min_{\mathbf{y} \in V} E(\mathbf{x}, \mathbf{y})$$

$$g(\bar{\mathbf{y}}) = \min_{\mathbf{y} \in V} g(\mathbf{y}) = \min_{\mathbf{y} \in V} \max_{\mathbf{x} \in U} E(\mathbf{x}, \mathbf{y})$$

Our question becomes: Are there optimal mixed strategies for A and/or B? The fundamental theorem for matrix games gives us an affirmative answer.

**Theorem C.** In any two-person matrix game, there are optimal mixed strategies for both players. Moreover, if $\bar{\mathbf{x}}$ and $\bar{\mathbf{y}}$ are optimal mixed strategies for players A and B, then

$$f(\bar{\mathbf{x}}) = \max_{\mathbf{x} \in U} \min_{\mathbf{y} \in V} E(\mathbf{x}, \mathbf{y}) = \min_{\mathbf{y} \in V} \max_{\mathbf{x} \in U} E(\mathbf{x}, \mathbf{y}) = g(\bar{\mathbf{y}})$$

and $(\bar{\mathbf{x}}, \bar{\mathbf{y}})$ is a saddle point for $E(\mathbf{x}, \mathbf{y})$.

## 52. Minimax Theorems and the Theory of Games

**Proof.** Apply Theorem B with $K(\mathbf{x}, \mathbf{y}) = E(\mathbf{x}, \mathbf{y})$ and

$$U = \left\{ \mathbf{x} \in \mathbf{R}^m : x_i \geq 0, \sum_1^m x_i = 1 \right\}$$

$$V = \left\{ \mathbf{y} \in \mathbf{R}^n : y_i \geq 0, \sum_1^n y_i = 1 \right\}$$

Note that $E$, being linear in both $\mathbf{x}$ and $\mathbf{y}$, is certainly concave in $\mathbf{x}$, convex in $\mathbf{y}$. In addition, $U$ is convex, being the intersection of the hyperplane $\sum_1^m x_i = 1$ with the nonnegative orthant, and it is closed and bounded, hence compact. Similar considerations apply to $V$. Thus $E$ has a saddle point $(\bar{\mathbf{x}}, \bar{\mathbf{y}})$, and $f(\bar{\mathbf{x}})$ is a maximum for $f$, $g(\bar{\mathbf{y}})$ is a minimum for $g$. The first assertion is proved.

Now let $\bar{\mathbf{x}}$ and $\bar{\mathbf{y}}$ be any two optimal mixed strategies for A and B. If we examine the proof of Theorem B, we see that there is a mixed strategy $\mathbf{x}_0$ such that $(\mathbf{x}_0, \bar{\mathbf{y}})$ is a saddle point of $E(\mathbf{x}, \mathbf{y})$, and this in turn implies that $f(\mathbf{x}_0) = g(\bar{\mathbf{y}})$. Now trivially $f(\bar{\mathbf{x}}) \leq g(\bar{\mathbf{y}})$ [formula (1) of Theorem A]. But if $f(\bar{\mathbf{x}}) < g(\bar{\mathbf{y}})$, then $f(\bar{\mathbf{x}}) < f(\mathbf{x}_0)$ and $\bar{\mathbf{x}}$ is not optimal. We conclude that $f(\bar{\mathbf{x}}) = g(\bar{\mathbf{y}})$. We leave the last step, showing that $(\bar{\mathbf{x}}, \bar{\mathbf{y}})$ is a saddle point, to the reader (Problem H). ●

The common value $\bar{v} = f(\bar{\mathbf{x}}) = g(\bar{\mathbf{y}})$ is called the **value** of the game. If $\bar{v} = 0$, the game is fair; if $\bar{v} > 0$, it is biased in favor of A; if $\bar{v} < 0$, it is biased in favor of B.

We return once again to our two examples. Both have optimal strategies for each player. Though it is one thing to know they exist and another to find them, we shall pass over the latter question for a moment, supposing that by some cleverness, it has been found that $\bar{\mathbf{x}} = (\frac{1}{2}, 0, \frac{1}{2})$ and $\bar{\mathbf{y}} = (\frac{10}{11}, 0, \frac{1}{11})$ are optimal mixed strategies in the first game; similarly $\bar{\mathbf{x}} = (0, 0, 1)$ and $\bar{\mathbf{y}} = (0, 1, 0)$ are optimal for the second game. (Verification that these are correct is left as Problem A. See also Problem B.) For the first game, $\bar{v} = 0$, while for the second $\bar{v} = 1$. The first game is fair; the second is biased in favor of A. In the first game, player A may expect to come out at least even in the long run if he plays his penny and dime at random, each with a probability of 1/2. In fact, suppose B, failing to make the proper analysis, chooses mixed strategy $\mathbf{y} = (y_1, y_2, y_3)$. A simple calculation gives $E(\bar{\mathbf{x}}, \mathbf{y}) = 2y_2$. To the extent that B plays his nickel, A will win in the long run.

Now let us return to the question of how to find the optimal strategies that we know exist. In general we hope for procedures that rely more on method than on inspiration. The references mentioned at the end of the section indicate that various techniques have been devised, but we shall only discuss the possibility of expressing the game problem in the form of a problem in linear programming. Toward this end we shall rephrase our problem, which is to find an $\bar{\mathbf{x}}$ that maximizes $f$ on $U$ and a $\bar{\mathbf{y}}$ that minimizes $g$ on $V$.

Note that $V = \{\mathbf{y} \in \mathbf{R}^n : y_j \geq 0, \sum_1^n y_j = 1\}$ is a compact convex set which has as its extreme points the $n$ standard unit vectors $\mathbf{e}_1, \ldots, \mathbf{e}_n$ (Problem J). Also, for fixed $\mathbf{x}$,

$$E(\mathbf{x}, \mathbf{y}) = \sum_{i=1}^{m} \sum_{j=1}^{n} x_i p_{ij} y_j$$

is linear (thus concave) in $\mathbf{y}$, and so by Theorem 51D, it attains its minimum at one of these extreme points. Thus,

$$f(\mathbf{x}) = \min_{\mathbf{y} \in V} E(\mathbf{x}, \mathbf{y}) = \min_{1 \leq j \leq n} E(\mathbf{x}, \mathbf{e}_j)$$

To maximize $f$ over $U$ is therefore to maximize a variable $z$ subject to

$$z \leq E(\mathbf{x}, \mathbf{e}_1), \quad \ldots, \quad z \leq E(\mathbf{x}, \mathbf{e}_n)$$

with $\mathbf{x}$ allowed to vary over $U$. Since $E(\mathbf{x}, \mathbf{e}_j) = \sum_{i=1}^{m} x_i p_{ij}$, our problem is to maximize $F(z) = z$ subject to the constraints

$$z - p_{11} x_1 - \cdots - p_{m1} x_m \leq 0$$
$$\vdots$$
$$z - p_{1n} x_1 - \cdots - p_{mn} x_m \leq 0$$
$$x_1 + \cdots + x_m = 1$$
$$x_i \geq 0, \quad i = 1, \ldots, m$$

This is, as we shall see in Section 53, a problem in linear programming.

In similar fashion we find that the problem of minimizing $g$ on $V$ is equivalent to minimizing $G(w) = w$ subject to

$$w - p_{11} y_1 - \cdots - p_{1n} y_n \geq 0$$
$$\vdots$$
$$w - p_{m1} y_1 - \cdots - p_{mn} y_n \geq 0$$
$$y_1 + \cdots + y_n = 1$$
$$y_j \geq 0, \quad j = 1, 2, \ldots, n$$

Not only have we expressed the problem of finding an optimal strategy for player B as a linear programming problem, but it turns out to be what is known as the dual to the problem for player A (Problem 53I).

## 52. Minimax Theorems and the Theory of Games

As we shall see, a procedure can be developed that solves a problem and its dual simultaneously.

### PROBLEMS AND REMARKS

**A.** The optimal mixed strategies given at the end of Section 52 for the two illustrative matrix games of that section may be verified by showing that the given values of $(\bar{\mathbf{x}}, \bar{\mathbf{y}})$ define a saddle point for the appropriate function $E(\mathbf{x}, \mathbf{y})$.

**B.** Suppose that $(\bar{\imath}, \bar{\jmath})$ is a saddle point for the $m \times n$ payoff matrix $P(i,j)$. Then $(\bar{\mathbf{x}}, \bar{\mathbf{y}})$, where $\bar{\mathbf{x}}$ and $\bar{\mathbf{y}}$ are probability vectors with 1's in the $\bar{\imath}$th and $\bar{\jmath}$th positions respectively, is a saddle point for

$$E(\mathbf{x}, \mathbf{y}) = \sum_{i=1}^{m} \sum_{j=1}^{n} x_i P(i,j) y_j$$

**C.** Analyze the familiar matching pennies game. If the players display like faces, A wins a penny; if unlike faces, B wins a penny.

**D.** Solve the games with the following payoff matrices, finding the optimal mixed strategies for each player and the value of the game.

(1) $\begin{bmatrix} 1 & -3 & -2 \\ 2 & 5 & -4 \\ 2 & 3 & 2 \end{bmatrix}$ (2) $\begin{bmatrix} -1 & 1 & 1 \\ 2 & -2 & 2 \\ 3 & 3 & -3 \end{bmatrix}$

(3) $\begin{bmatrix} a & 0 & 0 \\ 0 & b & 0 \\ 0 & 0 & c \end{bmatrix}$ (4) $\begin{bmatrix} 1 & 2 & 3 & 4 \\ 2 & 3 & 4 & 1 \\ 3 & 4 & 1 & 2 \\ 4 & 1 & 2 & 3 \end{bmatrix}$

(5) $P(i,j) = i - j$.

**E.** *Colonel Blotto Game* [Dresher, 1961, p. 7]. Colonel Blotto and his enemy each try to occupy two posts by properly distributing their forces. Let us assume that Colonel Blotto has four regiments and that the enemy has three regiments that are to be divided between the two posts. Define the payoff to Colonel Blotto at each post as follows. If Colonel Blotto has more regiments than the enemy at the post, he receives the enemy's regiments plus one (the occupation of the post being equivalent to capturing an extra regiment). If the enemy has more regiments than Colonel Blotto at the post, then Colonel Blotto loses one plus his regiments at the post. If each side places the same number of regiments at a post, it is a draw. The total payoff is the sum of the payoffs at the two posts. Find the 5 × 4 payoff matrix and show that the optimal mixed strategies for Colonel Blotto and the enemy are $(\frac{4}{9}, \frac{4}{9}, 0, 0, \frac{1}{9})$ and $(\frac{1}{18}, \frac{1}{18}, \frac{4}{9}, \frac{4}{9})$. What is the value of the game?

**F.** A square matrix $P$ is skew symmetric if $P(i,j) = -P(j,i)$ for all $i,j$. A game with a skew symmetric payoff matrix has value 0 and the players have the same optimal mixed strategies.

**G.** The set of saddle points for a function $K(\mathbf{x}, \mathbf{y})$ satisfying the hypothesis of Theorem B is convex in $\mathbf{R}^m \times \mathbf{R}^n$.

**H.** In a matrix game, $f(\bar{\mathbf{x}}) = g(\bar{\mathbf{y}})$ if and only if $(\bar{\mathbf{x}}, \bar{\mathbf{y}})$ is a saddle point for $E(\mathbf{x}, \mathbf{y})$. Moreover, the sets $U_0$ and $V_0$ of optimal strategies for A and B are convex, as is the set of saddle points for $E$.

**I.** Let $\{g_\mathbf{x}\}$, $\mathbf{x} \in U$, be a family of continuous functions all defined on a compact set $V$, and let $g(\mathbf{y}) = \sup_{\mathbf{x} \in U} g_\mathbf{x}(\mathbf{y})$. Then $g$ is continuous on $V$ if

(1)   $U$ is a finite set,
(2)   $U$ is a compact set and $g_\mathbf{x}(\mathbf{y})$ is continuous in $\mathbf{x}$ for each $\mathbf{y} \in V$.

**J.** Let $V = \{\mathbf{y} \in \mathbf{R}^n : y_j \geqslant 0, \ \Sigma_1^n y_j = 1\}$. The extreme points of $V$ are the $n$ standard unit vectors $\mathbf{e}_1, \ldots, \mathbf{e}_n$.

**K.** Many variations of Theorem B have appeared in which the statement of equality is seen to follow from hypotheses weaker than we used. One variation, due to Fan [1953], has been proved in several ways and may be stated here. Let $f$ be defined on $U \times V$ so that for fixed $\mathbf{y} \in V$, $f$ is convex on $U$, and for fixed $\mathbf{x} \in U$, $f$ is concave on $V$. We suppose $U$ is a compact Hausdorff space and $V$ to be arbitrary (not necessarily with a topology). Suppose $f : U \times V \to \mathbf{R}$ is, for fixed $\mathbf{y} \in V$, lower semicontinuous on $U$. Then

$$\min_{\mathbf{x} \in U} \sup_{\mathbf{y} \in V} f(\mathbf{x}, \mathbf{y}) = \sup_{\mathbf{y} \in V} \min_{\mathbf{x} \in U} f(\mathbf{x}, \mathbf{y})$$

[Hirschfeld, 1958; Sion, 1958].

★ ★ ★ ★ ★

The theory of games originated with Borel [1921; 1927], developed under von Neumann [1928], and flowered in the book by von Neumann and Morgenstern [1947]. More recent expositions are given by Berge and Ghouila–Houri [1965], Blackwell and Girshick [1954], Dresher [1961], Karlin [1959], Owen [1968], and Williams [1954].

## 53. Linear Programming

In this section we focus attention on the so-called linear programming problem, in which we seek to find the maximum value of a linear function defined on a convex domain in $\mathbf{R}^n$, the domain being bounded by a finite number of hyperplanes. Though this seems to be a very special problem if one thinks in terms of the general problems discussed in Section 51, the theory developed here provides the theoretical basis (and indeed was developed) for solving many important problems arising in game theory, control theory, and economics.

We begin this section with a simple example of a linear programming problem. Then after a careful statement of the general problem, we develop the theoretic principles on which methods of solution are based. We use these principles to solve our example, at the same time illustrating the need for a more practical way to proceed. The development of such a computational algorithm is postponed, however, to the next

## 53. Linear Programming

section. We turn instead to discuss duality in linear programming, proving the basic theorem on this subject and illustrating it with one more appeal to our durable example.

The problem we shall use as our example is that of maximizing the linear function $f(x_1, x_2) = -3x_1 + 2x_2$ on the region $\mathscr{F}$ of the plane (Fig. 53.1) bounded by $2x_1 - x_2 = -2$, $x_1 - 2x_2 = 3$, $x_1 + 2x_2 = 11$,

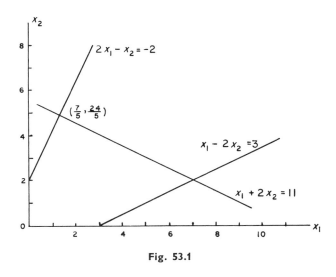

Fig. 53.1

$x_1 = 0$, $x_2 = 0$. Another way to describe $\mathscr{F}$ is to say it is the set of points $(x_1, x_2)$ that satisfy the following

$$\begin{align} 2x_1 - x_2 &\geqslant -2 \\ x_1 - 2x_2 &\leqslant 3 \\ x_1 + 2x_2 &\leqslant 11 \\ x_1 &\geqslant 0 \\ x_2 &\geqslant 0 \end{align} \tag{1}$$

These inequalities are called the **constraints** of the problem. The function $f(x_1, x_2) = -3x_1 + 2x_2$ is called the **objective function**. We now restate our problem.

**Example A.** Maximize the objective function

$$f(x_1, x_2) = -3x_1 + 2x_2$$

subject to the condition that $(x_1, x_2)$ be in the set

$$\mathscr{F} = \{(x_1, x_2): x_1, x_2 \text{ satisfy the linear constraints (1)}\}.$$

A linear expression in $n$ variables takes the form

$$a_{i1}x_1 + \cdots + a_{in}x_n$$

A system of $m$ such expressions, each $\leq$, $=$, $\geq b_i$, a constant, is said to be a set of linear constraints. It turns out that many problems that arise in practical applications include in the set of constraints the inequalities $x_1 \geq 0,\ldots, x_n \geq 0$ which restrict attention to the **nonnegative orthant** $\mathbf{R}_+^n$. Since this simplifies matters, we shall always assume these inequalities to be part of the constraints. This can actually be assumed without loss of generality, since if the constraints of a given problem do allow some negative variables, we can always replace such variables by the difference of two nonnegative variables (thereby increasing the number of variables in the problem, of course). The general problem in which we shall be interested, already illustrated by Example A, may now be described.

**The Linear Programming Problem.** Maximize the objective function

$$f(x_1,\ldots, x_n) = c_1 x_1 + \cdots + c_n x_n$$

subject to the condition that $(x_1,\ldots, x_n)$ be in

$$\mathscr{F} = \{\mathbf{x} \in \mathbf{R}_+^n: x_1,\ldots, x_n \text{ satisfy a finite set of linear constraints}\}$$

A vector $\mathbf{x} \in \mathscr{F}$ is called a **feasible solution**; it is an **optimal solution** if it is a feasible solution at which $f$ assumes a maximum on $\mathscr{F}$. Problem C suggests a number of possibilities about the form of the feasible set $\mathscr{F}$. It is easily seen, even in the case of two variables, that $\mathscr{F}$ may be unbounded, in which case the linear programming problem may not have an optimal solution. On the other hand, we may have $\mathscr{F} = \varnothing$, again precluding any solution. Although these difficulties are theoretically possible, most of them either are not encountered or else may be anticipated and appropriately handled when the constraints arise from an actual problem in which the variables represent physical entities. At any rate, allowing for unbounded sets, the empty set, or any other situation, the set $\mathscr{F}$ of feasible solutions always exhibits certain geometrical properties we shall now discuss.

## 53. Linear Programming

Note that any constraint expressed as an equality of the form $a_1x_1 + \cdots + a_nx_n = \langle \mathbf{a}, \mathbf{x} \rangle = b$ can be replaced by the two inequalities $\langle \mathbf{a}, \mathbf{x} \rangle \leq b$ and $\langle \mathbf{a}, \mathbf{x} \rangle \geq b$. Thus, any system of linear constraints can be thought of as a set of inequalities. Since an inequality determines a half-space in $\mathbf{R}^n$, the $\mathbf{x}$ satisfying a system of inequalities must lie in the intersection of a finite number of (closed) half-spaces. The restriction of $\mathbf{x}$ to $\mathbf{R}_+^n$ means that $\mathscr{F}$ contains no lines, so we have the following result.

**Theorem A.** *The set $\mathscr{F}$ of feasible solutions to the linear programming problem is a closed polyhedral set containing no lines.*

As an immediate consequence of Theorem 51E, we know (when $\mathscr{F} \neq \varnothing$) which of the feasible solutions to examine for optimal solutions to our problem, that is, for maximum points of the (convex) linear objective function $f$.

**Theorem B.** *If the objective function is bounded above on $\mathscr{F} \neq \varnothing$, then there are optimal solutions to the linear programming problem, at least one of which is an extreme point of $\mathscr{F}$.*

We are now interested in determining the extreme points of the set $\mathscr{F}$ of feasible solutions. In cases like Example A where $n = 2$, this can be done by visual inspection of the graph, but where more variables are involved, we clearly need other methods. Toward this end, we shall find it helpful to express our problem in a canonical form.

We have already observed that the constraints of a linear programming problem can be expressed as a set of inequalities, and since multiplication by $-1$ changes the direction of an inequality, we can make certain conventions about their directions. We therefore agree to state any linear programming problem in the following **canonical form**.

Maximize
$$f(x_1, \ldots, x_n) = c_1 x_1 + \cdots + c_n x_n$$

subject to the constraints

(P)
$$a_{11}x_1 + \cdots + a_{1n}x_n \leq b_1$$
$$\vdots$$
$$a_{m1}x_1 + \cdots + a_{mn}x_n \leq b_m$$
$$x_1 \geq 0, \ldots, x_n \geq 0$$

For vectors **y** and **z** with respective coordinates $(y_1,\ldots,y_n)$ and $(z_1,\ldots,z_n)$ we understand $\mathbf{y} \leq \mathbf{z}$ to mean $y_i \leq z_i$ for each $i$. Then recalling the convention whereby **x** is understood to represent a column vector when used in an expression involving matrix multiplication, we may with obvious notation state our problem as follows.

Maximize
$$f(\mathbf{x}) = \langle \mathbf{c}, \mathbf{x} \rangle$$

subject to the constraints

$$A\mathbf{x} \leq \mathbf{b}$$
$$\mathbf{x} \geq \mathbf{O}$$

To solve the problem $P$ we indulge in a common practice of mathematicians by converting this new problem (a system of linear inequalities) to an old problem (a system of linear equations) that we know how to solve. In this case, the trick is to note that corresponding to each of the first $m$ inequalities in the system $(P)$, we can define a nonnegative variable $x_{n+i}$ by the equation

$$a_{i1}x_1 + \cdots + a_{in}x_n + x_{n+i} = b_i$$

With the introduction of these so-called **slack variables**, we are led to what we shall call the related problem $P_R$.

Maximize
$$F(x_1,\ldots,x_{n+m}) = c_1x_1 + \cdots + c_nx_n + 0x_{n+1} + \cdots + 0x_{n+m}$$

subject to the constraints

$$(P_R) \quad \begin{bmatrix} a_{11} & \cdots & a_{1n} & 1 & & \bigcirc \\ \vdots & & \vdots & & \ddots & \\ a_{m1} & \cdots & a_{mn} & \bigcirc & & 1 \end{bmatrix} \begin{bmatrix} x_1 \\ \vdots \\ x_n \\ x_{n+1} \\ \vdots \\ x_{n+m} \end{bmatrix} = \begin{bmatrix} b_1 \\ \vdots \\ b_m \end{bmatrix}$$

$$x_1 \geq 0,\ldots, x_{n+m} \geq 0$$

The problems $P$ and $P_R$ are related in the following way. There is a

## 53. Linear Programming

one-to-one correspondence between the members of $\mathscr{F}$ and the members of $\mathscr{F}_R$ (the feasible set of $P_R$) described by

$$\mathbf{x} = \begin{bmatrix} x_1 \\ \vdots \\ x_n \end{bmatrix} \leftrightarrow \begin{bmatrix} x_1 \\ \vdots \\ x_n \\ x_{n+1} \\ \vdots \\ x_{n+m} \end{bmatrix} = \begin{bmatrix} \mathbf{x} \\ \mathbf{b} - A\mathbf{x} \end{bmatrix} = \mathbf{x}_R$$

Since this correspondence preserves convex combinations of points, extreme points of $\mathscr{F}$ correspond to extreme points of $\mathscr{F}_R$. It is clear that $f$ is maximized at $\bar{\mathbf{x}} \in \mathscr{F}$ if and only if $F$ is maximized at the corresponding $\bar{\mathbf{x}}_R \in \mathscr{F}_R$. We have much to gain then in considering the related problem $P_R$, for as we shall see, it is possible to say explicitly how to determine the extreme points of $\mathscr{F}_R$.

Let the matrix of $(P_R)$ be $A_R$, and denote the $(m + n)$ column vectors by $\mathbf{a}_1, \ldots, \mathbf{a}_{m+n}$. The submatrix formed by the last $m$ columns

$$[\mathbf{a}_{n+1} \cdots \mathbf{a}_{n+m}] = \begin{bmatrix} 1 & 0 & \cdots & 0 \\ 0 & 1 & \cdots & 0 \\ \vdots & \vdots & & \vdots \\ 0 & 0 & \cdots & 1 \end{bmatrix}$$

is nonsingular. Choose any $m$ linearly independent column vectors and use the resulting nonsingular submatrix to determine the corresponding $m$ coordinates as the unique solutions to

$$[\mathbf{a}_{i_1} \cdots \mathbf{a}_{i_m}] \begin{bmatrix} x_{i_1} \\ \vdots \\ x_{i_m} \end{bmatrix} = \begin{bmatrix} b_1 \\ \vdots \\ b_m \end{bmatrix} \tag{2}$$

We claim that any solution to (2) with all $x_{i_j} \geq 0$ determines an extreme point $(0, \ldots, x_{i_1}, \ldots, x_{i_m}, \ldots, 0)$ of the set $\mathscr{F}_R$, the only nonzero entries to the indicated extreme point being those positive coordinates determined by (2). In fact we shall claim more; we shall claim that each extreme point of $\mathscr{F}_R$ is determined in this way as a solution to some system of the form (2). Before proving these assertions, however, we shall illustrate what has been said by returning to our two-variable example.

**Example A'.** After conforming to the convention regarding the direction of inequalities and introducing the three slack variables required by this problem, we have it expressed in the form of maximizing $F(x_1, ..., x_5) = -3x_1 + 2x_2$ subject to

$$\begin{bmatrix} -2 & 1 & 1 & 0 & 0 \\ 1 & -2 & 0 & 1 & 0 \\ 1 & 2 & 0 & 0 & 1 \end{bmatrix} \begin{bmatrix} x_1 \\ x_2 \\ x_3 \\ x_4 \\ x_5 \end{bmatrix} = \begin{bmatrix} 2 \\ 3 \\ 11 \end{bmatrix}$$

$$x_1 \geq 0, \quad ..., \quad x_5 \geq 0$$

Since each set of three column vectors in this $3 \times 5$ matrix is linearly independent, we have $\binom{5}{3} = 10$ systems of the form (2) to solve. Behold how quickly even our simple example has grown!

Let us begin with the system of equations obtained by selecting the first three columns $\mathbf{a}_1$, $\mathbf{a}_2$, $\mathbf{a}_3$. We solve

$$\begin{bmatrix} -2 & 1 & 1 \\ 1 & -2 & 0 \\ 1 & 2 & 0 \end{bmatrix} \begin{bmatrix} x_1 \\ x_2 \\ x_3 \end{bmatrix} = \begin{bmatrix} 2 \\ 3 \\ 11 \end{bmatrix} \tag{123}$$

obtaining $(x_1, x_2, x_3) = (7, 2, 14)$. Using next the columns $\mathbf{a}_1$, $\mathbf{a}_2$, $\mathbf{a}_4$, we solve the system

$$\begin{bmatrix} -2 & 1 & 0 \\ 1 & -2 & 1 \\ 1 & 2 & 0 \end{bmatrix} \begin{bmatrix} x_1 \\ x_2 \\ x_4 \end{bmatrix} = \begin{bmatrix} 2 \\ 3 \\ 11 \end{bmatrix} \tag{124}$$

and obtain $(x_1, x_2, x_4) = (\frac{7}{5}, \frac{24}{5}, \frac{56}{5})$.

When all 10 systems have been solved, the results are as shown in the accompanying tabulation. Note that in listing the extreme points of $\mathscr{F}_R$, we only use solutions having all nonnegative coordinates, and that we determine the correct position of the nonzero entries by noting which columns were selected from $A_R$.

| Columns selected | Solution | Extreme points of $\mathscr{F}_R \subseteq \mathbf{R}^5$ | Extreme points of $\mathscr{F} \subseteq \mathbf{R}^2$ |
|---|---|---|---|
| (1 2 3) | (7, 2, 14) | (7, 2, 14, 0, 0) | (7, 2) |
| (1 2 4) | (7/5, 24/5, 56/5) | (7/5, 24/5, 0, 56/5, 0) | (7/5, 24/5) |
| (1 2 5) | (−7/3, −8/3, 56/3) | | |
| (1 3 4) | (11, 24, −8) | | |
| (1 3 5) | (3, 8, 8) | (3, 0, 8, 0, 8) | (3, 0) |
| (1 4 5) | (−1, 4, 12) | | |
| (2 3 4) | (11/2, −7/2, 14) | | |
| (2 3 5) | (−3/2, −7/2, 14) | | |
| (2 4 5) | (2, 7, 7) | (0, 2, 0, 7, 7) | (0, 2) |
| (3 4 5) | (2, 3, 11) | (0, 0, 2, 3, 11) | (0, 0) |

## 53. Linear Programming

The extreme points thus determined for $\mathscr{F}$ are of course the answers that are obvious from Fig. 53.1. We complete our problem by evaluating $f$ at each extreme point of $\mathscr{F}$, finding the maximum to be $f(\frac{7}{5}, \frac{24}{5}) = \frac{27}{5}$.

We are now ready to prove a theorem that justifies the assertions we have made and illustrated. Since our theorem is really about extreme points of the set of nonnegative solutions to a system of equations, we shall simplify our notation by dropping the subscript $R$ and stating our result for matrix $A$, an $m \times (n + m)$ matrix of rank $m$,

$$\mathscr{F} = \{\mathbf{x} \in \mathbf{R}_+^{n+m} : A\mathbf{x} = \mathbf{b}\}$$

and the systems of linear equations formed by selecting $m$ linearly independent column vectors from $A$ and writing

$$B \begin{bmatrix} x_{i_1} \\ \vdots \\ x_{i_m} \end{bmatrix} = [\mathbf{a}_{i_1} \cdots \mathbf{a}_{i_m}] \begin{bmatrix} x_{i_1} \\ \vdots \\ x_{i_m} \end{bmatrix} = \begin{bmatrix} b_1 \\ \vdots \\ b_m \end{bmatrix} \tag{3}$$

**Theorem C.** Any solution to (3) with all $x_{i_j} \geq 0$ determines an extreme point $(0, \ldots, x_{i_1}, \ldots, x_{i_m}, \ldots, 0)$ of the set $\mathscr{F}$ in which at least $n$ coordinates are 0. Moreover, any extreme point of $\mathscr{F}$ is so determined for some selection of $m$ linearly independent column vectors of $A$.

**Proof.** For convenience of notation, let us suppose that $(x_1, \ldots, x_m)$ is a nonnegative solution to (3) where $B = [\mathbf{a}_1 \cdots \mathbf{a}_m]$. Then $\mathbf{x} = (x_1, \ldots, x_m, 0, \ldots, 0)$ is surely in $\mathscr{F}$. If it is not an extreme point, then there must be two other points, say $\mathbf{y}$ and $\mathbf{z}$ in $\mathscr{F}$, such that

$$\mathbf{x} = \tfrac{1}{2}(\mathbf{y} + \mathbf{z})$$

Since the coordinates of $\mathbf{y}$ and $\mathbf{z}$ are all nonnegative, it must be that the last $n$ coordinates of $\mathbf{y}$ and $\mathbf{z}$, like the last $n$ coordinates of $\mathbf{x}$, are 0. That is,

$$\mathbf{y} = (y_1, \ldots, y_m, 0, \ldots, 0), \quad \mathbf{z} = (z_1, \ldots, z_m, 0, \ldots, 0)$$

Moreover, both points satisfy $A\mathbf{x} = \mathbf{b}$, so

$$y_1 \mathbf{a}_1 + \cdots + y_m \mathbf{a}_m = \mathbf{b} = z_1 \mathbf{a}_1 + \cdots + z_m \mathbf{a}_m$$

From the linear independence of $\mathbf{a}_1, \ldots, \mathbf{a}_m$, we conclude that $y_1 = z_1, \ldots, y_m = z_m$, and in turn that $\mathbf{y} = \mathbf{z}$. This means $\mathbf{x}$ is an extreme point of $\mathscr{F}$.

Now let $\mathbf{x}$ be an extreme point of $\mathscr{F}$, and again for notational convenience assume all the positive entries occur first; that is, $\mathbf{x} = (x_1, \ldots, x_k, 0, \ldots, 0)$ where $x_i > 0$, $i = 1, \ldots, k$. Since $A\mathbf{x} = \mathbf{b}$,

$$x_1 \mathbf{a}_1 + \cdots + x_k \mathbf{a}_k = \mathbf{b}$$

We claim that $\mathbf{a}_1, \ldots, \mathbf{a}_k$ are linearly independent. If not, there are $k$ numbers $y_1, \ldots, y_k$, not all zero (but some perhaps negative) such that

$$y_1 \mathbf{a}_1 + \cdots + y_k \mathbf{a}_k = \mathbf{O}$$

Choose $\alpha > 0$ so that both

$$\mathbf{u} = (x_1 + \alpha y_1, \ldots, x_k + \alpha y_k, 0, \ldots, 0)$$

and

$$\mathbf{v} = (x_1 - \alpha y_1, \ldots, x_k - \alpha y_k, 0, \ldots, 0)$$

have nonnegative coordinates. Note now that both $\mathbf{u}$ and $\mathbf{v}$ belong to $\mathscr{F}$, and that $\mathbf{x} = \frac{1}{2}(\mathbf{u} + \mathbf{v})$. That is, $\mathbf{x}$ is not an extreme point. This contradiction shows that $\mathbf{a}_1, \ldots, \mathbf{a}_k$ are linearly independent, and since rank $A = m$, $k \leqslant m$. The extreme point $\mathbf{x}$ is of the form $(x_1, \ldots, x_m, 0, \ldots, 0)$, and since it must satisfy $A\mathbf{x} = \mathbf{b}$, the numbers $(x_1, \ldots, x_m)$ will appear as a solution to some system (3). Finally, we call attention to the fact that we have assumed in our argument that $\mathbf{x} \neq \mathbf{O}$, which will always be the case when $\mathbf{b} \neq \mathbf{O}$. When $\mathbf{b} = \mathbf{O}$, $\mathbf{x} = \mathbf{O}$ is surely an extreme point and the $m$-dimensional vector $\mathbf{O}$ satisfies any system of the form (3). ●

We now have a procedure for solving a linear programming problem stated in the canonical form $P$. We introduce slack variables and consider instead the related problem $P_R$ having the feasible set $\mathscr{F}_R$. The extreme points of $\mathscr{F}_R$ are determined using Theorem C. The maximum of $F$, if it exists, occurs at an extreme point $\bar{\mathbf{x}}_R \in \mathscr{F}_R$. The maximum of $f$ occurs at the corresponding extreme point $\bar{\mathbf{x}} \in \mathscr{F}$, and $f(\bar{\mathbf{x}}) = F(\bar{\mathbf{x}}_R)$.

For practical purposes, the procedure outlined so far has some serious drawbacks. For one thing, at no step is there any clue as to whether the objective function really does assume a maximum value on the feasible set, so we may be led into a lot of work for nothing. Secondly, while we have made the solution to $P$ depend on solving a finite number of systems of $m$ equations, that finite number may be as large as $\binom{m+n}{m}$, the number

## 53. Linear Programming

of ways of selecting $m$ column vectors from a matrix with $n + m$ columns. As was apparent from the simple problem considered in Example A, this leads to an intolerable amount of work even when the number of variables involved is very small. The number of variables involved in linear programming problems that arise in applications may be judged by some remarks made by George Dantzig at a conference in Cambridge, England in 1968. He is quoted as having said in response to a question [Beale, 1968, p. 15]

> I recommend that a special large-scale technique be used to solve linear programs of size greater than one thousand equations. It is my opinion that if this is done, the computation time will be a fraction of that needed using the general simplex method (particularly for problems of the size of 8000 equations).

The **simplex method** to which Dantzig here refers is an algorithm that he developed and popularized as an efficient method for solving linear programming problems in a reasonable number of steps. It is the object of our attention in Section 54.

We now turn our attention to the subject of duality in linear programming, an instance of the phenomenon discussed in the introduction to Section 15. We state our linear programming problem, here referred to as the **primal problem** in canonical form.

**Primal Problem** $P$

Maximize

$$f(\mathbf{x}) = c_1 x_1 + \cdots + c_n x_n$$

with constraints

$$\begin{bmatrix} a_{11} & \cdots & a_{1n} \\ \vdots & & \vdots \\ a_{m1} & \cdots & a_{mn} \end{bmatrix} \begin{bmatrix} x_1 \\ \vdots \\ x_n \end{bmatrix} \leq \begin{bmatrix} b_1 \\ \vdots \\ b_m \end{bmatrix}$$

$$\mathbf{x} \geq \mathbf{0}$$

**Dual Problem** $P^*$

Minimize

$$g(\mathbf{y}) = b_1 y_1 + \cdots + b_m y_m$$

with constraints

$$\begin{bmatrix} a_{11} & \cdots & a_{m1} \\ \vdots & & \vdots \\ a_{1n} & \cdots & a_{mn} \end{bmatrix} \begin{bmatrix} y_1 \\ \vdots \\ y_m \end{bmatrix} \geq \begin{bmatrix} c_1 \\ \vdots \\ c_n \end{bmatrix}$$

$$\mathbf{y} \geq \mathbf{0}$$

The dual problem can be posed in the canonical form of a linear programming problem. Set $h(\mathbf{y}) = -g(\mathbf{y})$, and state the constraints with reversed inequalities as necessary. With obvious notational conveniences our problems may then be stated as follows.

**Primal Problem**

Maximize

$$f(\mathbf{x}) = \langle \mathbf{c}, \mathbf{x} \rangle$$

with constraints

$$A\mathbf{x} \leqslant \mathbf{b}$$
$$\mathbf{x} \geqslant \mathbf{O}$$

**Dual Problem**

Maximize

$$h(\mathbf{y}) = \langle -\mathbf{b}, \mathbf{y} \rangle$$

with constraints

$$-A^t \mathbf{y} \leqslant -\mathbf{c}$$
$$\mathbf{y} \geqslant \mathbf{O}$$

With the dual problem now stated in the form of the primal problem, we can write down the dual of the dual.

Minimize

$$H(\mathbf{x}) = \langle -\mathbf{c}, \mathbf{x} \rangle$$

with constraints

$$(-A^t)^t \mathbf{x} \geqslant -\mathbf{b}$$
$$\mathbf{x} \geqslant \mathbf{O}$$

Then restating this as a maximization problem for the function $-H(\mathbf{x}) = \langle \mathbf{c}, \mathbf{x} \rangle$ and constraints $A\mathbf{x} \leqslant \mathbf{b}$, we see that we have returned to the given primal problem, establishing one reason for calling these problems dual.

**Theorem D.** For a given primal linear programming problem, the dual of the dual problem is the primal problem.

We are now in position to state and prove the **duality theorem of linear programming** due to Gale *et al.* [1951].

**Theorem E.** Let $P$ be a (primal) linear programming problem with feasible set $\mathscr{F}$, and let $P^*$ be the dual problem with feasible set $\mathscr{F}^*$.

(a) If $\mathbf{x} \in \mathscr{F}$ and $\mathbf{y} \in \mathscr{F}^*$, then $f(\mathbf{x}) \leqslant g(\mathbf{y})$.

(b) If for some $\bar{\mathbf{x}} \in \mathscr{F}$ and some $\bar{\mathbf{y}} \in \mathscr{F}^*$, $f(\bar{\mathbf{x}}) = g(\bar{\mathbf{y}})$, then $\bar{\mathbf{x}}$ is an optimal solution to $P$ and $\bar{\mathbf{y}}$ is an optimal solution to $P^*$.

(c) If one of the problems $P$ or $P^*$ has an optimal solution $\bar{\mathbf{x}}$ or $\bar{\mathbf{y}}$, respectively, then so does the other, and $f(\bar{\mathbf{x}}) = g(\bar{\mathbf{y}})$.

## 53. Linear Programming

(d) If $f$ is bounded above on $\mathscr{F} \neq \varnothing$, or if $g$ is bounded below on $\mathscr{F}^* \neq \varnothing$, then both $P$ and $P^*$ have optimal solutions.

**Proof.** If $\mathbf{x} \in \mathscr{F}$, then $A\mathbf{x} \leqslant \mathbf{b}$ and for $\mathbf{y} \geqslant \mathbf{0}$, we may write

$$\langle A\mathbf{x}, \mathbf{y} \rangle \leqslant \langle \mathbf{b}, \mathbf{y} \rangle = g(\mathbf{y})$$

Similarly, if $\mathbf{y} \in \mathscr{F}^*$, $A^t\mathbf{y} \geqslant \mathbf{c}$ and for $\mathbf{x} \geqslant \mathbf{0}$, we may write

$$\langle \mathbf{x}, A^t\mathbf{y} \rangle \geqslant \langle \mathbf{x}, \mathbf{c} \rangle = f(\mathbf{x})$$

The two inequalities then give for $\mathbf{x} \in \mathscr{F}$, $\mathbf{y} \in \mathscr{F}^*$

$$f(\mathbf{x}) \leqslant \langle \mathbf{x}, A^t\mathbf{y} \rangle = \langle A\mathbf{x}, \mathbf{y} \rangle \leqslant g(\mathbf{y})$$

which establishes part (a). Now if $f(\bar{\mathbf{x}}) = g(\bar{\mathbf{y}})$, then for any $\mathbf{x} \in \mathscr{F}$,

$$f(\mathbf{x}) \leqslant g(\bar{\mathbf{y}}) = f(\bar{\mathbf{x}})$$

so $\bar{\mathbf{x}}$ is an optimal solution to $P$; and for any $\mathbf{y} \in \mathscr{F}^*$,

$$g(\mathbf{y}) \geqslant f(\bar{\mathbf{x}}) = g(\bar{\mathbf{y}})$$

meaning that $\bar{\mathbf{y}}$ is an optimal solution to $P^*$.

The duality of the two problems means that part (c) will be established if we show that the existence of an optimal solution $\bar{\mathbf{x}}$ for $P$ implies the existence of an optimal solution $\bar{\mathbf{y}}$ for $P^*$. But as we have already seen, problem $P$ is equivalent to the related problem $P_R$. We may start therefore with an optimal solution $\bar{\mathbf{x}}_R = (\bar{x}_1, \ldots, \bar{x}_{n+m})$ to $P_R$ which by Theorem B may be taken as an extreme point of $\mathscr{F}_R$. Thus,

$$F(\bar{\mathbf{x}}_R) = c_1 \bar{x}_1 + \cdots + c_n \bar{x}_n + 0\bar{x}_{n+1} + \cdots + 0\bar{x}_{n+m} = \langle \mathbf{c}_R, \bar{\mathbf{x}}_R \rangle$$

is a maximum. By Theorem C, we may select an $m \times m$ nonsingular matrix $B = [\mathbf{a}_{i_1}, \ldots, \mathbf{a}_{i_m}]$ of $A_R$ so that $B\mathbf{x}_B = \mathbf{b}$, and $\mathbf{x}_B$ contains all the nonzero coordinates of $\bar{\mathbf{x}}_R$. Let $\mathbf{c}_B = (c_{i_1}, \ldots, c_{i_m})$ and choose $\bar{\mathbf{y}}$ so that $B^t \bar{\mathbf{y}} = \mathbf{c}_B$. Then

$$g(\bar{\mathbf{y}}) = \langle \bar{\mathbf{y}}, \mathbf{b} \rangle = \langle \bar{\mathbf{y}}, B\mathbf{x}_B \rangle = \langle B^t \bar{\mathbf{y}}, \mathbf{x}_B \rangle$$
$$= \langle \mathbf{c}_B, \mathbf{x}_B \rangle = \langle \mathbf{c}_R, \bar{\mathbf{x}}_R \rangle = F(\bar{\mathbf{x}}_R) = f(\bar{\mathbf{x}})$$

By part (b), $\bar{\mathbf{x}}$ and $\bar{\mathbf{y}}$ are optimal solutions to $P$ and $P^*$, respectively.

Part (d) follows from Theorem B combined with part (c). ●

It is to be noted that we have not only proved Theorem E, but that we have actually given instructions for determining the optimal solution $\bar{\mathbf{y}}$ of the dual problem,

$$\bar{\mathbf{y}} = [B^t]^{-1} \mathbf{c}_B \tag{4}$$

We return once more to our example, this time to illustrate our discussion of duality.

### Example A″

**Primal Problem**

Maximize

$$f(\mathbf{x}) = -3x_1 + 2x_2$$

with constraints

$$\begin{bmatrix} -2 & 1 \\ 1 & -2 \\ 1 & 2 \end{bmatrix} \begin{bmatrix} x_1 \\ x_2 \end{bmatrix} \leqslant \begin{bmatrix} 2 \\ 3 \\ 11 \end{bmatrix}$$

$$\mathbf{x} \geqslant \mathbf{0}$$

**Dual Problem**

Minimize

$$g(\mathbf{y}) = 2y_1 + 3y_2 + 11y_3$$

with constraints

$$\begin{bmatrix} -2 & 1 & 1 \\ 1 & -2 & 2 \end{bmatrix} \begin{bmatrix} y_1 \\ y_2 \\ y_3 \end{bmatrix} \geqslant \begin{bmatrix} -3 \\ 2 \end{bmatrix}$$

$$\mathbf{y} \geqslant \mathbf{0}$$

The dual problem can be rephrased as a linear programming problem in which the function $h(\mathbf{y}) = -2y_1 - 3y_2 - 11y_3$ is to be maximized. The feasible set is an unbounded set in $\mathbf{R}^3$, making determination of extreme points by inspection of the graph an exercise in three-dimensional perspective drawing. Following the procedure outlined in Theorem C leads to $\binom{5}{2} = 10$ systems of equations again (though only in two unknowns this time).

The easiest way to solve the dual problem is to review the solution to the primal problem. We obtained $\bar{\mathbf{x}} = (\frac{7}{5}, \frac{24}{5})$ from $\bar{\mathbf{x}}_R = (\frac{7}{5}, \frac{24}{5}, 0, \frac{56}{5}, 0)$. Note that in an actual problem, the notational convenience by which we assumed all nonzero coordinates to appear first will probably not obtain. This is no problem, so long as in defining $B$ we remember to use the corresponding columns from the matrix $A_R$. In this case,

$$\bar{\mathbf{x}}_B = (x_1, x_2, x_4) = (\tfrac{7}{5}, \tfrac{24}{5}, \tfrac{56}{5})$$

$$\mathbf{c}_B = (-3, 2, 0)$$

$$B = \begin{bmatrix} -2 & 1 & 0 \\ 1 & -2 & 1 \\ 1 & 2 & 0 \end{bmatrix} \quad \text{so} \quad [B^t]^{-1} = \begin{bmatrix} -\tfrac{2}{5} & \tfrac{1}{5} & \tfrac{4}{5} \\ 0 & 0 & 1 \\ \tfrac{1}{5} & \tfrac{2}{5} & \tfrac{3}{5} \end{bmatrix}$$

## 53. Linear Programming

Then

$$\bar{y} = [B^t]^{-1} c_B = \begin{bmatrix} -\frac{2}{5} & \frac{1}{5} & \frac{4}{5} \\ 0 & 0 & 1 \\ \frac{1}{5} & \frac{2}{5} & \frac{3}{5} \end{bmatrix} \begin{bmatrix} -3 \\ 2 \\ 0 \end{bmatrix} = \begin{bmatrix} \frac{8}{5} \\ 0 \\ \frac{1}{5} \end{bmatrix}$$

The optimal solution to the dual problem occurs at $\bar{y} = (\frac{8}{5}, 0, \frac{1}{5})$, and we note that in accord with Theorem E, $g(\bar{y}) = \frac{27}{5} = f(\frac{7}{5}, \frac{24}{5})$.

## PROBLEMS AND REMARKS

**A.** Solve the dual problem of the example in this section by rephrasing it in the canonical form of a linear programming problem and using the method of Theorem C.

**B.** Consider the problem of maximizing

$$f(\mathbf{x}) = x_1 - 3x_2$$

subject to the constraints

$$2x_1 - x_2 \leq 4$$
$$x_1 + 2x_2 \leq 3$$
$$x_1 \geq 0, \quad x_2 \geq 0$$

(1) Solve, finding extreme points of the feasible set by geometric inspection.
(2) Solve by the method of Theorem C.
(3) State the dual; solve, finding extreme points by inspection.
(4) Solve the dual problem using the formula (4).

**C.** We wish to explore various difficulties that can occur in a linear programming problem.

(1) Maximize $f(x_1, x_2) = 4x_2 - x_1$ subject to the constraints

$$2x_1 - x_2 \geq 4$$
$$x_1 + 2x_2 \geq 3$$
$$x_1 \geq 0, \quad x_2 \geq 0$$

(2) Maximize $F(x_1, x_2) = -4x_2 - x_1$ subject to the constraints of part (1).
(3) Solve parts (1) and (2) again if we add the constraint $x_2 - x_1 \leq 0$.
(4) State and solve the dual problem for each of the four problems considered above.

**D.** Some authors say that the linear programming problem has no solution if the feasible set is empty, and that it has an **unbounded solution** if the objective function is unbounded on a nonempty feasible set.

(1) If the primal problem has an unbounded solution, then the dual problem has no solution.
(2) If the primal problem has no solution, we may not conclude that the dual problem has an unbounded solution; this may be demonstrated using a two-variable primal problem.

**E.** Find the maximum of $f(x_1, x_2) = 3x_1 - x_2$, constrained by

$$x_1 + x_2 \leq 6$$
$$2x_1 + x_2 \geq 4$$
$$x_1 - x_2 \leq 2$$
$$x_2 \leq 3$$
$$x_1 \geq 0, \quad x_2 \geq 0$$

by solving the dual problem.

**F.** In Problem H below we indicate an alternate proof of the duality theorem for linear programming. For this we need the **lemma of Farkas**, a result about systems of linear inequalities apparently first noticed by Minkowski and then proved by Farkas [1901]. The form we shall use states that for an $m \times n$ matrix $A$ and vector **b**, one and only one of the following two statements is true.

  (i) $A\mathbf{x} = \mathbf{b}$ has a nonnegative solution $\bar{\mathbf{x}} \geq 0$.
  (ii) There is a vector $\bar{\mathbf{y}}$ that satisfies $\mathbf{y}^t A \geq \mathbf{O}$ and $\langle \mathbf{y}, \mathbf{b} \rangle < 0$.

The result is proved in this form by Simonnard [1966, pp. 376–378].

**G.** A variety of theorems, identified as the **Minkowski-Farkas theorem**, appear in the literature. Show that the theorems listed below each imply the theorem stated in Problem F. Then decide which of them are implied by the theorem of Problem F.

(1) [Krekó, 1968] Let $A$ be an $m \times n$ matrix. Then

$$C = \{\mathbf{x} \in \mathbf{R}^m: \text{ There is a } \mathbf{w} \geq \mathbf{O} \in \mathbf{R}^n \text{ such that } A\mathbf{w} = \mathbf{x}\}$$

forms a convex cone. The set

$$C^+ = \{\mathbf{y} \in \mathbf{R}^m: \langle \mathbf{y}, \mathbf{x} \rangle \geq 0 \text{ for every } \mathbf{x} \in C\}$$

is called the **polar cone** of $C$. If there is a $\mathbf{z} \in \mathbf{R}^m$ such that $\langle \mathbf{z}, \mathbf{y} \rangle \geq 0$ for every $\mathbf{y} \in C^+$ then $\mathbf{z} \in C$; that is, $(C^+)^+ = C$.

(2) [Rockafellar, 1970a, p. 200] An inequality $\langle \mathbf{a}_0, \mathbf{x} \rangle \leq \alpha_0$ is said to be a **consequence** of the system $\langle \mathbf{a}_i, \mathbf{x} \rangle \leq \alpha_i$, $i = 1,\ldots, m$, if it is satisfied by every $\mathbf{x}$ that satisfies the system. (For example, $x_1 + x_2 \geq 0$ is a consequence of the system $x_1 \geq 0$, $x_2 \geq 0$). An inequality $\langle \mathbf{a}_0, \mathbf{x} \rangle \leq \alpha_0$ is a consequence of $A\mathbf{x} \leq \mathbf{O}$ if and only if there is a $\mathbf{y} \geq 0$ such that $\mathbf{y}^t A = \mathbf{a}_0$.

(3) [Berge and Ghouila-Houri, 1965, p. 67] Let $f, g_1, \ldots, g_m$ be concave on $\mathbf{R}^n$ and let $g_1, \ldots, g_k$, $k \leq m$, be affine. If the system

$$g_i(\mathbf{x}) \geq 0, \quad i = 1,\ldots, m, \quad f(\mathbf{x}) > 0$$

has no solution $\bar{\mathbf{x}} \in \mathbf{R}^n$, and if

$$g_i(\mathbf{x}) \begin{cases} > 0 & \text{for } i = 1,\ldots, k \\ \geq 0 & \text{for } i = k+1,\ldots, m \end{cases}$$

has a solution, then there is a $\mathbf{y} \geq \mathbf{O}$, $\mathbf{y} \neq \mathbf{O}$, such that for every $\mathbf{x} \in \mathbf{R}^n$,

$$f(\mathbf{x}) + \sum_{i=1}^m y_i g_i(\mathbf{x}) \leq 0$$

## 53. Linear Programming

**H.** An alternate approach to the duality theorem for linear programming starts with Theorem 52 A and makes use of the lemma of Farkas (Problem F). We outline the procedure, using the notation for the primal and dual problems as given above.

(1) Define $K(\mathbf{x}, \mathbf{y}) = \langle \mathbf{c}, \mathbf{x} \rangle + \langle \mathbf{b}, \mathbf{y} \rangle - \langle \mathbf{y}, A\mathbf{x} \rangle$. Observe that if $A\mathbf{x} \leq \mathbf{b}$, then

$$\inf_{\mathbf{y} \geq \mathbf{0}} K(\mathbf{x}, \mathbf{y}) = f(\mathbf{x})$$

and if $A^t \mathbf{y} \geq \mathbf{c}$, then

$$\sup_{\mathbf{x} \geq \mathbf{0}} K(\mathbf{x}, \mathbf{y}) = g(\mathbf{y})$$

(2) According to Theorem 52A,

$$\sup_{\mathbf{x} \in \mathcal{F}} f(\mathbf{x}) \leq \inf_{\mathbf{y} \in \mathcal{F}^*} g(\mathbf{y})$$

(3) If $\bar{\mathbf{x}}$ is an optimal solution to the primal problem, then the lemma of Farkas can be used to show that there is an optimal solution to the dual problem [Mangasarian, 1969, pp. 18–19].

The use of Theorem 52A on $K(\mathbf{x}, \mathbf{y})$ turns out to be the procedure that enables us to obtain a duality theorem for convex programming in Section 55. With this in mind, the argument above could be phrased so as to parallel more closely the argument to be used in the more general case [Karlin, 1959, pp. 118–122].

**I.** We have suggested several procedures (replacing unrestricted variables by the difference of two nonnegative variables, replacing an equality by a pair of inequalities, multiplying by $-1$ as necessary to get inequalities in the desired direction) for putting a linear programming problem into canonical form.

(1) Put into canonical form the problem, left at the end of Section 52, of finding the optimal strategy for player A in a matrix game.

(2) Similarly express the problem of finding an optimal strategy for player B.

(3) Verify that these two problems are dual in the sense of linear programming.

The subject of linear programming, to which we have given only a brief introduction, is large, relatively new as a defined area of mathematical activity, and fast growing. The simplex method, developed in 1947 by G. B. Dantzig while working on military logistic problems, first appeared in print four years later [Dantzig, 1951]. Of the growth and applicability of linear programming that followed his work, Dantzig later wrote in his book,

> In the summer of 1949 at the University of Chicago, a conference was held under the sponsorship of the Cowles Commission for Research in Economics; mathematicians, economists, statisticians from academic institutions and various government agencies presented research using the linear programming tool. The problems considered ranged from planning crop rotation to planning large scale miltary actions, from the routing of ships between harbors to the assessment of flow of commodities between industries of the economy. What was most surprising was that the research had taken place during the preceding two years.

In the same book [Dantzig, 1963], Dantzig devotes Chapter 2 to tracing factors that contributed to this growth. The reader looking for an introduction to the subject from the point of view of economics might begin with Dorfman, Samuelson, and Solow [1958] or Gale [1960]. Those wishing to see elementary applications to a variety of problems (24 to be exact) should consult Vajda [1958]; those looking for more sophisticated accounts of more recent applications are referred to Beale [1968]. The relationship of flows in networks to duality in linear programming was first pointed out by Ford and Fulkerson [1956] and further developed in the book [1962] by the same authors.

Besides the books already mentioned, we call attention to the texts of Karlin [1959] and Hadley [1962] which became standard references in this country, and to the text of Simonnard [1966] which in its original French version (1962) played somewhat the same role in Europe. We also recommend the book by Charnes *et al.* [1953] since it provides an excellent elementary introduction to linear programming, the simplex method, and applications to economics. Among many more recent texts providing an elementary introduction to the subject, that by Cooper and Steinberg [1970] might be suggested.

## 54. The Simplex Method

We now wish to describe an algorithm for solving the linear programming problem. This section can without loss of continuity be skipped by readers wishing to move on to the treatment of convex programming in Section 55.

It will be convenient in this section to consider an affine rather than a linear objective function. Thus, supposing that the slack variables have already been introduced, the linear programming problem to be considered may be written in the form:
Constraints

$$
\begin{aligned}
a_{11}x_1 + \cdots + a_{1S}x_S + \cdots + a_{1n}x_n - b_1 &= -x_{n+1} \\
&\vdots \\
a_{R1}x_1 + \cdots + a_{RS}x_S + \cdots + a_{Rn}x_n - b_R &= -x_{n+R} \\
&\vdots \\
a_{m1}x_1 + \cdots + a_{mS}x_S + \cdots + a_{mn}x_n - b_m &= -x_{n+m}
\end{aligned} \tag{1}
$$

Maximize $\hat{x}$ defined by

$$c_1 x_1 + \cdots + c_S x_S + \cdots + c_n x_n - d = \hat{x}$$

subject to the constraints (1) and all $x_i \geq 0$.

The system of constraints is said to be solved for $x_{n+1}, \ldots, x_{n+m}$ since if $x_1, \ldots, x_n$ are known, the others are immediately obtainable by substitution. Now if $a_{RS} \neq 0$, we may proceed with the usual elementary operations of linear algebra to write the above system in the equivalent form

## 54. The Simplex Method

$$\bar{a}_{11}x_1 + \cdots + 0x_S + \cdots + \bar{a}_{1n}x_n - \bar{b}_1 = -x_{n+1} + \frac{a_{1S}}{a_{RS}} x_{n+R}$$

$$\vdots$$

$$\frac{a_{R1}}{a_{RS}} x_1 + \cdots + x_S + \cdots + \frac{a_{Rn}}{a_{RS}} x_n - \frac{b_R}{a_{RS}} = -\frac{1}{a_{RS}} x_{n+R}$$

$$\vdots$$

$$\bar{a}_{m1}x_1 + \cdots + 0x_S + \cdots + \bar{a}_{mn}x_n - \bar{b}_m = -x_{n+m} + \frac{a_{mS}}{a_{RS}} x_{n+R}$$

$$\bar{c}_1 x_1 + \cdots + 0x_S + \cdots + \bar{c}_n x_n - \bar{d} = \hat{x} + \frac{c_S}{a_{RS}} x_{n+R}$$

Two comments about this new system are in order. First, we have placed bars on some coefficients to indicate that while their exact values are temporarily irrelevant, they are different from those in the original system (1). Second, we have carried along the equation defining $\hat{x}$ so that $\hat{x}$ can be expressed in terms of the same unknowns as appear on the left side of the constraint equations. Our problem is still to maximize $\hat{x}$ defined by the last equation, subject to the constraint equations above it (and the continuing understanding that all $x_i \geq 0$).

The system is now solved for $x_{n+1}, \ldots, x_S, \ldots, x_{n+m}$, where $x_S$ replaces $x_{n+R}$ by writing

$$a_{11}x_1 + \cdots + \frac{-a_{1S}}{a_{RS}} x_{n+R} + \cdots + \bar{a}_{1n}x_n - \bar{b}_1 = -x_{n+1}$$

$$\vdots$$

$$\frac{a_{R1}}{a_{RS}} x_1 + \cdots + \frac{1}{a_{RS}} x_{n+R} + \cdots + \frac{a_{Rn}}{a_{RS}} x_n - \frac{b_R}{a_{RS}} = -x_S \qquad (2)$$

$$\vdots$$

$$\bar{a}_{m1}x_1 + \cdots + \frac{-a_{mS}}{a_{RS}} x_{n+R} + \cdots + \bar{a}_{mn}x_n - \bar{b}_m = -x_{n+m}$$

$$\bar{c}_1 x_1 + \cdots + \frac{-c_S}{a_{RS}} x_{n+R} + \cdots + \bar{c}_n x_n - \bar{d} = \hat{x}$$

Using the so-called **tableau** notation, system (1) is summarized by (3).

| $x_1$ | | $x_S$ | | $x_n$ | $-1$ | | |
|---|---|---|---|---|---|---|---|
| $a_{11}$ | $\cdots$ | $a_{1S}$ | $\cdots$ | $a_{1n}$ | $b_1$ | $= -x_{n+1}$ | |
| $\vdots$ | | $\vdots$ | | $\vdots$ | $\vdots$ | | |
| $a_{R1}$ | $\cdots$ | $a_{RS}$ | $\cdots$ | $a_{Rn}$ | $b_R$ | $= -x_{n+R}$ | (3) |
| $\vdots$ | | $\vdots$ | | $\vdots$ | $\vdots$ | | |
| $a_{m1}$ | $\cdots$ | $a_{mS}$ | $\cdots$ | $a_{mn}$ | $b_m$ | $= -x_{n+m}$ | |
| $c_1$ | $\cdots$ | $c_S$ | $\cdots$ | $c_m$ | $d$ | $= \hat{x}$ | |

The new system is described by (4).

$$
\begin{array}{c|cccccc|c}
 & x_1 & & x_{n+R} & & x_n & -1 & \\
\hline
 & \bar{a}_{11} & \cdots & -\dfrac{a_{1S}}{a_{RS}} & \cdots & \bar{a}_{1n} & \bar{b}_1 & = -x_{n+1} \\
 & \vdots & & \vdots & & \vdots & \vdots & \\
 & \dfrac{a_{R1}}{a_{RS}} & \cdots & \dfrac{1}{a_{RS}} & \cdots & \dfrac{a_{Rn}}{a_{RS}} & \dfrac{b_R}{a_{RS}} & = -x_S \\
 & \vdots & & \vdots & & \vdots & \vdots & \\
 & \bar{a}_{mn} & \cdots & -\dfrac{a_{mS}}{a_{RS}} & \cdots & \bar{a}_{mn} & \bar{b}_m & = -x_{n+m} \\
 & \bar{c}_1 & \cdots & \dfrac{-c_S}{a_{RS}} & \cdots & \bar{c}_n & \bar{d} & = \hat{x}
\end{array}
\qquad (4)
$$

The process of going from tableau (3) to (4) is called **pivoting** about $a_{RS}$. The mechanics of going from (3) to (4) are described by the following flow chart.

**Pivoting**

> Replace $a_{RS}$ by $1/a_{RS}$.

↓

> Replace all other elements in column $S$ by their product with $-1/a_{RS}$;
> $$\bar{a}_{iS} = \dfrac{-a_{iS}}{a_{RS}}, \qquad i \neq R$$

↓

> Replace all other elements in row $R$ by their product with $1/a_{RS}$;
> $$\bar{a}_{Rj} = \dfrac{a_{Rj}}{a_{RS}}, \qquad j \neq S$$

↓

> Replace all other elements $t_{ij}$ of the tableau by the sum
> $$t_{ij} - t_{iS}\dfrac{a_{Rj}}{a_{RS}}, \qquad i \neq R,\ j \neq S.$$

## 54. The Simplex Method

Note that since pivoting leads to an equivalent system of equations, the candidates for maximizing $\hat{x}$ (the extreme points of the feasible set) remain the same, and the values of $\hat{x}$ obtained for these various candidates remain the same. Suppose now that after several pivoting steps, our tableau is of the form

|  | $x_{\bar{1}}$ | $x_{\bar{2}}$ | | $x_{\bar{n}}$ | $-1$ | |
|---|---|---|---|---|---|---|
|  | $\bar{a}_{11}$ | $\bar{a}_{12}$ | $\cdots$ | $\bar{a}_{1n}$ | $\bar{b}_1$ | $= -x_{\overline{n+1}}$ |
|  | $\vdots$ | $\vdots$ | | $\vdots$ | $\vdots$ | |
|  | $\bar{a}_{m1}$ | $\bar{a}_{m2}$ | $\cdots$ | $\bar{a}_{mn}$ | $\bar{b}_m$ | $= -x_{\overline{n+m}}$ |
|  | $\bar{c}_1$ | $\bar{c}_2$ | $\cdots$ | $\bar{c}_n$ | $\bar{d}$ | $= \hat{x}$ |

where $\bar{1}, \bar{2}, \ldots, \bar{n}, \overline{n+1}, \ldots \overline{n+m}$, is some permutation of $1, 2, \ldots, n+m$. If all $\bar{b}_i \geq 0$, then (Theorem 53 C) the point

$$(x_{\bar{1}}, \ldots, x_{\bar{n}}, x_{\overline{n+1}}, \ldots, x_{\overline{n+m}}) = (0, \ldots, 0, \bar{b}_1, \ldots, \bar{b}_m) \tag{5}$$

is one of the extreme points at which

$$\hat{x} = -\bar{d} + \bar{c}_1 x_{\bar{1}} + \cdots + \bar{c}_n x_{\bar{n}}$$

may take its maximum. If in addition, all $\bar{c}_j \leq 0$, then the fact that all $x_j \geq 0$ means that $\hat{x}$ cannot possibly get any larger than $-\bar{d}$, giving an optimal value at the point (6). Finally, if all $\bar{c}_j < 0$, no other point can make $\hat{x}$ as large as $-\bar{d}$, and we have a unique optimal solution. We summarize our results as follows.

> **Theorem A.** Suppose that a linear programming problem is represented by the tableau above.
> (a) If $\bar{b}_i \geq 0$ for every $i$, then (5) determines an extreme point of the feasible set.
> (b) If $\bar{b}_i \geq 0$ for all $i$, $\bar{c}_j \leq 0$ for all $j$, then (5) determines an optimal solution.
> (c) If $\bar{b}_i \geq 0$ for all $i$, $\bar{c}_j < 0$ for all $j$, then (5) determines a unique optimal solution.

Given a linear programming problem, we can always introduce the appropriate slack variables and write the problem in a tableau. Theorem A suggests the form to which we would like to move by a prudent choice of pivoting steps. Since not all linear programming problems have optimal solutions, we will not always be able to obtain the desired form, but when we can, the procedure we are about to describe, known as the **simplex method**, brings us to that form in a reasonable number of steps. Moreover, as is indicated on the flow diagrams outlining the procedure, the simplex method will alert us and bring us to a stop when we encounter a problem not having an optimal solution.

The simplex method is most conveniently described in two phases. In phase I, we try to move to an equivalent tableau in which all the constants $b_i \geqslant 0$. Geometrically, this corresponds, as we have seen, to determining an extreme point of the set of feasible solutions to the problem. We say a tableau in this form is in **I-feasible form**.

In phase II, we try to move to an equivalent tableau in which all coefficients $c_j \leqslant 0$. Taking the liberty of over-simplifying (see [Klee, 1964, pp. 147–150]) and ignoring possible degeneracies, we might describe phase II in geometric terms as follows. Beginning at an extreme point of the feasible set, examine the adjacent extreme points. If there is one that yields a greater value of the objective function, "move" to it (the algebraic process of pivoting). Continue the process; that is, look for a neighbor of the extreme point that provides a still greater value for the objective function. Since there are only finitely many extreme points, any sequence that continues to move to untried points must terminate at an extreme point **p** for which the adjacent extreme points do not provide any further increase in the value of the objective function. Either **p** is the desired optimal solution, or there issues from **p** an unbounded ray in the feasible region along which the objective function is unbounded from above. We shall see that the simplex method makes it computationally easy to distinguish between these two possibilities.

Since many applications lead immediately to a system of constraints in which all $b_i \geqslant 0$, phase II is often sufficient for solving practical problems. It also turns out to be easier to describe phase II first. For these reasons, we begin our discussion of the simplex method by showing how to procede if for all $i$, $b_i \geqslant 0$.

**Phase II** (getting all $c_j \leqslant 0$). To fix notation, we begin by assuming that we have a tableau in the form (6) where $\overline{1},\ldots,\overline{n+m}$ is a permutation of $1,\ldots, n+m$, and in which for all $i$, $b_i \geqslant 0$.

## 54. The Simplex Method

|  $x_{\bar{1}}$ | | $x_S$ | | $x_{\bar{n}}$ | $-1$ | |
|---|---|---|---|---|---|---|
| $a_{11}$ | $\cdots$ | $a_{1S}$ | $\cdots$ | $a_{1n}$ | $b_1$ | $= -x_{\overline{n+1}}$ |
| $\vdots$ | | $\vdots$ | | $\vdots$ | $\vdots$ | |
| $a_{R1}$ | $\cdots$ | $a_{RS}$ | $\cdots$ | $a_{Rn}$ | $b_R$ | $= -x_{\overline{n+R}}$ |
| $\vdots$ | | $\vdots$ | | $\vdots$ | $\vdots$ | |
| $a_{m1}$ | $\cdots$ | $a_{mS}$ | $\cdots$ | $a_{mn}$ | $b_m$ | $= -x_{\overline{n+m}}$ |
| $c_1$ | $\cdots$ | $c_S$ | $\cdots$ | $c_n$ | $d$ | $= \hat{x}$ |

(6)

Since the solution may be read off directly if all $c_j \leqslant 0$, we need only consider the case in which at least one $c_j > 0$. In such a case, we select $S$ so that $c_S \geqslant c_j$ for $j = 1,\ldots, n$.

If all the entries in column $S$ of the tableau, excepting $c_S$, are nonpositive, we note that for any $\lambda > 0$, the point

$$(x_{\bar{1}}, \ldots, x_{\bar{S}}, \ldots, x_{\bar{n}}, x_{\overline{n+1}}, \ldots, x_{\overline{n+m}})$$
$$= (0, \ldots, \lambda, \ldots, 0, b_1 - a_{1S}\lambda, \ldots, b_m - a_{mS}\lambda)$$

is feasible. That is, it satisfies the constraint equations, and all entries are nonnegative. But at this point,

$$\hat{x} = -d + c_S \lambda$$

Letting $\lambda \to \infty$, we see that the objective function has no maximum.

We now turn our attention to the case in which some entries in column $S$ are positive. Among those with $a_{iS} > 0$, choose $R$ such that

$$\frac{b_R}{a_{RS}} \leqslant \frac{b_i}{a_{iS}}$$

Consider the effect of pivoting about $a_{RS}$, bringing us to the equivalent tableau (7).

| $x_{\bar{1}}$ | $x_{\overline{n+R}}$ | $x_{\bar{n}}$ | $-1$ | |
|---|---|---|---|---|
| $\bar{a}_{11}$ | $\bar{a}_{1S}$ | $\bar{a}_{1n}$ | $\bar{b}_1$ | $= -x_{\overline{n+1}}$ |
| $\dfrac{a_{R1}}{a_{RS}}$ | $\dfrac{1}{a_{RS}}$ | $\dfrac{a_{Rn}}{a_{RS}}$ | $\dfrac{b_R}{a_{RS}}$ | $= -x_S$ |
| $\bar{a}_{m1}$ | $\bar{a}_{mS}$ | $\bar{a}_{mn}$ | $\bar{b}_m$ | $= -x_{\overline{n+m}}$ |
| $\bar{c}_1$ | $\bar{c}_S$ | $\bar{c}_n$ | $\bar{d}$ | $= \hat{x}$ |

(7)

We wish to show that this tableau is again I-feasible. As indicated, the constant term in row $R$ becomes $b_R/a_{RS} > 0$. In other rows, we have $\bar{b}_i = b_i - (a_{iS}/a_{RS}) b_R$, which is surely nonnegative if $a_{iS} \leqslant 0$, and for $a_{iS} > 0$, we have

$$\bar{b}_i = b_i - \frac{a_{iS}}{a_{RS}} b_R = a_{iS} \left( \frac{b_i}{a_{iS}} - \frac{b_R}{a_{RS}} \right) \geqslant 0.$$

We conclude that the new tableau is again I-feasible.

The pivoting about $a_{RS}$ aims at more, of course, than simply to produce another I-feasible tableau. Note that

$$\bar{d} = d - \frac{c_S}{a_{RS}} b_R,$$

and since $c_S > 0$, $a_{RS} > 0$, and $b_R \geqslant 0$, we see that $\bar{d} \leqslant d$. The new tableau yields a feasible solution, actually an extreme point of the feasible set, with a corresponding value of $\hat{x}$ greater than or equal to the previous one.

We are unable to argue, on the basis of what we have said here, that we will actually have $\bar{d} < d$. It is clear that this will be the case, eventually, unless by following the rules outlined above, we run into an infinite sequence of pivots in which the $b_R$ in the pivot row is 0. It has been shown by a specially constructed example that this phenomenon, called **cycling**, can occur. It has also been shown that with perturbation techniques, the simplex method can be modified to cope with this possibility [Charnes, Cooper, and Henderson, 1953]. Since cycling does not seem to occur in applications of the simplex method to problems arising in natural ways, we shall not concern ourselves with it here.

Barring cycling, we do move in finitely many steps from each tableau with some $c_j > 0$ to a new tableau in which $\bar{d} < d$. Since we can only encounter a finite number of different values for $d$ (no more than the number of extreme points of the feasible set), the process must terminate. That is, we must come to a tableau in which $c_j \leqslant 0$ for all $j$.

## 54. The Simplex Method

**Phase II** (getting all $c_j \leq 0$)

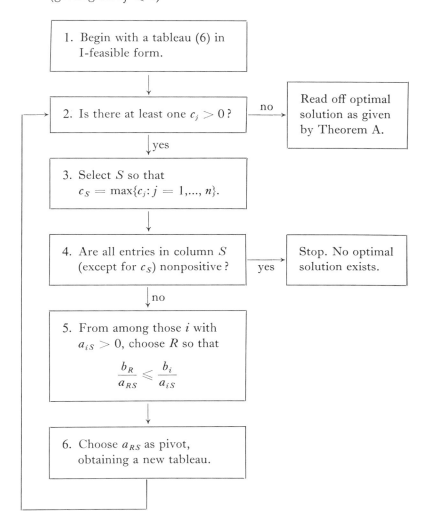

**Example A.** We return to the optimization problem considered in Example 53A. The data given in that problem is exhibited in the tableau, Stage 1.

### Stage 1

| $x_1$ | $x_2$ | $-1$ | |
|---|---|---|---|
| $-2$ | ① | $2$ | $= -x_3$ |
| $1$ | $-2$ | $3$ | $= -x_4$ |
| $1$ | $2$ | $11$ | $= -x_5$ |
| $-3$ | $2$ | $0$ | $= \hat{x}$ |

This tableau is already in I-feasible form. Selecting the circled entries as pivots, we successively move to Stages 2 and 3.

### Stage 2

| $x_1$ | $x_3$ | $-1$ | |
|---|---|---|---|
| $-2$ | $1$ | $2$ | $= -x_2$ |
| $-3$ | $2$ | $7$ | $= -x_4$ |
| ⑤ | $-2$ | $7$ | $= -x_5$ |
| $1$ | $-2$ | $-4$ | $= \hat{x}$ |

### Stage 3

| $x_5$ | $x_3$ | $-1$ | |
|---|---|---|---|
| $2/5$ | $1/5$ | $24/5$ | $= -x_2$ |
| $3/5$ | $4/5$ | $56/5$ | $= -x_4$ |
| $1/5$ | $-2/5$ | $7/5$ | $= -x_1$ |
| $-1/5$ | $-8/5$ | $-27/5$ | $= \hat{x}$ |

The three stages nicely illustrate the geometric description we gave of phase II. In Section 53 we listed the extreme points of the feasible set $\mathscr{F}_R \subseteq R^5$. In the solution just presented, we started (Stage 1) at the extreme point $(0, 0, 2, 3, 11)$, and we moved successively to $(0, 2, 0, 7, 7)$ and then $(7/5, 24/5, 0, 56/5, 0)$. The coefficients $c_j < 0$, $j = 1, 2$ in Stage 3, so we know we have found the unique optimal solution. In terms of the problem as originally stated (without the slack variables), we have the maximum of $\hat{x} = 27/5$ occuring at $(7/5, 24/5)$.

**Phase I** (getting all $b_i \geq 0$). We now consider the problem of what to do if at least one $b_i < 0$. Geometrically, we are asking how to locate an extreme point of the feasible set. This is a nontrivial question. Following a method due to Rockafellar [1964], we shall introduce an additional variable $x_0$, transform the corresponding augmented linear programming problem to I-feasible form, and then relate the result to the original problem.

## 54. The Simplex Method

For convenience, we repeat the original problem.

$$a_{11}x_1 + \cdots + a_{1n}x_n - b_1 = -x_{n+1}$$
$$\vdots \quad (8)$$
$$a_{m1}x_1 + \cdots + a_{mn}x_n - b_m = -x_{n+m}$$
$$\text{Maximize} \quad c_1 x_1 + \cdots + c_n x_n - d = \hat{x}$$

subject to the constraints (8) and all $x_i \geq 0$.

The augmented problem we wish to consider is

$$a_{11}x_1 + \cdots + a_{1n}x_n - (b_1 + x_0) = -x_{n+1}$$
$$\vdots$$
$$a_{m1}x_1 + \cdots + a_{mn}x_n - (b_m + x_0) = -x_{n+m} \quad (9)$$
$$\text{Maximize} \quad -x_0 = \hat{\hat{x}}$$

subject to the constraints (9) and $x_i \geq 0$, $i = 0, 1,\ldots, n + m$.

We note several things about the augmented problem. First, its feasible set is nonempty; let $x_0 \geq \max\{-b_1, \ldots, -b_m\}$, choose $x_1 = \cdots = x_n = 0$, and determine $x_{n+1}, \ldots, x_{n+m}$. Second, since $\hat{\hat{x}} \leq 0$, the augmented problem has an optimal solution (Theorem 53E); that is, $\hat{\hat{x}}$ achieves a maximum that is at most 0. Third, the constraints (9) reduce to the constraints (8) when $x_0 = 0$. Consequently, the original constraints are satisfied if and only if the maximum value of $\hat{\hat{x}} = -x_0 = 0$. In particular, if the maximum value of $\hat{\hat{x}}$ is less than 0, the original problem has an empty feasible set. Finally, it is easy to transform the augmented problem to one in which all the $b_i \geq 0$. Consider the tableau (10) of the augmented problem.

| | $x_0$ | $x_1$ | $\cdots$ | $x_n$ | $-1$ | |
|---|---|---|---|---|---|---|
| | $-1$ | $a_{11}$ | $\cdots$ | $a_{1n}$ | $b_1$ | $= -x_{n+1}$ |
| | $\vdots$ | $\vdots$ | | $\vdots$ | $\vdots$ | |
| | $-1$ | $a_{R1}$ | $\cdots$ | $a_{Rm}$ | $b_R$ | $= -x_{n+R}$ |
| | $\vdots$ | $\vdots$ | | $\vdots$ | $\vdots$ | |
| | $-1$ | $a_{m1}$ | $\cdots$ | $a_{mn}$ | $b_m$ | $= -x_{n+m}$ |
| | $-1$ | $0$ | | $0$ | $0$ | $= \hat{\hat{x}}$ |

(10)

Select $R$ so that $b_R \leq b_i$, $i = 1,\ldots, m$ and use the $-1$ in the first column of row $R$ as a pivot, obtaining the tableau (11).

$$
\begin{array}{c|cccc|c|l}
 & x_{n+R} & x_1 & & x_n & -1 & \\
\hline
 & -1 & a_{11} - a_{R1} & \cdots & a_{1n} - a_{Rn} & b_1 - b_R & = -x_{n+1} \\
 & \vdots & \vdots & & \vdots & \vdots & \\
 & -1 & -a_{R1} & \cdots & -a_{Rn} & -b_R & = -x_0 \quad (11)\\
 & \vdots & \vdots & & \vdots & \vdots & \\
 & -1 & a_{m1} - a_{R1} & \cdots & a_{mn} - a_{Rn} & b_m - b_R & = -x_{n+m} \\
\hline
 & -1 & -a_{R1} & & -a_{Rn} & -b_R & = \hat{x}
\end{array}
$$

The elements in the last column are clearly nonnegative. In one step we have moved to a tableau that is in I-feasible form. The solution to the augmented problem may now be completed using phase II of the simplex method, bringing us after a finite number of pivoting steps to the tableau (12).

$$
\begin{array}{c|cccc|c|l}
 & x_{\bar{0}} & x_{\bar{1}} & & x_{\bar{n}} & -1 & \\
\hline
 & \bar{a}_{10} & \bar{a}_{11} & \cdots & \bar{a}_{1n} & \bar{b}_1 \geqslant 0 & = -x_{\overline{n+1}} \\
 & \vdots & \vdots & & \vdots & \vdots & \\
 & \bar{a}_{m0} & \bar{a}_{m1} & \cdots & \bar{a}_{mn} & \bar{b}_m \geqslant 0 & = -x_{\overline{n+m}} \quad (12)\\
\hline
 & \bar{c}_0 \leqslant 0 & \bar{c}_1 \leqslant 0 & & \bar{c}_n \leqslant 0 & \bar{d} & = \hat{x}
\end{array}
$$

If $\bar{d} > 0$ (so $\hat{x} = -\bar{d} < 0$), we have seen that the original problem has no feasible solutions. We have also seen that the original problem has feasible solutions if $\bar{d} = 0$. Moreover, in this case, we may without loss of generality suppose that $x_0$ appears somewhere along the top of tableau (12). For if it appears on the right, say in row $R$, a simple argument shows that not all elements in row $R$ can be 0, and any nonzero entry in this row may be used as a pivot to bring $x_0$ to the top.

With $x_0$ on the top of tableau (12), say in column $S$ so $x_0 = x_S$, we have a point $(x_0, \ldots, x_S, \ldots, x_{n+m})$ that must satisfy the equivalent system (9). And since $x_0 = -\hat{x} = 0$, the remaining $n + m$ coordinates must satisfy (8), giving us the desired feasible solution to the original problem. All that we lack now is an expression that would give us the original objective function in terms of the permuted coordinates $x_{\bar{1}}, \ldots, x_{\overline{n+m}}$. We could have kept track of the effect of the pivoting in solving the augmented problem if we had simply carried along in our tableau one more line corresponding to the original expression for $\hat{x}$. Thus, phase I would begin with tableau (13).

## 54. The Simplex Method

$$\begin{array}{c|cccc|c|l}
 & x_0 & x_1 & & x_n & -1 & \\
\hline
 & -1 & a_{11} & \cdots & a_{1n} & b_1 & = -x_{n+1} \\
 & \vdots & \vdots & & \vdots & \vdots & \\
 & -1 & a_{m1} & \cdots & a_{mn} & b_m & = -x_{n+m} \\
\hline
 & 0 & c_1 & \cdots & c_n & d & = \hat{x} \\
\hline
 & -1 & 0 & \cdots & 0 & 0 & = \hat{\hat{x}}
\end{array} \qquad (13)$$

Solution of the augmented problem terminates in a tableau of the form (14).

$$\begin{array}{c|cccccc|c|l}
 & x_{\bar{0}} & x_{\bar{1}} & & x_{\bar{S}} & & x_{\bar{n}} & -1 & \\
\hline
 & \bar{a}_{10} & \bar{a}_{11} & \cdots & \bar{a}_{1S} & \cdots & \bar{a}_{1n} & \bar{b}_1 \geq 0 & = -x_{\overline{n+1}} \\
 & \vdots & \vdots & & \vdots & & \vdots & \vdots & \\
 & \bar{a}_{m0} & \bar{a}_{m1} & \cdots & \bar{a}_{mS} & \cdots & \bar{a}_{mn} & \bar{b}_m \geq 0 & = -x_{\overline{n+m}} \\
\hline
 & \bar{c}_0 & \bar{c}_1 & \cdots & \bar{c}_S & \cdots & \bar{c}_n & \bar{d} & = \hat{x} \\
\hline
 & \bar{\bar{c}}_0 & \bar{\bar{c}}_1 & \cdots & \bar{\bar{c}}_S & \cdots & \bar{\bar{c}}_n & \bar{\bar{d}} \geq 0 & = \hat{\hat{x}}
\end{array} \qquad (14)$$

If $\bar{\bar{d}} > 0$, the original problem has no feasible solution. If $\bar{\bar{d}} = 0$, the original problem has feasible solutions and we may take $x_{\bar{S}}$, some variable along the top of tableau, to be $x_0 = 0$. This amounts to deleting column $S$ from the tableau and since we are no longer interested in $\hat{\hat{x}}$, the last row may be similarly deleted. The remaining tableau is equivalent to the tableau for the original problem, and it is in I-feasible form. We may proceed with phase II to obtain the maximum of $\hat{x}$.

**Phase I** (getting all $b_i \geq 0$)

> 1. Begin with a linear programming problem which, when written in canonical form, has at least one $b_i < 0$. Set up the tableau (13) of the augmented problem.

> 2. Select $R$ so that $b_R \leq b_i$ for $i = 1,\ldots, m$.

## Phase I *(continued)*

3. Using the $-1$ in column 1 of row $R$ as a pivot, obtain an I-feasible tableau equivalent to (13).

4. Use phase II to solve the augmented problem, terminating with a tableau in the form (14).

5. Is $\bar{d} > 0$? — yes → Stop. The problem has no feasible solutions.

no ↓

6. Is $x_0$ one of the variables along the top of tableau (14)? — yes → Go to step 9.

no ↓

7. Designate by I the row in which $x_0$ appears on the right. Choose $a_{IS} \neq 0$ in row I.

8. Using $a_{IS}$ as a pivot, obtain a tableau (14) having $x_0 = x_S$ as one of the variables along the top.

9. Delete from the tableau (14) the column headed by $x_0$ and the last row.

10. The remaining tableau is, for the given problem, an equivalent tableau in I-feasible form. Go to phase II.

## 54. The Simplex Method

**Example B.** Given the constraints

$$4x_1 + x_2 \leq 16$$
$$-x_1 - x_2 \leq -7$$
$$2x_1 - x_2 = -4$$

we shall maximize $3x_1 - 2x_2 = \hat{x}$. The augmented tableau is shown as Stage 1. Since $-7$ is the smallest

|  | $x_0$ | $x_1$ | $x_2$ | $-1$ |  |
|---|---|---|---|---|---|
|  | $-1$ | 4 | 1 | 16 | $= -x_3$ |
|  | (−1) | $-1$ | $-1$ | $-7$ | $= -x_4$ |
|  | $-1$ | 2 | $-1$ | $-4$ | $= -x_5$ |
|  | 0 | 3 | $-2$ | 0 | $= \hat{x}$ |
|  | $-1$ | 0 | 0 | 0 | $= \hat{x}$ |

Stage 1

|  | $x_4$ | $x_1$ | $x_2$ | $-1$ |  |
|---|---|---|---|---|---|
|  | $-1$ | 5 | 2 | 23 | $= -x_3$ |
|  | $-1$ | 1 | (1) | 7 | $= -x_0$ |
|  | $-1$ | 3 | 0 | 3 | $= -x_5$ |
|  | 0 | 3 | $-2$ | 0 | $= \hat{x}$ |
|  | $-1$ | 1 | 1 | 7 | $= \hat{x}$ |

Stage 2

of the constants, we choose the $-1$ in the second row as our pivot, thus moving to Stage 2. We next focus on maximizing $\hat{x}$. Noting the 1 in the last row, third column, we examine the ratios $7/1$, $23/2$, and select the indicated 1 as our pivot for moving to Stage 3.

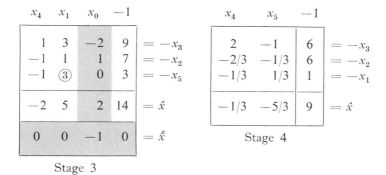

|  | $x_4$ | $x_1$ | $x_0$ | $-1$ |  |
|---|---|---|---|---|---|
|  | 1 | 3 | $-2$ | 9 | $= -x_3$ |
|  | $-1$ | 1 | 1 | 7 | $= -x_2$ |
|  | $-1$ | (3) | 0 | 3 | $= -x_5$ |
|  | $-2$ | 5 | 2 | 14 | $= \hat{x}$ |
|  | 0 | 0 | $-1$ | 0 | $= \hat{x}$ |

Stage 3

|  | $x_4$ | $x_5$ | $-1$ |  |
|---|---|---|---|---|
|  | 2 | $-1$ | 6 | $= -x_3$ |
|  | $-2/3$ | $-1/3$ | 6 | $= -x_2$ |
|  | $-1/3$ | $1/3$ | 1 | $= -x_1$ |
|  | $-1/3$ | $-5/3$ | 9 | $= \hat{x}$ |

Stage 4

We have accomplished our purpose; all coefficients in the expression for $\hat{x}$ are nonpositive. Since the corresponding maximum value for $\hat{x} = -x_0 = 0$, the given problem has a nonempty feasible set. Moreover, the tableau at Stage 3 happens to have $x_0$ along the top, so we may immediately delete the last row and the column headed by $x_0$. In the remaining tableau which is now an I-feasible tableau equivalent to the given problem, we set about maximizing $\hat{x}$ using the procedures of Phase II. The indicated pivot is a 3. From the resulting tableau (Stage 4), we are able to read off the optimal solution. The objective function $\hat{x} = 3x_1 - 2x_2$ takes its maximum value of $-9$ at $x_1 = 1$, $x_2 = 6$.

## THE SIMPLEX METHOD AND THE DUAL PROBLEM

As we know, associated with every linear programming problem, there is a dual problem. The dual problem is easily incorporated into our tableau for the given problem (15).

$$
\begin{array}{c|ccc|c|l}
 & x_1 & \cdots & x_n & -1 & \\
\hline
y_1 & a_{11} & \cdots & a_{1n} & b_1 & = -x_{n+1} \\
\vdots & \vdots & & \vdots & \vdots & \\
y_m & a_{m1} & \cdots & a_{mn} & b_m & = -x_{n+m} \\
\hline
-1 & c_1 & \cdots & c_n & d & = \hat{x} \\
\hline
 & \| & & \| & \| & \\
 & y_{m+1} & & y_{m+n} & \hat{y} &
\end{array}
\tag{15}
$$

We understand the new notation to mean

$$
\begin{aligned}
a_{11} y_1 + \cdots + a_{m1} y_m - c_1 &= y_{m+1} \\
&\vdots \\
a_{1n} y_1 + \cdots + a_{mn} y_m - c_n &= y_{m+n}
\end{aligned}
\tag{16}
$$

where we are to minimize

$$b_1 y_1 + \cdots + b_m y_m - d = \hat{y}$$

subject to the constraints (16) and $y_j \geq 0$, $j = 1,\ldots, m + n$.

It will be noted that this differs in form from (1) only in that the slack variables on the right side are not negative, owing to the reversed inequality of the dual problem. This turns out to be exactly what we

## 54. The Simplex Method

need, however, to assure us that the rules for exchanging $x_S$ and $x_{n+R}$ about the pivot $a_{RS}$ are the rules to be followed in exchanging $y_R$ and $y_{m+S}$ about the pivot $a_{RS}$. Thus, if the original problem has a solution so that a sequence of pivoting steps brings us to the tableau (17),

|     | $x_{\bar{1}}$ | $\cdots$ | $x_{\bar{n}}$ | $-1$ |     |
| --- | --- | --- | --- | --- | --- |
| $y_{\bar{1}}$ | $\bar{a}_{11}$ | $\cdots$ | $\bar{a}_{1n}$ | $\bar{b}_1 \geq 0$ | $= -x_{\overline{n+1}}$ |
|     | $\vdots$ |     | $\vdots$ | $\vdots$ |     |
| $y_{\bar{m}}$ | $\bar{a}_{m1}$ | $\cdots$ | $\bar{a}_{mn}$ | $\bar{b}_m \geq 0$ | $= x_{\overline{n+m}}$ |
| $-1$ | $\bar{c}_1 \leq 0$ | $\cdots$ | $\bar{c}_n \leq 0$ | $\bar{d}$ | $= \hat{x}$ |
|     | $\parallel$ |     | $\parallel$ | $\parallel$ |     |
|     | $y_{\overline{m+1}}$ |     | $y_{\overline{m+n}}$ | $\hat{y}$ |     |

(17)

we are similarly brought to minimizing

$$\hat{y} = -\bar{d} + \bar{b}_1 y_{\bar{1}} + \cdots + \bar{b}_m y_{\bar{m}}$$

subject to the condition $y_{\bar{j}} \geq 0$, $j = 1,\ldots, m+n$. Reasoning as before, it is clear that $\hat{y}$ is minimized by choosing $y_1 = \cdots = y_m = 0$ and setting $y_{\overline{m+1}} = -\bar{c}_1, \ldots, y_{\overline{m+n}} = -\bar{c}_n$.

**Example C.** At the close of Section 53, we considered the dual problems:

**Primal Problem**

Maximize

$$f(\mathbf{x}) = -3x_1 + 2x_2$$

with constraints

$$\begin{bmatrix} -2 & 1 \\ 1 & -2 \\ 1 & 2 \end{bmatrix} \begin{bmatrix} x_1 \\ x_2 \end{bmatrix} \leq \begin{bmatrix} 2 \\ 3 \\ 11 \end{bmatrix}$$

$$\mathbf{x} \geq 0$$

**Dual Problem**

Minimize

$$g(\mathbf{y}) = 2y_1 + 3y_2 + 11y_3$$

with constraints

$$\begin{bmatrix} -2 & 1 & 1 \\ 1 & -2 & 2 \end{bmatrix} \begin{bmatrix} y_1 \\ y_2 \\ y_3 \end{bmatrix} \geq \begin{bmatrix} -3 \\ 2 \end{bmatrix}$$

$$\mathbf{y} \geq 0$$

We solved the primal problem by the simplex method in Example $A$ of this section. If we add the variables $y_i$ to the tableaus indicated in that solution, we have at the concluding stage the tableau (18)

|       | $x_5$ | $x_3$ | $-1$   |         |
|-------|-------|-------|--------|---------|
| $y_5$ | 2/5   | 1/5   | 24/5   | $=-x_2$ |
| $y_2$ | 3/5   | 4/5   | 56/5   | $=-x_4$ |
| $y_4$ | 1/5   | $-2/5$ | 7/5   | $=-x_1$ |
| $-1$  | $-1/5$ | $-8/5$ | $-27/5$ | $=\hat{x}$ |
|       | $\|\|$ | $\|\|$ | $\|\|$ |         |
|       | $y_3$ | $y_1$ | $\hat{y}$ |     |

(18)

Thus, $\hat{y}$ is minimized by choosing $y_5 = y_2 = y_4 = 0$, $y_3 = 1/5$, and $y_1 = 8/5$. In terms of the original variables, we have our optimal solution at $\bar{y} = (8/5, 0, 1/5)$ where $\hat{y} = 27/5$.

## PROBLEMS AND REMARKS

**A.** In Problem 53C, we used a simple example to illustrate various difficulties that can occur in a linear programming problem. Set up the various possibilities in a tableau and use the simplex method as a tool for identifying the difficulties.

**B.** Use the simplex method to simultaneously solve the example of Problem 53E and its dual.

**C.** In saying that a pivoting step could always be used in phase I to bring $x_0$ to the top of tableau (12), we said a simple argument shows that the row $R$ containing $x_0$ could not have all zero entries. Complete the argument.

★ ★ ★ ★ ★

Most of the material in this section is presented in any text that treats the simplex method. In following Rockafellar [1964] in our treatment of phase I, however, we deviate from conventional procedures. We believe the procedure here is conceptually more simple in that one pivoting step in the augmented tableau serves to return us to an I-feasible form.

## 55. Convex Programming

Conditioned by now to think in terms of convex functions, it would be natural, in choosing between several equivalent formulations, to state the convex programming problem as one in which we seek to minimize a convex function. In order to parallel our development of linear programming in Section 53, however, we shall choose instead to maximize a concave function. It is clearly a matter of terminology, since to minimize the convex function $F(\mathbf{x})$ is to maximize the concave function $f(\mathbf{x}) = -F(\mathbf{x})$.

## 55. Convex Programming

***The Convex Programming Problem P.*** Let $U \subseteq \mathbf{R}^n$ be a convex set and suppose $f: U \to \mathbf{R}$ is concave. Maximize $f$ on $U$, subject to the constraints

$$\phi_1(\mathbf{x}) \leqslant 0$$
$$\vdots$$
$$\phi_m(\mathbf{x}) \leqslant 0$$

where each of the functions $\phi_i: U \to \mathbf{R}$ is convex.

The set $U$ is often taken to be the nonnegative orthant $\mathbf{R}_+^n$, in which case the problem may be stated by omitting reference to $U$ and adding the familiar constraint $\mathbf{x} \geqslant \mathbf{O}$. If for notational convenience we define $\Phi: U \to \mathbf{R}^m$ by the coordinate functions

$$\Phi: \begin{array}{l} y_1 = \phi_1(\mathbf{x}) \\ \vdots \\ y_m = \phi_m(\mathbf{x}) \end{array}$$

then the **feasible set** for $P$ is defined by

$$\mathscr{F} = \{\mathbf{x} \in U: \Phi(\mathbf{x}) \leqslant \mathbf{O}\}$$

The problem asks us to find a point $\bar{\mathbf{x}} \in \mathscr{F}$, if such a point exists, such that for all other $\mathbf{x} \in \mathscr{F}, f(\mathbf{x}) \leqslant f(\bar{\mathbf{x}})$.

**Theorem A.** *The set $\mathscr{F}$ of feasible solutions is convex.*

**Proof.** $\mathscr{F} = \bigcap_1^m \{\mathbf{x} \in U: \phi_i(\mathbf{x}) \leqslant 0\}$. This is the intersection of a finite number of convex sets. ●

If all the functions involved in our problem were differentiable, the technique of elementary calculus using Lagrange multipliers would suggest itself. That is, we would consider

$$K(\mathbf{x}, \mathbf{y}) = f(\mathbf{x}) - y_1\phi_1(\mathbf{x}) - \cdots - y_m\phi_m(\mathbf{x}) \tag{1}$$

Although we have not made any differentiability assumptions, it is still useful to consider this function. Note that for fixed $\mathbf{y} \geqslant \mathbf{O}$, the function $K(\mathbf{x}, \mathbf{y})$ is a concave function of $\mathbf{x}$ on $U$, and for fixed $\mathbf{x} \in U$, it is a convex function of $\mathbf{y}$. For such a function, the point $(\bar{\mathbf{x}}, \bar{\mathbf{y}}) \in U \times \mathbf{R}_+^m$ is a saddle point if for all $\mathbf{x} \in U, \mathbf{y} \in \mathbf{R}_+^m$,

$$K(\mathbf{x}, \bar{\mathbf{y}}) \leqslant K(\bar{\mathbf{x}}, \bar{\mathbf{y}}) \leqslant K(\bar{\mathbf{x}}, \mathbf{y})$$

The question of whether or not the convex programming problem has a solution is closely related to whether or not the function $K(\mathbf{x}, \mathbf{y})$ defined by (1) has a saddle point. To state precisely the connection between these problems, we need first to define one more term. A point $\mathbf{x} \in \mathscr{F}$ is called a **strictly feasible solution** to the convex programming problem if $\phi_i(\mathbf{x}) < 0$ for each $i = 1, \ldots, m$.

> **Theorem B.** We refer to the convex programming problem $P$ and the function $K(\mathbf{x}, \mathbf{y})$ defined on $U \times \mathbf{R}_+^m$ by (1).
>
> (a) If $K$ has a saddle point $(\bar{\mathbf{x}}, \bar{\mathbf{y}}) \in U \times \mathbf{R}_+^m$, then $\bar{\mathbf{x}}$ is an optimal solution to $P$.
>
> (b) Suppose that $P$ has a strictly feasible solution $\mathbf{s}$. Then if $P$ has an optimal solution $\bar{\mathbf{x}}$, there is a $\bar{\mathbf{y}} \in \mathbf{R}_+^m$ such that $(\bar{\mathbf{x}}, \bar{\mathbf{y}})$ is a saddle point for $K$.

**Proof.** If $(\bar{\mathbf{x}}, \bar{\mathbf{y}}) \in U \times \mathbf{R}_+^m$ is a saddle point for $K$, then from the definition of $K$,

$$f(\mathbf{x}) - \langle \bar{\mathbf{y}}, \Phi(\mathbf{x}) \rangle \leqslant f(\bar{\mathbf{x}}) - \langle \bar{\mathbf{y}}, \Phi(\bar{\mathbf{x}}) \rangle \leqslant f(\bar{\mathbf{x}}) - \langle \mathbf{y}, \Phi(\bar{\mathbf{x}}) \rangle \qquad (2)$$

for all $\mathbf{x} \in U$, $\mathbf{y} \in \mathbf{R}_+^m$. From the right-hand inequality,

$$\langle \bar{\mathbf{y}}, \Phi(\bar{\mathbf{x}}) \rangle \geqslant \langle \mathbf{y}, \Phi(\bar{\mathbf{x}}) \rangle \qquad (3)$$

Since this must hold for all $\mathbf{y} \geqslant \mathbf{O}$, it must be that $\Phi(\bar{\mathbf{x}}) \leqslant 0$, hence that $\bar{\mathbf{x}} \in \mathscr{F}$. Also, since $\Phi(\bar{\mathbf{x}}) \leqslant \mathbf{O}$, $\langle \bar{\mathbf{y}}, \Phi(\bar{\mathbf{x}}) \rangle \leqslant 0$; but if we choose $\mathbf{y} = \mathbf{O}$ in (3), we get the opposite inequality. We conclude that $\langle \bar{\mathbf{y}}, \Phi(\bar{\mathbf{x}}) \rangle = 0$, so that the left inequality of (2) really says

$$f(\mathbf{x}) - \langle \bar{\mathbf{y}}, \Phi(\mathbf{x}) \rangle \leqslant f(\bar{\mathbf{x}})$$

for all $\mathbf{x} \in U$. Now if $\mathbf{x} \in \mathscr{F}$ where $\Phi(\mathbf{x}) \leqslant \mathbf{O}$, we have $-\langle \bar{\mathbf{y}}, \Phi(\mathbf{x}) \rangle \geqslant 0$. It follows that $f(\mathbf{x}) \leqslant f(\bar{\mathbf{x}})$, concluding the proof of part (a).

To prove part (b), we first define two sets $M$ and $N$ in $\mathbf{R} \times \mathbf{R}^m$ as follows.

$$M = \{(y_0, \mathbf{y}): \text{ for some } \mathbf{x} \in U, y_0 \leqslant f(\mathbf{x}), \mathbf{y} \geqslant \Phi(\mathbf{x})\}$$

$$N = \{(y_0, \mathbf{y}): y_0 > f(\bar{\mathbf{x}}), y_i < 0 \text{ for } i \geqslant 1\}$$

The set $M$ is convex as can be seen by considering a convex combination of points in $M$ and making use of the fact that $f$ is concave and the $\phi_i$ are convex. Set $N$, being the intersection of $m + 1$ half-spaces, is also convex. Both sets have nonempty interiors. Finally, we wish to show

## 55. Convex Programming

that $M \cap N = \varnothing$. Consider a point $(y_0, y_1, \ldots, y_m) \in M$. If the corresponding $\mathbf{x}$ is not in $\mathscr{F}$, then for some $i = 1, \ldots, m$, it must be that $\phi_i(\mathbf{x}) > 0$, thus that $y_i > 0$, which excludes the point from $N$. And if the corresponding $\mathbf{x} \in \mathscr{F}$, then from the optimality of $\bar{\mathbf{x}}$,

$$y_0 \leqslant f(\mathbf{x}) \leqslant f(\bar{\mathbf{x}})$$

again excluding the point from $N$.

We are now able to assert that there is a hyperplane $H$ that separates $M$ and $N$ (Theorem 32B). That is, there is a nonzero vector

$$(a_0, \mathbf{a}) \in \mathbf{R} \times \mathbf{R}^m$$

and a real number $\alpha$ such that for all $(y_0, \mathbf{y}) \in M$, $(z_0, \mathbf{z}) \in N$,

$$a_0 y_0 - \langle \mathbf{a}, \mathbf{y} \rangle \leqslant \alpha \leqslant a_0 z_0 - \langle \mathbf{a}, \mathbf{z} \rangle \tag{4}$$

Since the right-hand inequality holds for all $z_0 > f(\bar{\mathbf{x}})$ and all $(z_1, \ldots, z_m)$ with $z_i < 0$, it follows that $a_0 \geqslant 0$ and $\mathbf{a} \geqslant \mathbf{O}$. Since for any $\mathbf{x} \in U$, the vector $(f(\mathbf{x}), \Phi(\mathbf{x})) \in M$ while the vector

$$\mathbf{z}_\varepsilon = (f(\bar{\mathbf{x}}) + \varepsilon, -\varepsilon, \ldots, -\varepsilon) = (f(\bar{\mathbf{x}}) + \varepsilon, -\boldsymbol{\varepsilon})$$

is in $N$, we may use (4) to write

$$a_0 f(\mathbf{x}) - \langle \mathbf{a}, \Phi(\mathbf{x}) \rangle \leqslant a_0 f(\bar{\mathbf{x}}) + \langle \mathbf{a}, \boldsymbol{\varepsilon} \rangle$$

After letting $\varepsilon \downarrow 0$, we get

$$a_0 f(\mathbf{x}) - \langle \mathbf{a}, \Phi(\mathbf{x}) \rangle \leqslant a_0 f(\bar{\mathbf{x}}) \tag{5}$$

The latter holds for all $\mathbf{x} \in U$, and in particular for the given strictly feasible solution $\mathbf{s}$. Now $a_0 > 0$, for if $a_0 = 0$, then $-\langle \mathbf{a}, \Phi(\mathbf{s}) \rangle \leqslant 0$, which would imply $\mathbf{a} = \mathbf{O}$, violating the fact that hyperplane $H$ is determined by a nonzero vector. Thus we may multiply (5) by $1/a_0$, let $\bar{\mathbf{y}} = (1/a_0)\mathbf{a}$, and obtain

$$f(\mathbf{x}) - \langle \bar{\mathbf{y}}, \Phi(\mathbf{x}) \rangle \leqslant f(\bar{\mathbf{x}}) \tag{6}$$

for all $\mathbf{x} \in U$. Since this is surely true for $\mathbf{x} = \bar{\mathbf{x}}$, $\langle \bar{\mathbf{y}}, \Phi(\bar{\mathbf{x}}) \rangle \geqslant 0$. The opposite inequality always holds for $\mathbf{x} \in \mathscr{F}$, so $\langle \bar{\mathbf{y}}, \Phi(\bar{\mathbf{x}}) \rangle = 0$. We may therefore extend (6) to say

$$f(\mathbf{x}) - \langle \bar{\mathbf{y}}, \Phi(\mathbf{x}) \rangle \leqslant f(\bar{\mathbf{x}}) - \langle \bar{\mathbf{y}}, \Phi(\bar{\mathbf{x}}) \rangle \leqslant f(\bar{\mathbf{x}}) - \langle \mathbf{y}, \Phi(\bar{\mathbf{x}}) \rangle$$

for all $\mathbf{x} \in U$, $\mathbf{y} \in \mathbf{R}_+^m$, concluding our proof. ●

The conclusions of Theorem B can, of course, be stated without specific reference to the function $K(\mathbf{x}, \mathbf{y})$. In proving part (a), we saw that if $(\bar{\mathbf{x}}, \bar{\mathbf{y}})$ was a saddle point, then

$$f(\mathbf{x}) - \bar{y}_1\phi_1(\mathbf{x}) - \cdots - \bar{y}_m\phi_m(\mathbf{x})$$

has a minimum at $\bar{\mathbf{x}}$, and

$$\bar{y}_1\phi_1(\bar{\mathbf{x}}) + \cdots + \bar{y}_m\phi_m(\bar{\mathbf{x}}) = 0$$

These conditions were seen to be sufficient to imply that $P$ has an optimal solution. In part (b) we saw that if $P$ had a strictly feasible solution, then the existence of an optimal solution at $\bar{\mathbf{x}}$ implied the conditions above as necessary consequences. Since $\bar{\mathbf{y}} \geq 0$ and $\Phi(\bar{\mathbf{x}}) \leq \mathbf{O}$, the second condition means that if $\phi_i(\bar{\mathbf{x}}) < 0$, then $y_i = 0$. We may now restate Theorem B in the following form.

**Theorem C.** Suppose the convex programming problem $P$ has a strictly feasible solution. Then for $P$ to have an optimal solution at $\bar{\mathbf{x}}$, it is both necessary and sufficient that there exist a $\bar{\mathbf{y}} \in \mathbf{R}_+^m$ satisfying the **Kuhn–Tucker conditions**: $f(\mathbf{x}) - [\bar{y}_1\phi_1(\mathbf{x}) + \cdots + \bar{y}_m\phi_m(\mathbf{x})]$ has a maximum on the set $U$ at $\bar{\mathbf{x}}$, and if $\phi_i(\bar{\mathbf{x}}) < 0$, then $\bar{y}_i = 0$.

We began our discussion of the convex programming problem with the comment that if the objective function $f$ and the constraint functions $\phi_i$ were differentiable, it would be natural to think of using Lagrange multipliers to attack the maximization problem. Although we have proceeded without any assumptions about differentiability, we point out that the original paper in this area [Kuhn–Tucker, 1951] made just such an assumption. Their result follows easily from our approach (Problem B).

It is now but a short step to the duality theorem for convex programming. Let us restate the convex programming problem.

**Primal Problem P.** Maximize the concave function $f: U \to \mathbf{R}$, subject to the constraints $\phi_i(\mathbf{x}) \leq 0$, where $\phi_i: U \to \mathbf{R}$ is convex, $i = 1, \ldots, m$.

We define $g: \mathbf{R}_+^m \to \mathbf{R}$ by

$$g(\mathbf{y}) = \sup_{\mathbf{x} \in U} K(\mathbf{x}, \mathbf{y}) = \sup_{\mathbf{x} \in U} [f(\mathbf{x}) - y_1\phi_1(\mathbf{x}) - \cdots - y_m\phi_m(\mathbf{x})]$$

For fixed $\mathbf{x}$, $K(\mathbf{x}, \mathbf{y})$ is an affine function of $\mathbf{y}$, and $g$ is therefore the supremum of a family of convex (affine) functions. Therefore (Problem

## 55. Convex Programming

41 B) $g$ is convex. Let $\mathscr{F}^*$ be the set on which $g$ is finite. Call $\mathscr{F}^*$ the **feasible set** for the function $g$.

**Dual Problem $P^*$.** If $\mathscr{F}^* \neq \varnothing$, minimize the convex function $g: \mathbf{R}_+^m \to \mathbf{R}$ on $\mathscr{F}^*$.

The linear programming problem is a special case of a convex programming problem in which $U = \mathbf{R}_+^n$, the (concave) objective function is

$$f(\mathbf{x}) = c_1 x_1 + \cdots + c_n x_n = \langle \mathbf{c}, \mathbf{x} \rangle$$

and the (convex) constraint functions are

$$\begin{aligned} \phi_1(\mathbf{x}) &= a_{11} x_1 + \cdots + a_{1n} x_n - b_1 \\ &\vdots \\ \phi_m(\mathbf{x}) &= a_{m1} x_1 + \cdots + a_{mn} x_n - b_m \end{aligned} \quad \text{or} \quad \Phi(\mathbf{x}) = A\mathbf{x} - \mathbf{b} \quad (7)$$

The function $g: \mathbf{R}_+^m \to \mathbf{R}$ is defined by

$$\begin{aligned} g(\mathbf{y}) &= \sup_{\mathbf{x} \in \mathbf{R}_+^n} \{\langle \mathbf{c}, \mathbf{x} \rangle - \langle \mathbf{y}, A\mathbf{x} - \mathbf{b} \rangle\} \\ &= \sup\{\langle \mathbf{b}, \mathbf{y} \rangle - \langle A^t \mathbf{y} - \mathbf{c}, \mathbf{x} \rangle\} \\ &= \begin{cases} \infty & \text{if } A^t\mathbf{y} - \mathbf{c} < 0 \\ \langle \mathbf{b}, \mathbf{y} \rangle & \text{if } A^t\mathbf{y} - \mathbf{c} \geq 0 \end{cases} \end{aligned}$$

The set $\mathscr{F}^* \subseteq \mathbf{R}_+^m$ where $g$ is finite is the set on which $A^t\mathbf{y} \geq \mathbf{c}$. The dual problem is therefore to minimize $\langle \mathbf{b}, \mathbf{y} \rangle$ on $\mathbf{R}_+^m$ subject to the constraint $A^t\mathbf{y} \geq \mathbf{c}$. We recognize this to be the dual problem of linear programming.

As might be expected for more general functions, we obtain a similar but weaker form of the **Kuhn–Tucker duality theorem**.

> **Theorem D.** Let $P$ be a primal convex programming problem and let $P^*$ be the dual problem as stated above.
> (a) If $\mathbf{x}$ is a feasible solution to $P$ and $\mathbf{y}$ is a feasible solution to $P^*$, then $f(\mathbf{x}) \leq g(\mathbf{y})$.
> (b) Suppose $P$ has a strictly feasible solution. Then if problem $P$ has an optimal solution $\bar{\mathbf{x}}$, Problem $P^*$ has an optimal solution $\bar{\mathbf{y}}$ and $f(\bar{\mathbf{x}}) = g(\bar{\mathbf{y}})$.

**Proof.** Since $g$ is defined as a supremum, the problem of minimizing this function calls to mind the minimax-type theorems of Section 52.

If we define $F: U \to \mathbf{R}$ by

$$F(\mathbf{x}) = \inf_{\mathbf{y} \in \mathbf{R}_+^m} K(\mathbf{x}, \mathbf{y}) = \begin{cases} f(\mathbf{x}), & \mathbf{x} \in \mathscr{F} \\ -\infty, & \mathbf{x} \in U \setminus \mathscr{F} \end{cases}$$

then according to Theorem 52A

$$F(\mathbf{x}) \leqslant \sup_{\mathbf{x} \in U} F(\mathbf{x}) \leqslant \inf_{\mathbf{y} \in \mathbf{R}_+^m} g(\mathbf{y}) \leqslant g(\mathbf{y})$$

for all $\mathbf{x} \in U$, $\mathbf{y} \in \mathbf{R}_+^m$. This in turn means that $f(\mathbf{x}) \leqslant g(\mathbf{y})$ for all $\mathbf{x} \in \mathscr{F}$, $\mathbf{y} \in \mathscr{F}^*$, establishing (a).

From Theorem B, we know that the existence of an optimal solution $\bar{\mathbf{x}}$ for $P$ implies the existence of a $\bar{\mathbf{y}} \in \mathbf{R}_+^m$ such that $(\bar{\mathbf{x}}, \bar{\mathbf{y}})$ is a saddle point for $K$. Another appeal to Theorem 52A (see also its proof) gives us $F(\bar{\mathbf{x}}) = g(\bar{\mathbf{y}})$, or equivalently, $f(\bar{\mathbf{x}}) = g(\bar{\mathbf{y}})$. This combined with (a) makes $\bar{\mathbf{y}}$ an optimal solution to $P^*$ ●

## PROBLEMS AND REMARKS

**A.** The example in part (1) below illustrates the necessity of requiring in Theorems B and C some condition like the existence of a strictly feasible solution.

(1) For functions defined on $\mathbf{R}$, maximize $f(x) = x$ subject to the constraint $\phi(x) = x^2 \leqslant 0$. There is an optimal solution, but $K(x, y) = x - yx^2$ has no saddle point.

(2) The condition that there be a strictly feasible solution is often called **Slater's condition**. Mangasarian [1969, p. 78] and others discuss other conditions that may be used in Theorems B and C for the same purpose. For example, Slater's condition is satisfied ⇔ Karlin's condition is satisfied. **Karlin's condition**: There is no vector $\mathbf{a} \in \mathbf{R}_+^m$ such that $\langle \mathbf{a}, \Phi(\mathbf{x}) \rangle \geqslant 0$ for all $\mathbf{x} \in \mathscr{F}$, the feasible set for $P$.

*(3) If the constraints of $P$ are affine, taking the form (7), Theorem B can be proved without (one of the equivalent) assumptions about a strictly feasible solution. (See, for example, [Karlin, 1959, p. 203].)

**B.** As a special case of Theorem C, we have the version originally proved by Kuhn and Tucker [1951]. Let the concave objective function $f(\mathbf{x})$ and the convex constraint functions $\phi_i(\mathbf{x})$ be differentiable on $\mathbf{R}_+^n$. Suppose also that there is a strictly feasible solution. Then for the convex programming problem to have an optimal solution at $\bar{\mathbf{x}}$, it is both necessary and sufficient that there exist a $\bar{\mathbf{y}} \in \mathbf{R}_+^m$ such that for $K(\mathbf{x}, \mathbf{y}) = f(\mathbf{x}) - \langle \mathbf{y}, \Phi(\mathbf{x}) \rangle$,

$$\frac{\partial K}{\partial \mathbf{x}}(\bar{\mathbf{x}}, \bar{\mathbf{y}}) \leqslant 0, \qquad \left\langle \frac{\partial K}{\partial \mathbf{x}}(\bar{\mathbf{x}}, \bar{\mathbf{y}}), \bar{\mathbf{x}} \right\rangle = 0$$

$$\frac{\partial K}{\partial \mathbf{y}}(\bar{\mathbf{x}}, \bar{\mathbf{y}}) \geqslant 0, \qquad \left\langle \frac{\partial K}{\partial \mathbf{y}}(\bar{\mathbf{x}}, \bar{\mathbf{y}}), \bar{\mathbf{y}} \right\rangle = 0$$

*C. The following problems are adapted from Berge and Ghouila-Houri [1965,

## 55. Convex Programming

pp. 76–84]. We employ our usual notation with $U = \mathbf{R}^n$; thus $f: \mathbf{R}^n \to \mathbf{R}$ and $\Phi: \mathbf{R}^n \to \mathbf{R}^m$ where $\Phi$ is described by the coordinate functions

$$\Phi: \begin{array}{l} y_1 = \phi_1(\mathbf{x}) \\ \vdots \\ y_m = \phi_m(\mathbf{x}) \end{array}$$

The feasible set $\mathscr{F}$ is defined by $\mathscr{F} = \{\mathbf{x} \in \mathbf{R}^n : \Phi(\mathbf{x}) \leqslant \mathbf{O}\}$ and $K(\mathbf{x}, \mathbf{y}) = f(\mathbf{x}) - \langle \mathbf{y}, \Phi(\mathbf{x}) \rangle$. Note, however, that we are not making any initial assumptions about functions being convex or concave.

**Problem 1**  Maximize $f$ on $\mathscr{F}$.

**Problem 2**  Assuming all functions are differentiable, find $\bar{\mathbf{x}} \in \mathscr{F}$ such that $f'(\bar{\mathbf{x}})\mathbf{u} \leqslant 0$ for every $\mathbf{u} \in \mathbf{R}^n$ for which $\phi_i(\bar{\mathbf{x}}) = 0$ implies $\phi_i'(\bar{\mathbf{x}})\mathbf{u} \leqslant 0$.

**Problem 3**  Assuming all functions are differentiable, find $\bar{\mathbf{x}} \in \mathscr{F}$ and $\bar{\mathbf{y}} \in \mathbf{R}_+^m$ such that $\langle \bar{\mathbf{y}}, \Phi(\bar{\mathbf{x}}) \rangle = 0$ and

$$\frac{\partial f}{\partial x_i}(\bar{\mathbf{x}}) - \sum_{j=1}^m \bar{y}_j \frac{\partial \phi_j(\bar{\mathbf{x}})}{\partial x_i} = 0 \quad \text{for} \quad i = 1,\ldots, n$$

**Problem 4**  Find $(\bar{\mathbf{x}}, \bar{\mathbf{y}}) \in \mathbf{R}^n \times \mathbf{R}_+^m$ that is a saddle point for $K$.

(1) If $(\bar{\mathbf{x}}, \bar{\mathbf{y}})$ is a solution to Problem 4 (P4), then $\bar{\mathbf{x}}$ is a solution to (P1).
(2) (P2) and (P3) are equivalent problems.
(3) If the functions are all differentiable, every solution to (P4) is a solution to (P3).
(4) If $f$ is concave and the coordinate functions of $\Phi$ are all convex, and if there exists at least one $\mathbf{s} \in \mathscr{F}$ such that $\phi_i(\mathbf{s}) \neq 0$ for all $\phi_i$ that are not affine, then (P1) and (P4) are equivalent. If, in addition, all the functions are differentiable, then all the problems are equivalent.

**D.**  Let $S$ be an $n \times n$ symmetric matrix, and let $\mathbf{c} \in \mathbf{R}^n$. Let

$$f(\mathbf{x}) = \langle \mathbf{c}, \mathbf{x} \rangle + \langle \mathbf{x}, S\mathbf{x} \rangle$$

When $f$ is concave, the convex programming problem of maximizing $f$ subject to linear constraints is called the **quadratic programming problem**. That is, we consider the problem of maximizing the concave function $f$ defined above, subject to the constraints $\mathbf{x} \in \mathbf{R}_+^n$ and $A\mathbf{x} \leqslant \mathbf{b}$, where $A$ is an $m \times n$ matrix.

(1) $f$ is concave [strictly concave] if matrix $-S$ is nonnegative definite [positive definite].

*(2) By introducing slack variables as in the case of the linear programming problem, we obtain an equivalent problem in which the constraints are stated as equalities. To avoid cumbersome notation, we shall assume this has been done, so that we now have the problem,

$$\text{maximize} \quad f(\mathbf{x}) = \langle \mathbf{c}, \mathbf{x} \rangle + \langle \mathbf{x}, S\mathbf{x} \rangle$$
$$\text{subject to the constraints} \quad A\mathbf{x} = \mathbf{b}, \quad \mathbf{x} \geqslant 0$$

We may show that for $\bar{\mathbf{x}} \in \mathbf{R}_+^n$ to be an optimal solution, it is necessary and sufficient that we find vectors $\bar{\mathbf{u}} \in \mathbf{R}^m$, $\bar{\mathbf{v}} \in \mathbf{R}_+^n$ such that

$$2S\bar{\mathbf{x}} - A^t\bar{\mathbf{u}} + \bar{\mathbf{v}} = -\mathbf{c}$$
$$\bar{x}_i \bar{v}_i = 0, \quad i = i,\ldots, n$$

*(3) With the quadratic programming problem stated as in part (2) above, the dual problem is defined in the literature as follows. Find $\bar{\mathbf{y}} \in \mathbf{R}^n$, $\bar{\mathbf{u}} \in \mathbf{R}^m$ so as to

$$\text{minimize} \quad g(\mathbf{y}) = \langle \mathbf{b}, \mathbf{u} \rangle - \langle \mathbf{y}, S\mathbf{y} \rangle$$
$$\text{subject to the constraints} \quad -2S\mathbf{y} + A^t\mathbf{u} \geqslant \mathbf{c}, \quad \mathbf{y} \geqslant 0$$

How does this relate to the dual as we have stated it for an arbitrary convex programming problem?

*(4) If one of the two programming problems, the primal problem of part (2) or the dual problem of (3), has an optimal solution, so does the other.

We have followed Hadley [1964, Chapter 7] in this formulation of the quadratic programming problem. Additional information on quadratic programming is to be found in most of the books mentioned at the end of this section.

**E.** The subject of duality for convex programs is intimately related to that of conjugacy for convex functions (Section 15). To illustrate this relationship, let $f: U \to \mathbf{R}$ be concave on $U \subseteq \mathbf{R}^n$, and let $\phi: U \to \mathbf{R}$ be a convex constraint function.

**Primal Problem P**

Maximize $f(\mathbf{x})$
on $\mathscr{F} = \{\mathbf{x} \in U: \phi(\mathbf{x}) \leqslant 0\}$

**Dual Problem P\***

Minimize
$g(y) = \sup_{\mathbf{x} \in U} [f(\mathbf{x}) - y\,\phi(\mathbf{x})]$
on $\mathscr{F}^* = \{y \in \mathbf{R}_+: g(y) < \infty\}$

(1) Introduce the **perturbation function** for **P** [Rockafellar, 1970a, p. 276] defined by

$$p(v) = \sup\{f(\mathbf{x}): \mathbf{x} \in U, \phi(\mathbf{x}) + v \leqslant 0\}$$

Clearly, it is the behavior of $p$ near 0 that is critical for our problem. In particular, if $f$ has a maximum at $\bar{\mathbf{x}}$, then $f(\bar{\mathbf{x}}) = p(0)$.

(2) Let $I = \{v \in \mathbf{R}: \phi(\mathbf{x}) + v \leqslant 0 \text{ for some } \mathbf{x} \in U, p(v) < \infty\}$. Then $I$ is an interval and for $y \in \mathscr{F}^*$,

$$g(y) = \sup_{v \in I} [yv + p(v)]$$

(3) On $\mathscr{F}^*$, $g$ agrees with the conjugate of $-p$.

This can all be extended to the case of $m$ constraints, but then we need the theory of conjugate convex functions developed for functions of several variables, a topic to which we have only briefly alluded (Section 43). The best reference for further study of this subject is by Rockafellar [1970a], but see also Karlin [1959, chapter 7], Berge and Ghouila–Houri [1965, Chapter 5], and Stoer and Witzgall [1970, Chapter 5].

★ ★ ★ ★ ★

Most of the work on nonlinear programming builds in one way or another on the paper of Kuhn and Tucker [1951]. Our development is influenced by Karlin [1959] and lecture notes from R. T. Rockafellar. The reader wishing to see a fuller treatment of this topic is also referred to books by Hadley [1964], and Mangasarian [1969], and for a number of solved examples using the methods of this section, to Bracken and McCormick [1968].

## 56. Approximation

Most students interested in any kind of scientific analysis encounter the following problem. Data of some sort have been used to plot points in the plane. According to some theory, the points should lie on a straight line. However, they fall in a pattern that only approximates a line. Assuming that circumstances preclude the time-honored custom of moving a few points, the student's problem is to find the line that best fits the points. In order to have something specific to talk about, we pose a simple problem of this type.

**Example A.** We seek the line $y = mx + b$ that best fits the points (2, 7), (7, 12), and (12, 15).

For any line $y = mx + b$, the three values $x_1 = 2$, $x_2 = 7$, and $x_3 = 12$ determine corresponding $y$ values $y_1 = 2m + b$, $y_2 = 7m + b$, and $y_3 = 12m + b$ (Fig. 56.1). The differences by which the line misses the desired $y$ values for each of the $x_i$ are

$$t_1 = y_1 - 7, \qquad t_2 = y_2 - 12, \qquad t_3 = y_3 - 15$$

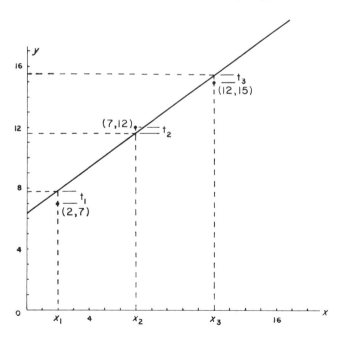

Fig. 56.1

Our first problem is to decide how to measure the fit of the line to the three points. Presumably such a measure should be a function of $(t_1, t_2, t_3)$, which is large for a bad fit, small for a good one. We quickly reject the idea of using the sum $t_1 + t_2 + t_3$ since it allows negative differences to cancel positive differences, making the sum small for lines that badly violate our intuitive sense of a good fit (Problem A). Three measures that do seem reasonable are $|t_1| + |t_2| + |t_3|$, $\max\{|t_1|, |t_2|, |t_3|\}$, and $t_1^2 + t_2^2 + t_3^2$. Once the measure has been selected, the problem is to find the line that minimizes it.

Each of the three proposed measures is a nonnegative convex function. Of these, the last one is often used because it has a derivative and therefore lends itself to the methods of elementary calculus. It should not be expected that the line that best fits the points with respect to one of the measures above will necessarily be best with respect to another (Problem B). In fact, it is well to note here that we do not even know answers to the following questions.

1a. With respect to a given criterion for determining "goodness of fit," is there necessarily a best line (or might we get a sequence of lines for which the selected measure of the fit forms a corresponding sequence converging to a value never actually realized for any line)?

2a. If we get a "best fitting line" that minimizes some measure of goodness of fit, will this line be unique, or might there be other lines for which the same minimum is achieved?

In Example A we tried to fit a line to a finite set of points. Suppose instead that we try to find a line that best approximates the graph of a continuous curve on [0, 1]. Again we pose a specific problem.

**Example B.** We seek an affine function $A(x) = mx + b$ that best approximates the function $f(x) = x/(x + 1)$ on [0, 1].

How shall we measure the "distance" between $f$ and $A$? We faced the question once before and proposed two answers (Example 21D').

$$\|f - A\| = \max_{x \in [0,1]} |f(x) - A(x)|$$

$$|f - A| = \left\{ \int_0^1 [f(x) - A(x)]^2 \, dx \right\}^{1/2}$$

Again we call attention to the two questions raised after Example A. Will there be a best affine function? Will such a function, if it exists, be unique?

## 56. Approximation

The general approximation problem, of which Examples A and B are special cases, may be stated as follows.

**The Approximation Problem.** Let $\mathbf{L}$ be a normed linear space and suppose $K \subseteq \mathbf{L}$. For an arbitrary $\mathbf{z} \in \mathbf{L}$, find an element $\bar{\mathbf{z}} \in K$, called a **best approximation** to $\mathbf{z}$ in $K$, such that

$$\text{dist}(\mathbf{z}, K) = \inf_{\mathbf{y} \in K} \|\mathbf{y} - \mathbf{z}\| = \|\bar{\mathbf{z}} - \mathbf{z}\|$$

Two questions are to be considered.

1. Does $K$ contain a best approximation to $\mathbf{z}$?
2. If a best approximation exists, is it unique?

The approximation problem is now seen as a problem in minimizing a translate of the norm function, or if $\mathbf{z} = \mathbf{O}$, as a problem in minimizing the norm function itself. We wish to state both of our examples in the form of the general problem.

**Example A'** In the linear space $\mathbf{L}^3$ of points $\mathbf{y} = (y_1, y_2, y_3)$, we have the convex set (actually a subspace)

$$K = \{(2m + b, 7m + b, 12m + b): \quad m \text{ and } b \text{ real}\}$$

and the point $\mathbf{z} = (7, 12, 15)$ that we wish to approximate by a point $\bar{\mathbf{z}} \in K$. We consider the problem for each of the norms

$$N_1(\mathbf{y}) = |y_1| + |y_2| + |y_3|$$
$$N_2(\mathbf{y}) = \max\{|y_1|, |y_2|, |y_3|\}$$
$$N_3(\mathbf{y}) = (y_1^2 + y_2^2 + y_3^2)^{1/2}$$

Two things should be noted about this reformulation of Example A. First, it turns our attention to finding a point $\bar{\mathbf{z}} \in K$ so as to minimize $\|\mathbf{y} - \mathbf{z}\|$ rather than to minimize the norm of the differences $\mathbf{t} = (t_1, t_2, t_3)$ with which we originally worked. Secondly, we have slightly altered our third measure of "goodness of fit" by taking the square root, thus defining a norm. It is clear that minimization of $y_1^2 + y_2^2 + y_3^2$ is equivalent to minimization of $(y_1^2 + y_2^2 + y_3^2)^{1/2}$. The former is more amenable to computational procedures, the latter to theoretical discussions.

**Example B'.** The space is $C[0, 1]$, the space of continuous functions on $[0, 1]$; $K$ is the subset of all affine functions $A(x) = mx + b$. The point $z$ to be approximated is $f(x) = x/(x+1)$, and we consider the problem for each of the norms

$$N_1(f) = \|f\| = \max_{x \in [0,1]} |f(x)|$$

$$N_2(f) = |f| = \left[\int_0^1 f^2(x)\, dx\right]^{1/2}$$

Before obtaining any formal results, a number of geometric observations may be helpful. The set $M_K(z)$ of best approximations to $z$ in $K$ (Fig. 56.2) is always convex if $K$ is convex, since it is the intersection

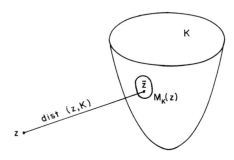

**Fig. 56.2**

of the closed ball $N_\rho(z)$ of radius $\rho = \mathrm{dist}(z, K)$ with the set $K$. Being convex, $M_K(z)$ is either empty, a single point, or an infinite set.

As a further guide to our intuition, consider the situation in the plane. Denoting points by $\mathbf{x} = (r, s)$ to avoid subscripts, we defined three different norms in Section 21,

$$N_1(\mathbf{x}) = |r| + |s|$$
$$N_2(\mathbf{x}) = \max\{|r|, |s|\}$$
$$N_3(\mathbf{x}) = (r^2 + s^2)^{1/2}$$

They are, except for the dimension of the space, exactly the norms that came up in a natural way in Example A. Let $K = \{(r, s): r + s \geq 2\}$ and take $\mathbf{z} = \mathbf{O}$ as the point to be approximated. Imagine a ball about the origin with variable radius $\rho$. Let $\rho$ increase from 0 until the expanding ball approaches the set $K$. The two questions associated with the approximation problem become the following:

1b.  Is there a ball that contains points of $K$ on its boundary $\|\mathbf{x}\| = \rho$ but no points of $K$ in its interior $\|\mathbf{x}\| < \rho$?

## 56. Approximation

2b.  If such a ball exists, does it contain just one point of $K$?

The situation for each of the three norms is indicated in Fig. 56.3.

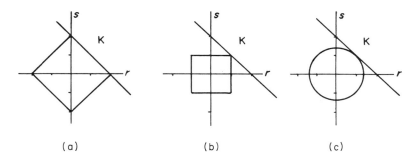

**Fig. 56.3.**  (a) $|r| + |s| = 2$.  (b) $\max(|r|, |s|) = 1$.  (c) $(r^2 + s^2)^{1/2} = \sqrt{2}$.

The distances from $K$ to the origin are, respectively, 2, 1, $\sqrt{2}$. In all three cases, the answer to question 1b is yes. But for question 2b, it is no for the first, yes for the other two. If the set $K$ were the open half-space $\{(r, s):\ r + s > 2\}$, then question 1b would have a negative answer in all three cases. Other sets $K$ are considered in Problem C.

Some reflection makes it clear that answers to the approximation questions depend both on properties of the set $K$ (closed? compact? convex? subspace?) and on the geometry of the unit sphere. Our first theorem gives us conditions under which we can be certain that a best approximation exists.

**Theorem A.**  Let $K$ be a finite-dimensional closed subset of a normed linear space $\mathbf{L}$. Then there is at least one point of $K$ at a minimum distance from a given fixed point $\mathbf{z}$.

**Proof.**  There is no loss in generality if we take $\mathbf{z}$ to be $\mathbf{O}$. Choose any point $\mathbf{x}_0 \in K$ and form the set

$$M = \{\mathbf{x} \in K:\ \|\mathbf{x}\| \leqslant \|\mathbf{x}_0\|\}$$

$M$ is closed and bounded in a finite-dimensional subspace of $\mathbf{L}$; hence it is compact. The continuous norm function $N(\mathbf{x}) = \|\mathbf{x}\|$ takes a minimum value at some $\mathbf{x}_1 \in M$, and for any $\mathbf{x} \in K$, we have $\|\mathbf{x}\| \geqslant \|\mathbf{x}_1\|$. ●

This theorem settles the question of existence for both Examples A′

and B'. In the first example, $K$ is the two-dimensional subspace spanned by $(2, 7, 12)$ and $(1, 1, 1)$; in the second, it is the two-dimensional subspace spanned by $p_0(x) = 1$ and $p_1(x) = x$. Since finite-dimensional subspaces are always closed, each $K$ meets the requirements of the theorem; a best approximation exists.

The question of uniqueness remains to be investigated. Toward this end, not surprisingly, we must define a concept for norms that is really a condition that prevents "flat" spots on the unit sphere. We say a normed linear space **L** is a **strictly convex space** if its unit sphere contains no line segments. More precisely, we require that

$$\| \mathbf{x} \| = \| \mathbf{y} \| = \| \tfrac{1}{2}(\mathbf{x} + \mathbf{y}) \| = 1 \quad \text{implies} \quad \mathbf{x} = \mathbf{y}$$

All inner product spaces (hence all Euclidean spaces $\mathbf{R}^n$) are strictly convex (Problem E). We emphasize that strict convexity is a property not preserved by topological isomorphism, as is apparent from considering the (topologically equivalent) norms represented in Fig. 56.3.

> **Theorem B.** Let $K$ be a finite-dimensional closed convex set in a strictly convex normed linear space **L**. Then there is a unique point of $K$ at a minimum distance from a given fixed point $\mathbf{z} \in \mathbf{L}$.

**Proof.** The existence of a point of $K$ at a minimum distance from $\mathbf{z}$ is guaranteed by Theorem A; only the uniqueness needs to be established. Assume as usual that $\mathbf{z}$ is the origin, and let $\mathbf{x}$ and $\mathbf{y}$ be two points of $K$ at a minimum distance $\alpha > 0$ from **O**. Since $K$ is convex, $\tfrac{1}{2}(\mathbf{x} + \mathbf{y}) \in K$, and

$$\alpha \leqslant \| \tfrac{1}{2}(\mathbf{x} + \mathbf{y}) \| \leqslant \tfrac{1}{2} \| \mathbf{x} \| + \tfrac{1}{2} \| \mathbf{y} \| = \tfrac{1}{2}\alpha + \tfrac{1}{2}\alpha = \alpha$$

Thus, $\| \tfrac{1}{2}(\mathbf{x} + \mathbf{y}) \| = \alpha$ and so $\mathbf{x}/\alpha$, $\mathbf{y}/\alpha$, and $\tfrac{1}{2}(\mathbf{x} + \mathbf{y})/\alpha$ all have norm 1. It follows from the strict convexity that $\mathbf{x}/\alpha = \mathbf{y}/\alpha$, hence that $\mathbf{x} = \mathbf{y}$. ●

Both of the theorems proved in this section are essentially finite dimensional even though set in general normed linear spaces. True infinite-dimensional results (that is, where $K$ is infinite dimensional) exist in many forms. We mention one such result in Problem K where the space **L** is a so-called uniformly convex space.

Another characteristic of the theorems in this section is that both are existence theorems. They assert that under certain conditions, solutions (or unique solutions) exist. They give no clue as to how to find them. Constructive results (theorems and algorithms telling how to find a best

## 56. Approximation

approximation) abound in approximation theory, but again we restrict ourselves to mentioning a few such results as problems. For instance, Problem F gives a formula for the unique best approximation in the case where $K$ is a finite-dimensional subspace of an inner product space **L**.

It is clear from the comments of the last two paragraphs that we have given only the most elementary introduction to approximation theory. Our goal was modest. We wished to show that approximation involves the minimization of a convex function on a normed linear space, hence that techniques already discussed in this chapter become tools for approximation. (See particularly Problems L, M.) We also wished to illustrate the role of such notions as strict convexity, uniform convexity, and other geometric properties of the unit sphere. More sophisticated existence theorems together with many constructive techniques are treated extensively (though never exhaustively) in a variety of good books on approximation theory. We refer the reader to the enduring book by Jackson [1930], the expository paper of Buck [1959], and the recent texts by Cheney [1966], Lorentz [1966], and Rice [1964, 1969].

### PROBLEMS AND REMARKS

**A.** Consider using the sum $t_1 + t_2 + t_3$ as a measure of the goodness of the fit of a line to the points of Example A. Calculate this sum for the lines $y = 3x - 10$ and $y = x + 5$. Draw these lines on a graph on which the three given points have been plotted.

**B.** We have seen that there is a unique line that is the best fit to the points of Example A when we minimize $t_1^2 + t_2^2 + t_3^2$.

(1) Use the methods of elementary calculus to show that the line is described by $y = \frac{4}{5} x + \frac{86}{15}$.

(2) For a given line $l$, determine $t_1$, $t_2$, $t_3$ as in Example A, and let $m(l) = |t_1| + |t_2| + |t_3|$. If $l_1$ is the line determined in part (1) and $l_2$ is the line described by $y = \frac{4}{5} x + \frac{27}{5}$, show for the points of Example A that $m(l_2) < m(l_1)$. This shows that the best fitting line as measured by minimizing $t_1^2 + t_2^2 + t_3^2$ is not the best fitting line as measured by minimizing $|t_1| + |t_2| + |t_3|$.

(3) Find a best fitting line for the three points of Example A as determined by minimizing $m(l)$ in part (2). Is the answer unique?

**C.** We define the following sets in the plane.

$$K_1 = \{(r, s): rs \geq 2\}$$
$$K_2 = \{(r, s): rs > 2\}$$
$$K_3 = \{(r, s): r = 2\}$$
$$K_4 = \{(r, s): r^2 + s^2 \geq 2\}$$

Answer the two questions associated with the approximation problem, using **O** as the fixed point and $N_1(\mathbf{x}) = |r| + |s|$, for each set $K_i$. Do the same for the norms $N_2(\mathbf{x}) = \max\{|r|, |s|\}$ and $N_3(\mathbf{x}) = (r^2 + s^2)^{1/2}$.

**D.** Find the unique function $A(x) = mx + b$ that minimizes $|f(x) - A(x)|$ in Example B by minimizing

$$R(m, b) = \int_0^1 \left[\frac{x}{x+1} - (mx + b)\right]^2 dx$$

**E.** Many important normed linear spaces are strictly convex.

(1) The parallelogram law (Problem 21J) may be used to show that all inner product spaces are strictly convex.

(2) The space $l_n^p = \{\mathbf{x} = (x_1, \ldots, x_n)\}$ with norm

$$\|\mathbf{x}\| = \left(\sum_{k=1}^n |x_k|^p\right)^{1/p}$$

for $1 < p < \infty$ is strictly convex (cf. Theorem 62C). What about the case $p = 1$?

**F.** Let $K$ be the subspace spanned by an orthonormal set $\{\mathbf{u}_1, \ldots, \mathbf{u}_n\}$ in an inner product space $\mathbf{L}$. Given $\mathbf{z} \in \mathbf{L}$, there is a unique point $\bar{\mathbf{z}} \in K$ closest to $\mathbf{z}$ given by the formula $\bar{\mathbf{z}} = \sum_1^n \langle \mathbf{z}, \mathbf{u}_k \rangle \mathbf{u}_k$.

**G.** Problem F may be used to solve Examples A' and B' for the inner product norms.

(1) In Example A' with norm $N_3$, take

$$\mathbf{u}_1 = \frac{1}{\sqrt{3}}(1, 1, 1), \quad \mathbf{u}_2 = \frac{1}{\sqrt{2}}(-1, 0, 1)$$

Find $\bar{\mathbf{z}} = (\frac{22}{3}, \frac{34}{3}, \frac{46}{3})$. The desired line passes through $(2, \frac{22}{3})$, $(7, \frac{34}{3})$, and $(12, \frac{46}{3})$.

(2) In Example B' with norm $N_2$, take $\mathbf{u}_1 = 1$, $\mathbf{u}_2 = (1/\sqrt{5})(2x - 1)$.

**H.** In the space $C[-\pi, \pi]$ of continuous functions on $[-\pi, \pi]$ with inner product $\langle f, g \rangle = \int_{-\pi}^{\pi} f(x) g(x) \, dx$, let $K$ be the subspace spanned by the orthonormal set

$$\left\{\frac{1}{\sqrt{2\pi}}, \frac{1}{\sqrt{\pi}}\cos x, \frac{1}{\sqrt{\pi}}\sin x, \ldots, \frac{1}{\sqrt{\pi}}\cos nx, \frac{1}{\sqrt{\pi}}\sin nx\right\}$$

Find the best approximations to (1) $f(x) = x$, (2) $f(x) = x^2$, (3) $f(x) = |x|$. These approximations are the $n$th-order Fourier series corresponding to the given functions.

**I.** The system of equations

$$3r + s = 2$$
$$r + 4s = 4$$
$$2r + 3s = 3$$

is inconsistent, so no point $(\bar{r}, \bar{s})$ will satisfy all three equations. Discuss, in light of this section, the problem of finding a point $(\bar{r}, \bar{s})$ that comes the closest to satisfying all three. (For an extensive discussion of this kind of problem, see Cheney [1966, Chapter 2].)

**J.** An infinite-dimensional subspace $K$ of a normed linear space $\mathbf{L}$ need not contain a best approximation to a given point $\mathbf{z} \in \mathbf{L}$. Let $\mathbf{L} = \{\mathbf{x} = (x_1, x_2, \ldots): x_n \to 0\}$ with $\|\mathbf{x}\| = \max_n |x_n|$, and let $K = \{\mathbf{x} \in \mathbf{L}: \sum_1^\infty 2^{-n} x_n = 0\}$ [Cheney, 1966, p. 21].

**K.** Call a normed linear space $\mathbf{L}$ **uniformly convex** if for each $\varepsilon > 0$ there is a $\delta > 0$ such that $\|\mathbf{x} - \mathbf{y}\| < \varepsilon$ whenever $\|\mathbf{x}\| = \|\mathbf{y}\| = 1$ and $\|\frac{1}{2}(\mathbf{x} + \mathbf{y})\| > 1 - \delta$.

## 56. Approximation

(1) Every uniformly convex normed linear space **L** is strictly convex, and if **L** is finite dimensional, the two notions are equivalent.

(2) A nonempty closed convex set in a uniformly convex Banach space possesses a unique point closest to a given point [Cheney, 1966, p. 22].

(3) An inner product space is uniformly convex. It follows from part (2) that every nonempty closed convex set in a Hilbert space has a unique point of smallest norm. Rudin [1966, p. 78] proves this directly and obtains a number of important consequences.

**L.** We may obtain a general solution for problems like the one considered in Example A. We seek a line $y = mx + b$ that best fits the $n$ points $(x_1, y_1),\ldots, (x_n, y_n)$ in the sense of least squares, that is, in the sense of minimizing $t_1^2 + \cdots + t_n^2$ where $t_i = y_i - mx_i - b$. We know (Theorem B) there is a unique solution. The methods of calculus (Problem 51I) can be used to show that the solution is $y - \bar{y} = m(x - \bar{x})$ where

$$\bar{x} = \frac{1}{n}\sum_1^n x_i, \qquad \bar{y} = \frac{1}{n}\sum_1^n y_i, \qquad m = \frac{(1/n\,\sum_1^n x_i y_i) - \bar{x}\bar{y}}{(1/n\,\sum_1^n x_i^2) - \bar{x}^2}$$

See Buck [1965, pp. 356–358].

**M.** Consider Problem L, that of finding a best fitting line for $n$ given points, for each of the norms $N_1(t) = |t_1| + \cdots + |t_n|$ and $N_2(t) = \max\{|t_1|,\ldots, |t_n|\}$. In both cases we may formulate the problem as minimization of a convex function (the norm) on the affine set

$$A = \{(t_1,\ldots, t_n): t_i = y_i - mx_i - b,\ i = 1,\ldots, n\}$$

Both are nonlinear programming problems in the sense of Section 55. Both can be transformed into linear programming problems. Replace $t_i$ by $u_i - v_i$ where $u_i = \max\{t_i, 0\}$, $v_i = \max\{-t_i, 0\}$, so that $|t_i| = u_i + v_i$. Our two problems become

(1) minimize $g(u_i,\ldots, u_n, v_1,\ldots, v_n) = \sum_1^n (u_i + v_i)$ subject to

$$\begin{aligned} u_i - v_i &= y_i - mx_i - b \\ u_i &\geq 0,\ v_i \geq 0 \end{aligned} \qquad i = 1,\ldots, n$$

(2) minimize $f(z) = z$ subject to

$$\begin{aligned} z &\geq u_i + v_i \\ u_i - v_i &= y_i - mx_i - b \qquad i = 1,\ldots, n \\ u_i &\geq 0,\ v_i \geq 0 \end{aligned}$$

**N.** A subset $K$ of a normed linear space **L** is called a **Chebyshev set** if each point in **L** has a unique nearest point in $K$. From Theorem B, we conclude that a nonempty closed convex set in $\mathbf{R}^n$ is a Chebyshev set.

(1) The converse is true; that is, in $\mathbf{R}^n$ a nonempty closed set is convex if and only if it is a Chebyshev set.

(2) Convex sets may be characterized in the same way in any finite-dimensional normed linear space $\mathbf{L}^n$ that is strictly convex and has a smooth (Problem 44E) unit sphere.

This result, due to Motzkin and others, is discussed by Valentine [1964, p. 94]. Research continues on attempts to extend the characterization of convex sets to infinite-dimensional spaces [Klee, 1961; Brønsted, 1965, 1966a,b; Klee, 1967].

# VI

*Inequalities*

One cannot escape the feeling that these mathematical formulas have an independent existence and an intelligence of their own, that they are wiser than we are, wiser even than their discoverers, that we get more out of them than was originally put into them.

HEINRICH HERTZ

... it should be emphasized that the theory of convexity..., taken together with a few elementary devices, can be used to derive a large number of the most familiar and important inequalities of analysis.

D. S. MITRINOVIĆ

## 60. Introduction

It has been said that analysis is primarily the study of inequalities. If this is an overstatement, it is nevertheless true that inequalities play an important role in analysis, applied mathematics, and even algebra and geometry. The classic work, "Inequalities" by Hardy, Littlewood, and Polya, [1952], has been supplemented with more recent books by Beckenbach and Bellman [1965], Marcus and Minc [1964], Kazarinoff [1961], and Mitrinović [1970]. These books provide handy references and extended bibliographies for the reader wishing to explore the topic in depth. Our purpose here is to show that the theory of convex functions affords a unified treatment of some of the most important inequalities in mathematics.

Our basic inequality is that which defines a convex function, namely

$$f(\alpha x + \beta y) \leqslant \alpha f(x) + \beta f(y)$$

for $\alpha > 0$, $\beta > 0$, $\alpha + \beta = 1$. From it follows (as we shall see in Section 61) the inequality

$$x^\alpha y^\beta \leqslant \alpha x + \beta y \tag{1}$$

for $x \geqslant 0$, $y \geqslant 0$, and this in turn can be used to establish the inequalities of Hölder and Minkowski. Inequality (1) is itself a special case of the geometric mean–arithmetic mean inequality. Its extensions and ramifications are the subject of Section 62. Finally in Section 63, we apply convexity theory to the study of matrices, obtaining among others the inequalities of Hadamard and Minkowski for determinants.

## 61. The Classical Inequalities

Our first result, Jensen's inequality, has already been proved in Section 40. Here we give a particularly simple proof which, however, works best on open sets since it depends on the existence of support for a convex function at each point of the set.

**Theorem A (Jensen's Inequality).** Let $f$ be convex on the open (possibly infinite) interval $(a, b)$ and let $x_i \in (a, b)$. If $\alpha_i > 0$ and $\sum_{i=1}^n \alpha_i = 1$, then

$$f\left(\sum_1^n \alpha_i x_i\right) \leqslant \sum_1^n \alpha_i f(x_i) \tag{1}$$

**Proof.** Recall from Theorem 12D that $f$ has support at each $x_0 \in (a, b)$. That is, for each $x_0$ there is a number $m$ (depending on $x_0$) such that $f(x) \geq f(x_0) + m(x - x_0)$. In particular, If $x_0 = \sum_1^n \alpha_i x_i$, then for $i = 1, \ldots, n$,

$$f(x_i) \geq f(x_0) + m(x_i - x_0)$$

If we multiply both sides by $\alpha_i$, sum, and simplify, we obtain (1). ●

The **geometric mean–arithmetic mean** (GM–AM) inequality in its most familiar form is

$$(x_1 x_2 \cdots x_n)^{1/n} \leq \frac{1}{n}(x_1 + \cdots + x_n) \tag{2}$$

which holds for $x_i \geq 0$ and $n$ a positive integer. We prove a more general form which includes this by taking $\alpha_i = 1/n$.

**Theorem B (GM–AM Inequality).** If $x_i \geq 0$, $\alpha_i > 0$, and $\sum_1^n \alpha_i = 1$, then

$$x_1^{\alpha_1} x_2^{\alpha_2} \cdots x_n^{\alpha_n} \leq \alpha_1 x_1 + \cdots + \alpha_n x_n \tag{3}$$

**Proof.** We need to prove (3) only when $x_i > 0$ for all $i$, in which case we may set $y_i = \log x_i$. Then

$$x_i^{\alpha_i} = \exp(\alpha_i \log x_i) = \exp(\alpha_i y_i)$$

Since $f(t) = e^t$ is convex on $(-\infty, \infty)$, we may appeal to Theorem A to write

$$\prod_1^n x_i^{\alpha_i} = \exp\left(\sum \alpha_i y_i\right) = f\left(\sum \alpha_i y_i\right)$$

$$\leq \sum \alpha_i f(y_i) = \sum \alpha_i \exp(y_i) = \sum_1^n \alpha_i x_i \quad \bullet$$

The special case where $n = 2$, $\alpha_1 = 1/p$, $\alpha_2 = 1/q$, $x_1 = x^p$, and $x_2 = y^q$ is fundamental to what follows. It is

$$xy \leq \frac{1}{p} x^p + \frac{1}{q} y^q \tag{4}$$

## 61. The Classical Inequalities

This inequality also happens to be a special case of Young's inequality that appeared in connection with conjugate convex functions (Section 15). The functions on the right of (4) are conjugate convex functions.

**Theorem C (Hölder's Inequality).** If $x_i \geq 0$, $y_i \geq 0$, $p > 1$, and $1/p + 1/q = 1$, then

$$\sum_1^n x_i y_i \leq \left(\sum_1^n x_i^p\right)^{1/p} \left(\sum_1^n y_i^q\right)^{1/q} \tag{5}$$

In particular we have the CBS inequality

$$\sum_1^n x_i y_i \leq \left(\sum_1^n x_i^2\right)^{1/2} \left(\sum_1^n y_i^2\right)^{1/2} \tag{6}$$

**Proof.** We may and do suppose that at least one of the $x_i$'s and one of the $y_i$'s are greater than zero. Then both $u = (\sum x_i^p)^{1/p}$ and $v = (\sum y_i^q)^{1/q}$ are positive. Appealing to (4) with $x = x_i/u$, $y = y_i/v$, we obtain

$$\frac{x_i}{u} \cdot \frac{y_i}{v} \leq \frac{1}{p}\left(\frac{x_i}{u}\right)^p + \frac{1}{q}\left(\frac{y_i}{v}\right)^q$$

or, after summing

$$\frac{\sum_1^n x_i y_i}{uv} \leq \frac{1}{p} + \frac{1}{q} = 1$$

which is equivalent to (5). ●

**Theorem D (Minkowski's Inequality).** If $x_i \geq 0$, $y_i \geq 0$, and $p \geq 1$, then

$$\left[\sum_1^n (x_i + y_i)^p\right]^{1/p} \leq \left[\sum_1^n x_i^p\right]^{1/p} + \left[\sum_1^n y_i^p\right]^{1/p} \tag{7}$$

**Proof.** When $p = 1$, we have equality. If $p > 1$, we may choose $q$ so that $1/p + 1/q = 1$, in which case $(p-1)q = p$. Thus if we write

$$\sum_1^n (x_i + y_i)^p = \sum_1^n x_i(x_i + y_i)^{p-1} + \sum_1^n y_i(x_i + y_i)^{p-1}$$

and apply Hölder's inequality to both terms on the right, we get

$$\sum_{1}^{n} (x_i + y_i)^p \leq \left[\sum_{1}^{n} x_i^p\right]^{1/p} \left[\sum_{1}^{n} (x_i + y_i)^p\right]^{1/q}$$

$$+ \left[\sum_{1}^{n} y_i^p\right]^{1/p} \left[\sum_{1}^{n} (x_i + y_i)^p\right]^{1/q}$$

$$= \left[\sum_{1}^{n} (x_i + y_i)^p\right]^{1/q} \left[\left(\sum_{1}^{n} x_i^p\right)^{1/p} + \left(\sum_{1}^{n} y_i^p\right)^{1/p}\right]$$

This is equivalent to (7). ●

A somewhat less standard inequality, but one that we will need later, is the following.

**Theorem E.** If $x_i \geq 0$, $y_i \geq 0$, and $k$ is a positive integer, then

$$\left[\prod_{i=1}^{k} (x_i + y_i)\right]^{1/k} \geq \left(\prod_{1}^{k} x_i\right)^{1/k} + \left(\prod_{1}^{k} y_i\right)^{1/k} \qquad (8)$$

**Proof.** We may suppose that $x_i + y_i > 0$ for $i = 1,\ldots, k$. Then

$$\frac{(\prod_{1}^{k} x_i)^{1/k} + (\prod_{1}^{k} y_i)^{1/k}}{[\prod_{1}^{k} (x_i + y_i)]^{1/k}} = \left(\prod_{1}^{k} \frac{x_i}{x_i + y_i}\right)^{1/k} + \left(\prod_{1}^{k} \frac{y_i}{x_i + y_i}\right)^{1/k}$$

$$\leq \frac{1}{k} \sum_{1}^{k} \frac{x_i}{x_i + y_i} + \frac{1}{k} \sum_{1}^{k} \frac{y_i}{x_i + y_i} = 1$$

Here we used the GM–AM inequality in its simplest form, namely (2). ●

## PROBLEMS AND REMARKS

**A.** Suppose in Theorem A that we require $f$ to be strictly convex. Then there is equality in (1) $\Leftrightarrow$ all the $x_i$'s are equal.

**B.** It is often of interest to be able to state precise conditions under which an inequality can be replaced by an equality. For the following inequalities, the conditions are indicated.

(1) GM–AM inequality: all $x_i$'s are equal.
(2) Hölder's inequality: there are nonnegative numbers $A$ and $B$ (not both zero) such that $Ax_i^p = By_i^q$, $i = 1,\ldots, n$.
(3) Minkowski's inequality: same as (2) if $p > 1$.

## 61. The Classical Inequalities

**C.** If $x_i \geq 0$, $\alpha_i > 0$, $\sum_{i=1}^{n} \alpha_i = 1$, and $p \geq 1$, then

$$(\alpha_1 x_1 + \cdots + \alpha_n x_n)^p \leq \alpha_1 x_1^p + \cdots + \alpha_n x_n^p$$

Consider also the case where $p < 1$.

**D.** From Hölder's inequality, it may be shown that if $x_i \geq 0$, $p > 1$,

$$\left( \sum_{i=1}^{n} x_i^p \right)^{1/p} = \max_{\mathbf{y} \in M} \sum_{i=1}^{n} x_i y_i$$

where

$$M = \left\{ \mathbf{y} \in \mathbf{R}_+^n : \sum_{i=1}^{n} y_i^q = 1 \right\}$$

Thus the nonlinear function on the left is an envelope of linear functions. Beckenbach and Bellman [1965, p. 23] call this a quasilinearization.

**E.** The representation in Problem D may be used to prove Minkowski's inequality.

**\*F.** From the convexity of $\log(1 + e^x)$, it can be shown that for $s_i \geq 0$, $t_i \geq 0$, $\alpha_i > 0$, and $\sum_{i=1}^{n} \alpha_i = 1$,

$$(s_1^{\alpha_1} \cdots s_n^{\alpha_n}) + (t_1^{\alpha_1} \cdots t_n^{\alpha_n}) \leq (s_1 + t_1)^{\alpha_1} \cdots (s_n + t_n)^{\alpha_n}$$

[*Hint*: Let $x_j = \log(t_j/s_j)$; use Jensen's inequality.]

**\*G.** From the convexity of $(1 + x^p)^{1/p}$ for $p > 1$, it can be shown that for $s_i \geq 0$, $t_i \geq 0$,

$$\{(s_1 + \cdots + s_n)^p + (t_1 + \cdots + t_n)^p\}^{1/p} \leq (s_1^p + t_1^p)^{1/p} + \cdots + (s_n^p + t_n^p)^{1/p}$$

(*Hint*: Let $x_j = t_j/s_j$ and $\alpha_j = s_j / \sum_{i=1}^{n} s_i$; use Jensen's inequality.)

**\*H.** If $x_{ij} \geq 0$, $\alpha_i > 0$, and $\sum_{1}^{n} \alpha_i = 1$, then

$$\sum_{j=1}^{m} \prod_{i=1}^{n} x_{ij}^{\alpha_i} \leq \prod_{i=1}^{n} \left( \sum_{j=1}^{m} x_{ij} \right)^{\alpha_i}$$

This extends Problem F.

**\*I.** If $x_{ij} \geq 0$ and $p > 1$, then

$$\left\{ \sum_{i=1}^{m} \left( \sum_{j=1}^{n} x_{ij} \right)^p \right\}^{1/p} \leq \sum_{j=1}^{n} \left( \sum_{i=1}^{m} x_{ij}^p \right)^{1/p}$$

This extends Problem G.

**\*J.** Theorem A may be generalized as follows. Let $f$ be convex on $(a, b)$ and let $x: [c, d] \to \mathbf{R}$ be integrable with $a < x(t) < b$. If $\alpha: [c, d] \to \mathbf{R}$ is nonnegative, $\int_c^d \alpha(t) \, dt = 1$, and $\alpha x$ is integrable on $[c, d]$, then

$$f\left( \int_c^d \alpha(t) x(t) \, dt \right) \leq \int_c^d \alpha(t) f(x(t)) \, dt$$

The result may also be formulated in terms of a Stieltjes integral.

*K.  Hölder's and Minkowski's inequalities also hold for integrals.

(1) If $x(t) \geqslant 0$, $y(t) \geqslant 0$, $p > 1$, $1/p + 1/q = 1$, and the functions $x^p$ and $y^q$ are integrable, then $xy$ is integrable and

$$\int_a^h x(t)y(t)\, dt \leqslant \left(\int_a^b x^p(t)\, dt\right)^{1/p} \left(\int_a^b y^q(t)\, dt\right)^{1/q}$$

(2) If $x(t) \geqslant 0$, $y(t) \geqslant 0$, $p > 1$, and $x^p$ and $y^p$ are integrable, then $(x+y)^p$ is integrable and

$$\left[\int_a^b (x(t)+y(t))^p\, dt\right]^{1/p} \leqslant \left[\int_a^b x^p(t)\, dt\right]^{1/p} + \left[\int_a^b y^p(t)\, dt\right]^{1/p}$$

Actually, this is still only part of the story. Both results are valid for arbitrary abstract integrals [Hewitt and Stromberg, 1965, pp. 190, 191].

*L.  The convexity of $x^p$ for $p \geqslant 1$ can be used to show

(1) $u \geqslant 0$, $v \geqslant 0 \Rightarrow (u+v)^p \leqslant 2^{p+1}(u^p + v^p)$,
(2) $f(t) \geqslant 0 \Rightarrow [\int_a^b f(t)\, dt]^p \leqslant (b-a)^{p-1} \int_a^b f^p(t)\, dt$.

M.  Theorem E can be used to show that $f: \mathbf{R}_+^n \to \mathbf{R}$ defined by $f(x_1,...,x_n) = (\prod_{j=1}^k x_j)^{1/k}$ is concave, $k \leqslant n$.

*N.  Let $x_i > 0$, $A_n = \frac{1}{n}\sum_1^n x_i$, and $G_n = (\prod_1^n x_i)^{1/n}$. Then

(1) $n(A_n - G_n) \geqslant (n-1)(A_{n-1} - G_{n-1})$ (**Rado's inequality**),
(2) $(A_n/G_n)^n \geqslant (A_{n-1}/G_{n-1})^{n-1}$ (**Popoviciu's inequality**).

For generalizations, see Bullen [1971b].

O.  If $f: [0,1] \to \mathbf{R}$ is convex and $f'(0) = 0$, then $f$ is increasing. For $x_i > 0$, $y_i > 0$,

$$f(t) = \left(\sum_1^n x_i^{1+t} y_i^{1-t}\right)\left(\sum_1^n x_j^{1-t} y_j^{1+t}\right)$$

$$= \sum_{i,j=1} x_i y_i x_j y_j \left(\frac{x_i y_j}{x_j y_i}\right)^t$$

is such a function. In particular, $f(0) \leqslant f(1)$, which is the CBS inequality. This technique is used to prove and generalize many familiar inequalities in Daykin and Eliezer [1969] and Eliezer and Mond [1972].

## 62. The Generalized Geometric Mean—Arithmetic Mean Inequality and Norms

Let $\mathbf{x} = (x_1,...,x_n)$, $\boldsymbol{\alpha} = (\alpha_1,...,\alpha_n)$ where $x_i > 0$, $\alpha_i > 0$, and $\sum_1^n \alpha_i = 1$. We define a **mean of order** $t$ by

$$M_t(\mathbf{x}, \boldsymbol{\alpha}) = \left(\sum_{i=1}^n \alpha_i x_i^t\right)^{1/t}, \qquad t \neq 0$$

## 62. The Generalized GM-AM Inequality and Norms

and
$$M_0(\mathbf{x}, \boldsymbol{\alpha}) = \lim_{t \to 0} M_t(\mathbf{x}, \boldsymbol{\alpha}) = \prod_{i=1}^{n} x_i^{\alpha_i}$$

The second equality above is verified by considering the limit of $\log M_t(\mathbf{x}, \boldsymbol{\alpha})$, computed with the aid of L'Hospital's rule. If $t > 0$ and $x_k = \max(x_1, \ldots, x_n)$, then

$$\alpha_k^{1/t} x_k \leqslant M_t(\mathbf{x}, \boldsymbol{\alpha}) \leqslant x_k$$

and since $\alpha_k^{1/t} \to 1$ as $t \to \infty$, we may define

$$M_\infty(\mathbf{x}, \boldsymbol{\alpha}) = \lim_{t \to \infty} M_t(\mathbf{x}, \boldsymbol{\alpha}) = \max(x_1, \ldots, x_n)$$

Similarly,
$$M_{-\infty}(\mathbf{x}, \boldsymbol{\alpha}) = \lim_{t \to -\infty} M_t(\mathbf{x}, \boldsymbol{\alpha}) = \min(x_1, \ldots, x_n)$$

Then for fixed $\mathbf{x}$ and $\boldsymbol{\alpha}$, $M_t(\mathbf{x}, \boldsymbol{\alpha})$ may be thought of as a real-valued continuous function defined on $[-\infty, \infty]$. A number of expressions commonly used in mathematics are obtained as special cases of $M_t(\mathbf{x}, \boldsymbol{\alpha})$. Specifically, the values $t = -1, 0, 1, 2$ give expressions known, respectively, as the harmonic mean, geometric mean, arithmetic mean, and root mean square of $x_1, \ldots, x_n$.

**Theorem A (Generalized GM–AM Inequality).** If $M_t(\mathbf{x}, \boldsymbol{\alpha})$ is the mean of order $t$ defined above, then

$$s \leqslant t \quad \text{implies} \quad M_s(\mathbf{x}, \boldsymbol{\alpha}) \leqslant M_t(\mathbf{x}, \boldsymbol{\alpha}) \tag{1}$$

Moreover, $g(t) = t \log M_t(\mathbf{x}, \boldsymbol{\alpha})$ is convex on $(-\infty, \infty)$.

**Proof.** Let $D$ denote differentiation with respect to $t$, an operation we apply to the identity

$$t \log M_t(\mathbf{x}, \boldsymbol{\alpha}) = \log \sum_{1}^{n} \alpha_i x_i^t \tag{2}$$

For $t \neq 0$, we obtain after a bit of algebra

$$\left( t^2 \sum_{1}^{n} \alpha_i x_i^t \right) DM_t(\mathbf{x}, \boldsymbol{\alpha})$$
$$= M_t(\mathbf{x}, \boldsymbol{\alpha}) \left[ \sum_{1}^{n} \alpha_i x_i^t \log x_i^t - \left( \sum_{1}^{n} \alpha_i x_i^t \right) \log \left( \sum_{1}^{n} \alpha_i x_i^t \right) \right] \tag{3}$$

Now $y \log y$ is convex on $(0, \infty)$, so by Jensen's inequality,

$$\sum_1^n \alpha_i y_i \log y_i \geq \left(\sum_1^n \alpha_i y_i\right) \log \left(\sum_1^n \alpha_i y_i\right)$$

When this is applied with $y_i = x_i^t$ to the expression in brackets, we find that the right side of (3) is nonnegative. Hence, $DM_t(\mathbf{x}, \boldsymbol{\alpha}) \geq 0$, at least for $t \neq 0$. This combined with the continuity of $M_t(\mathbf{x}, \boldsymbol{\alpha})$ at $t = 0$ yields (1).

To establish the convexity of $g$, one might compute $g''(t)$ with the intent of showing it to be nonnegative. This is easy to do for $t \neq 0$, but difficulties near the origin discourage this frontal assault on the problem. Theorem 13F on the closure of log-convex functions under addition provides a beautiful alternative. Note that (2) is true even at $t = 0$ and that

$$\log \alpha_i x_i^t = \log \alpha_i + t \log x_i$$

is convex on $(-\infty, \infty)$. It follows that $\log \sum_1^n \alpha_i x_i^t$ is convex. ●

Of perhaps more interest than $M_t(\mathbf{x}, \boldsymbol{\alpha})$ is the closely related **sum of order** $t$, $S_t(\mathbf{x})$, defined for $x_i > 0$ and $t \neq 0$ by

$$S_t(\mathbf{x}) = n^{1/t} M_t(\mathbf{x}, 1/n) = \left(\sum_{i=1}^n x_i^t\right)^{1/t} \tag{4}$$

Note that

$$\lim_{t \downarrow 0} \sum_1^n x_i^t = n,$$

so if $n \geq 2$,

$$S_{0+}(\mathbf{x}) = \lim_{t \downarrow 0} S_t(\mathbf{x}) = \infty$$

Also,

$$S_\infty(\mathbf{x}) = \lim_{t \to \infty} n^{1/t} M_t(\mathbf{x}, 1/n) = \max(x_1, \ldots, x_n)$$

**Theorem B.** *If $S_t(\mathbf{x})$ is the sum of order $t$ defined above and $n \geq 2$, then*

$$0 < r < t \leq \infty \quad \text{implies} \quad S_r(\mathbf{x}) > S_t(\mathbf{x}) \tag{5}$$

*Moreover, $t \log S_t(\mathbf{x})$, $\log S_t(\mathbf{x})$, and $S_t(\mathbf{x})$ are all convex as functions of $t$ on $(0, \infty)$.*

## 62. The Generalized GM-AM Inequality and Norms

**Proof.** For $n \geqslant 2$, we always have $0 < x_i < S_r(\mathbf{x})$, so for $r < t < \infty$,

$$\sum_1^n \left[\frac{x_i}{S_r(\mathbf{x})}\right]^t < \sum_1^n \left[\frac{x_i}{S_r(\mathbf{x})}\right]^r = 1$$

or

$$\sum_1^n x_i^t < [S_r(\mathbf{x})]^t$$

This is equivalent to (5). If $t = \infty$ and we again set $x_k = \max(x_1, \ldots, x_n)$, then

$$S_t(\mathbf{x}) = x_k = (x_k^r)^{1/r} < S_r(\mathbf{x})$$

as desired.

Next from (4),

$$t \log S_t(\mathbf{x}) = \log n + t \log M_t(\mathbf{x}, 1/n) \tag{6}$$

and is therefore convex by Theorem A.

The convexity of $\log S_t(\mathbf{x})$ is considerably more difficult to establish. For fixed $a > 0$, let $b = S_a(\mathbf{x})$, and then define on $(0, \hat{a})$ the functions

$$f(t) = t, \quad g(t) = t \log S_t(\mathbf{x}/b) = \log \sum_1^n (x_i/b)^t$$

Clearly $f$ is convex, positive, and decreasing on $(0, a)$, and we shall show shortly that $g$ has these same properties. Assuming this for the moment, it follows from Theorem 13C that $fg$ is convex on $(0, a)$. Then from the fact that $S_t(\mathbf{x}\, b) = S_t(\mathbf{x})\, b$, we have

$$f(t)g(t) = \log S_t(\mathbf{x}) - \log b$$

Thus $\log S_t(\mathbf{x})$ is convex on $(0, a)$, and since $a$ was arbitrary, it is convex on $(0, \infty)$. The convexity of $S_t(\mathbf{x})$ on $(0, \infty)$ follows from the general property (Section 13) that any log-convex function is convex.

As for the claimed properties of $g$, we note first that (6) with the fixed point $\mathbf{x}$ replaced by $\mathbf{x}\, b$ establishes its convexity. Since $t < a$, it follows from (5) that

$$S_t\left(\frac{\mathbf{x}}{b}\right) = \frac{S_t(\mathbf{x})}{b} = \frac{S_t(\mathbf{x})}{S_a(\mathbf{x})} > 1,$$

so $g$ is positive. Finally

$$g'(t) = \frac{\sum_1^n (x_i/b)^t \log(x_i/b)}{\sum_1^n (x_i/b)^t}$$

which is negative since $0 < x_i < b = S_a(\mathbf{x})$. This implies that $g$ is decreasing on $(0, a)$. ●

$S_t(\mathbf{x})$ is of interest partly because of its relation to an important set of norms on finite-dimensional linear spaces. For $\mathbf{x} = (x_1, \ldots, x_n)$ and $p \geqslant 1$, define

$$\|\mathbf{x}\|_p = \left(\sum_{i=1}^n |x_i|^p\right)^{1/p}$$

and

$$\|\mathbf{x}\|_\infty = \max\{|x_1|, \ldots, |x_n|\}$$

We recognize $\|\ \|_2$ as the ordinary Euclidean norm, and we have also met $\|\ \|_1$ and $\|\ \|_\infty$ several times before.

**Theorem C.** If $1 \leqslant p \leqslant \infty$, then $\|\ \|_p$ is a norm on the linear space of $n$-tuples $\mathbf{x} = (x_1, \ldots, x_n)$. If $1 \leqslant p < r < \infty$, then

$$\|\mathbf{x}\|_p \geqslant \|\mathbf{x}\|_r \geqslant \|\mathbf{x}\|_\infty \tag{7}$$

and if $q = \alpha p + \beta r$ where $\alpha > 0$, $\beta > 0$, and $\alpha + \beta = 1$, then

$$\|x\|_q \leqslant (\|\mathbf{x}\|_p)^\alpha (\|\mathbf{x}\|_r)^\beta \leqslant \alpha \|\mathbf{x}\|_p + \beta \|x\|_r \tag{8}$$

**Proof.** The only nontrivial part in showing that $\|\ \|_p$ is a norm is establishing the triangle inequality. For $1 \leqslant p < \infty$, this is a consequence of Minkowski's inequality; for $p = \infty$ it follows easily from the definition of $\|\ \|_\infty$.

If $n = 1$, then (7) and (8) are trivialities. For $n \geqslant 2$, (7) follows from (5) if $|x_i| > 0$ for all $i$'s, and then by taking limits if some $x_i$'s are zero. The first inequality of (8) is just a rewriting of the convexity of $\log S_t(|\mathbf{x}|)$ in the case where $|\mathbf{x}| = (|x_1|, \ldots, |x_n|)$ and all $x_i$'s are different from zero. If some $x_i$'s are zero, the same result follows by taking limits. The second inequality of (8) is an application of the GM–AM inequality. ●

## PROBLEMS AND REMARKS

**A.** Hardy, Littlewood, and Polya [1952, pp. 11–15] require only $x_i \geqslant 0$ in the definition of $M_t(\mathbf{x}, \boldsymbol{\alpha})$, but then for $t < 0$, they interpret $M_t(\mathbf{x}, \boldsymbol{\alpha}) = 0$ if any $x_i = 0$. Note that $\lim_{t \downarrow 0} M_t(\mathbf{x}, \boldsymbol{\alpha}) = 0$ if one of the $x_i$'s is 0, so that $M_t(\mathbf{x}, \boldsymbol{\alpha})$ is still continuous at $t = 0$. Moreover, the statement about the monotonicity of $M_t(\mathbf{x}, \boldsymbol{\alpha})$ is still valid as is

## 62. The Generalized GM-AM Inequality and Norms

the convexity of $t \log M_t(\mathbf{x}, \boldsymbol{\alpha})$ in the following form. If $s = \lambda r + \mu t$ where $\lambda > 0$, $\mu > 0$, and $\lambda + \mu = 1$, then

$$[M_s(\mathbf{x}, \boldsymbol{\alpha})]^s \leq [M_r(\mathbf{x}, \boldsymbol{\alpha})]^{\lambda r} [M_t(\mathbf{x}, \boldsymbol{\alpha})]^{\mu t}$$

**B.** For each $\mathbf{x} = (x_1, \ldots, x_n)$, $\lim_{p \to \infty} \|\mathbf{x}\|_p = \|\mathbf{x}\|_\infty$.

**C.** If $\mathbf{x} = (x_1, \ldots, x_n)$ and $1 \leq p \leq \infty$, then

$$\|\mathbf{x}\|_\infty \leq \|\mathbf{x}\|_p \leq \|\mathbf{x}\|_1 \leq n \|\mathbf{x}\|_\infty$$

Thus all $p$ norms are topologically equivalent, a fact we proved earlier in greater generality (Theorem 21E).

***D.** Let $l^p$ ($1 \leq p < \infty$) be the class of all sequences $\mathbf{x} = \{x_i\}$ of real numbers such that

$$\|\mathbf{x}\|_p = \left(\sum_{i=1}^\infty |x_i|^p\right)^{1/p} < \infty$$

and let $l^\infty$ be the class of sequences for which

$$\|\mathbf{x}\|_\infty = \sup_i |x_i| < \infty$$

(Note that $l^\infty$ has already been encountered in Example 21F'.)

(1) For $1 \leq p \leq \infty$, the class $l^p$ is a linear space.
(2) $\|\ \|_p$ defines a norm on $l^p$, $1 \leq p \leq \infty$.
(3) If $1 \leq p < q \leq \infty$ and $\mathbf{x} \in l^p$, then $\mathbf{x} \in l^q$ and $\|\mathbf{x}\|_p \geq \|\mathbf{x}\|_q$, but the norms are not topologically equivalent.
(4) If $\mathbf{x} \in l^p$ for some $p$, then $\mathbf{x} \in l^p$ for all larger $p$ and $\lim_{p \to \infty} \|\mathbf{x}\|_p = \|\mathbf{x}\|_\infty$.
(5) If $q = \alpha p + \beta r$ where $\alpha > 0$, $\beta > 0$, and $\alpha + \beta = 1$, and if $1 \leq p < r < \infty$, then

$$\|\mathbf{x}\|_q \leq (\|\mathbf{x}\|_p)^\alpha (\|\mathbf{x}\|_r)^\beta \leq \alpha \|\mathbf{x}\|_p + \beta \|\mathbf{x}\|_r$$

***E.** Let $C[a, b]$ be the class of continuous functions on $[a, b]$, and define

$$\|f\|_p = \left(\int_a^b |f(t)|^p \, dt\right)^{1/p}, \quad 1 \leq p < \infty$$

$$\|f\|_\infty = \max_{t \in [a, b]} |f(t)|$$

(1) $\|\ \|_p$ defines a norm on $C[a, b]$, $1 \leq p \leq \infty$.
(2) If $1 \leq p < q < \infty$, then $\|f\|_p \leq \|f\|_q (b - a)^{1/p - 1/q}$, but $\|\ \|_p$ and $\|\ \|_q$ are not topologically equivalent.
(3) $\|f\|_\infty = \lim_{p \to \infty} \|f\|_p$.
(4) If $q = \alpha p + \beta r$ where $\alpha > 0$, $\beta > 0$, and $\alpha + \beta = 1$, and if $1 \leq p < r < \infty$, then

$$\|f\|_q \leq (\|f\|_p)^\alpha (\|f\|_r)^\beta \leq \alpha \|f\|_p + \beta \|f\|_r$$

There are versions of these results for abstract integrals as well [Hardy, Littlewood, and Polya 1952, pp. 126–171; Hewitt and Stromberg, 1965, pp. 188–209].

*F. Let $\alpha$, $x$ be positive continuous functions on $[a, b]$ with $\int_a^b \alpha(s)\, ds = 1$, and let

$$M_t(x, \alpha) = \left[\int_a^b \alpha(s)\, x^t(s)\, ds\right]^{1/t}, \qquad t \neq 0$$

$$M_0(x, \alpha) = \exp\left[\int_a^b \alpha(s)\, \log x(s)\, ds\right]$$

$$M_{-\infty}(x, \alpha) = \min_{s \in [a,b]} x(s), \qquad M_\infty(x, \alpha) = \max_{s \in [a,b]} x(s)$$

(1) $M_t(x, \alpha)$ is continuous in $t$ on $[-\infty, \infty]$.
(2) $r \leqslant t$ implies $M_r(x, \alpha) \leqslant M_t(x, \alpha)$.
(3) $t \log M_t(x, \alpha)$ is convex in $t$ on $(-\infty, \infty)$.

G. Let $\mathbf{x} = (x_1, x_2, \ldots, x_n)$, $\boldsymbol{\alpha} = (\alpha_1, \alpha_2, \ldots, \alpha_n)$ where $x_i > 0$, $\alpha_i > 0$, $\sum_1^n \alpha_i = 1$. For $F : (0, \infty) \to \mathbf{R}$ a strictly monotone function, we may introduce an associated generalized mean $\mathscr{F}(\mathbf{x}, \boldsymbol{\alpha})$ by

$$\mathscr{F}(\mathbf{x}, \boldsymbol{\alpha}) = F\left(\sum_1^n \alpha_i F^{-1}(x_i)\right)$$

The choice $F(u) = u^{1/t}$, for example, makes $\mathscr{F}(\mathbf{x}, \boldsymbol{\alpha}) = M_t(\mathbf{x}, \boldsymbol{\alpha})$. If $F$ is strictly monotonic, $G$ is strictly increasing, and $G^{-1} \circ F$ is convex, then $\mathscr{F}(\mathbf{x}, \boldsymbol{\alpha}) \leqslant \mathscr{G}(\mathbf{x}, \boldsymbol{\alpha})$ which generalizes Theorem A. (*Hint*: Write $F$ as $GG^{-1}F$ and apply Jensen's inequality.) Bullen [1971b] uses this result to prove and unify a large number of inequalities.

## 63. Matrix Inequalities

Our aim is to prove a very general inequality for functions of matrices (Theorem C) from which a number of classical results follow as special cases. We need two preliminary theorems that deal with the class $\Omega_n$ of $n \times n$ **doubly stochastic matrices**, that is, matrices with nonnegative entries in which the sum of the elements in each row and in each column is 1. The first theorem, due to Birkoff [1946], was anticipated in Problem 320 for the case $n = 2$. It identifies the extreme points of $\Omega_n$ as permutation matrices. A **permutation matrix** is a matrix obtained by permuting the rows of the identity matrix; it has exactly one 1 in each row and in each column, all other entries being 0.

**Theorem A.** *The set $\Omega_n$ of doubly stochastic matrices is a convex subset of the linear space of all $n \times n$ matrices. The extreme points of $\Omega_n$ are the permutation matrices, meaning that $\Omega_n$ is the convex hull of the permutation matrices.*

## 63. Matrix Inequalities

**Proof.** It is easy to verify that $\Omega_n$ is convex, and that any permutation matrix is an extreme point. The difficulty comes in showing that every extreme point is a permutation matrix. To do this, we transform the problem into one we have studied earlier.

The entries in any $n \times n$ matrix $X = [x_{ij}]$ may be written in a column vector $\mathbf{x}$ with $n^2$ entries by letting $x_k = x_{ij}$, $k = n(i-1) + j$, $i, j = 1, \ldots, n$. Then let $A$ be the $(2n - 1) \times n^2$ matrix

$$A = \begin{bmatrix} 1\ 1\ 1\ \cdots\ 1 & 0\ 0\ 0\ \cdots\ 0 & \cdots & 0\ 0\ 0\ \cdots\ 0 \\ 0\ 0\ 0\ \cdots\ 0 & 1\ 1\ 1\ \cdots\ 1 & \cdots & 0\ 0\ 0\ \cdots\ 0 \\ \vdots & \vdots & & \vdots \\ 0\ 0\ 0\ \cdots\ 0 & 0\ 0\ 0\ \cdots\ 0 & \cdots & 1\ 1\ 1\ \cdots\ 1 \\ 1\ 0\ 0\ \cdots\ 0 & 1\ 0\ 0\ \cdots\ 0 & \cdots & 1\ 0\ 0\ \cdots\ 0 \\ 0\ 1\ 0\ \cdots\ 0 & 0\ 1\ 0\ \cdots\ 0 & \cdots & 0\ 1\ 0\ \cdots\ 0 \\ \vdots & \vdots & & \vdots \\ 0\ 0\ 0\ \cdots\ 1\ 0 & 0\ 0\ 0\ \cdots\ 1\ 0 & \cdots & 0\ 0\ 0\ \cdots\ 1\ 0 \end{bmatrix}$$

and let $\mathbf{b}$ be the column vector of dimension $2n - 1$ having all entries 1. We now see that under the one-to-one correspondence (an isomorphism in the language of Section 21) we have described between $n \times n$ matrices $X$ and vectors $\mathbf{x} \in \mathbf{R}^{n^2}$, the convex set $\Omega_n$ corresponds to the convex set

$$\mathscr{F} = \{\mathbf{x} \in \mathbf{R}_+^{n^2} : A\mathbf{x} = \mathbf{b}\}$$

Extreme points of $\Omega_n$ correspond to extreme points of $\mathscr{F}$, and the happy fact is that we can identify the extreme points of $\mathscr{F}$ with the help of Theorem 53C. If $\mathbf{p} = (p_1, \ldots, p_{n^2})$ is such an extreme point, then at least $n^2 - (2n - 1) = (n - 1)^2$ of its entries are 0, and the others $(p_{k_1}, \ldots, p_{k_{2n-1}}) = \mathbf{p}^*$ satisfy $B\mathbf{p}^* = \mathbf{b}$ for some nonsingular $(2n - 1) \times (2n - 1)$ submatrix $B$ of $A$.

Now $\det B = \pm 1$, a fact that follows from the special form of $A$ (Problem A). Thus, $B^{-1}$ has integer entries, and then so does $\mathbf{p}^* = B^{-1}\mathbf{b}$. We conclude that vector $\mathbf{p}$ and the corresponding $n \times n$ matrix $P$ have integer entries. But since $P$ is doubly stochastic, it can only have integer entries if it is a permutation matrix. This establishes that every extreme point of $\Omega_n$ is a permutation matrix.

Finally we note that $\mathscr{F}$ is a closed bounded convex subset of $\mathbf{R}^{n^2}$, so we may appeal to Theorem 32D to conclude that it is the convex hull of its extreme points. The same is then true of the isomorphic convex set $\Omega_n$. ●

Next, for a fixed vector $\boldsymbol{\lambda} = (\lambda_1, \ldots, \lambda_n)$ in $\mathbf{R}^n$, let

$$K_\lambda = \{\mathbf{x} \in \mathbf{R}^n : \mathbf{x} = S\boldsymbol{\lambda}, S \in \Omega_n\} \tag{1}$$

and observe that $K_\lambda$ is a convex set. Also, let $\mathbf{s}_i$ denote the $i$th row of the matrix $S$.

**Theorem B.** If $f: U \to \mathbf{R}$ is convex on any set $U$ containing $K_\lambda$, then $g: \Omega_n \to \mathbf{R}$ defined by

$$g(S) = f(S\lambda) = f(\langle \mathbf{s}_1, \lambda \rangle, \ldots, \langle \mathbf{s}_n, \lambda \rangle)$$

is convex. Moreover, $g$ assumes its maximum value on $\Omega_n$ at a permutation matrix $P_\lambda$.

**Proof.** If $S, T \in \Omega_n$, $\alpha > 0$, $\beta > 0$, and $\alpha + \beta = 1$, then

$$g(\alpha S + \beta T) = f[(\alpha S + \beta T)\lambda] = f[\alpha S\lambda + \beta T\lambda]$$
$$\leqslant \alpha f(S\lambda) + \beta f(T\lambda) = \alpha g(S) + \beta g(T)$$

which means $g$ is convex. Now let $P_1, \ldots, P_m$ be the $m = n!$ permutation matrices in $\Omega_n$. For an arbitrary $S \in \Omega_n$, we may write (Theorem A)

$$S = \sum_1^m \alpha_i P_i$$

for some choices of $\alpha_i \geqslant 0$ with $\sum_1^m \alpha_i = 1$. Since $g$ is convex,

$$g(S) \leqslant \sum_1^m \alpha_i g(P_i) \leqslant \sum_1^m \alpha_i g(P_\lambda) = g(P_\lambda)$$

where $P_\lambda$ is a permutation matrix chosen so that $g(P_\lambda) = \max_{1 \leqslant i \leqslant m} g(P_i)$. We have used the notation $P_\lambda$ to emphasize that the matrix $P_\lambda$ will in general depend on $\lambda$. ●

Now we come to the theorem toward which we have been working, a general inequality for convex functions of symmetric matrices.

**Theorem C.** Let $A$ be a real symmetric $n \times n$ matrix and let $\lambda = (\lambda_1, \ldots, \lambda_n)$ be a vector consisting of the (necessarily real) characteristic values of $A$ arranged in some order. If $f: U \to \mathbf{R}$ is convex on a set $U$ containing the set $K_\lambda$ defined by (1), then for any orthonormal set $\{\mathbf{v}_1, \ldots, \mathbf{v}_n\}$ in $\mathbf{R}^n$,

$$f(\langle A\mathbf{v}_1, \mathbf{v}_1 \rangle, \ldots, \langle A\mathbf{v}_n, \mathbf{v}_n \rangle) \leqslant f(P_\lambda \lambda) = f(\lambda_{i_1}, \ldots, \lambda_{i_n})$$

$P_\lambda$ being the permutation matrix whose existence is asserted in Theorem B.

## 63. Matrix Inequalities

**Proof.** Let $\mathbf{u}_1, \ldots, \mathbf{u}_n$ be the normalized characteristic vectors of $A$ corresponding to the characteristic values $\lambda_1, \ldots, \lambda_n$ so that $A\mathbf{u}_i = \lambda_i \mathbf{u}_i$. Let $S$ be the matrix with entries $s_{ij} = \langle \mathbf{u}_i, \mathbf{v}_j \rangle^2$; $S$ is doubly stochastic since it is symmetric and

$$1 = \|\mathbf{u}_i\|^2 = \sum_{j=1}^n \langle \mathbf{u}_i, \mathbf{v}_j \rangle^2 = \sum_{j=1}^n s_{ij}$$

that is, its row sums are 1. Now

$$A\mathbf{v}_j = A \sum_{i=1}^n \langle \mathbf{u}_i, \mathbf{v}_j \rangle \mathbf{u}_i = \sum_{i=1}^n \langle \mathbf{u}_i, \mathbf{v}_j \rangle \lambda_i \mathbf{u}_i$$

Hence,

$$\langle A\mathbf{v}_j, \mathbf{v}_j \rangle = \sum_{i=1}^n \langle \mathbf{u}_i, \mathbf{v}_j \rangle^2 \lambda_i = \langle \mathbf{s}_j, \boldsymbol{\lambda} \rangle$$

where $\mathbf{s}_j$ is the $j$th column (equivalently the $j$th row) of $S$. Consequently,

$$\begin{aligned} f(\langle A\mathbf{v}_1, \mathbf{v}_1 \rangle, \ldots, \langle A\mathbf{v}_n, \mathbf{v}_n \rangle) \\ = f(\langle \mathbf{s}_1, \boldsymbol{\lambda} \rangle, \ldots, \langle \mathbf{s}_n, \boldsymbol{\lambda} \rangle) \\ = f(S\boldsymbol{\lambda}) \leqslant f(P_\lambda \boldsymbol{\lambda}) = f(\lambda_{i_1}, \ldots, \lambda_{i_n}) \end{aligned} \quad (2)$$

where $P_\lambda$ is the permutation matrix guaranteed by Theorem B. ●

We have actually proved a bit more than we have claimed. By choosing $\{\mathbf{v}_1, \ldots, \mathbf{v}_n\}$ as the correct permutation, namely $\{\mathbf{u}_{i_1}, \ldots, \mathbf{u}_{i_n}\}$, of the characteristic vectors, we obtain equality in (2). Thus,

$$\max f(\langle A\mathbf{v}_1, \mathbf{v}_1 \rangle, \ldots, \langle A\mathbf{v}_n, \mathbf{v}_n \rangle) = f(P_\lambda \boldsymbol{\lambda}) \quad (3)$$

the maximum being taken over all orthonormal sets in $\mathbf{R}^n$.

Inequalities of endless variety may now be derived by choosing different convex functions $f$ in Theorem C. Our first choice is, for fixed $k \leqslant n$,

$$f(t_1, \ldots, t_n) = \sum_{j=1}^k t_j \quad (4)$$

defined on all of $\mathbf{R}^n$.

**Theorem D.** *If $A$ is a real symmetric $n \times n$ matrix and $\{\mathbf{v}_1, \ldots, \mathbf{v}_n\}$ is an orthonormal set in $\mathbf{R}^n$, then*

$$\sum_{j=1}^k \lambda_{n-j+1} \leqslant \sum_{j=1}^k \langle A\mathbf{v}_j, \mathbf{v}_j \rangle \leqslant \sum_{j=1}^k \lambda_j \quad (5)$$

*where $\lambda_1 \geqslant \lambda_2 \geqslant \cdots \geqslant \lambda_n$ are the characteristic values of $A$.*

**Proof.** Note that $f$ defined in (4) is linear, so both $-f$ and $f$ are convex on $\mathbf{R}^n$. For the function $f$, Theorem C says there is a permutation $i_1,\ldots, i_n$ of $1,\ldots, n$ so that
$$\sum_{j=1}^{k} \langle A\mathbf{v}_j, \mathbf{v}_j \rangle \leq \sum_{j=1}^{k} \lambda_{i_j}$$

Due to the ordering of the $\lambda_i$, the sum on the right is surely less than or equal to $\sum_1^k \lambda_j$, giving the second inequality of (5). The first inequality follows in a similar way from considering the function $-f$. Note that if $\mathbf{v}_j$ is the normalized characteristic vector corresponding to $\lambda_{n-j+1}$, then there is equality at the left in (5). Similarly if $\mathbf{v}_j$ is the normalized characteristic vector corresponding to $\lambda_j$, the right side becomes an equality. •

Next we choose
$$f(t_1,\ldots, t_n) = \left(\prod_{j=1}^{k} t_j\right)^{1/k} \qquad (6)$$

defined for $t_j \geq 0$ and $k \leq n$. It follows from Theorem 61E (See Problem 61M) that $f$ is concave, and this leads to the following result.

**Theorem E.** If $A$ is a real nonnegative definite $n \times n$ matrix and $\{\mathbf{v}_1,\ldots, \mathbf{v}_n\}$ is an orthonormal set in $\mathbf{R}^n$, then
$$\prod_{j=1}^{k} \lambda_{n-j+1} \leq \prod_{j=1}^{k} \langle A\mathbf{v}_j, \mathbf{v}_j \rangle \leq \left(\frac{1}{k}\prod_{j=1}^{k} \lambda_j\right)^k \qquad (7)$$

where $\lambda_1 \geq \lambda_2 \geq \cdots \geq \lambda_n \geq 0$ are the characteristic values of $A$.

**Proof.** Since $f$ defined by (6) is concave, $-f$ is convex on the nonnegative orthant $\mathbf{R}_+^n$. Consequently, by Theorem C,
$$\left(\prod_{j=1}^{k} \lambda_{n-j+1}\right)^{1/k} \leq \left(\prod_{j=1}^{k} \langle A\mathbf{v}_j, \mathbf{v}_j \rangle\right)^{1/k},$$

which is equivalent to the left side of (7). Again there is equality if $\mathbf{v}_j$ is the normalized characteristic vector corresponding to $\lambda_{n-j+1}$. To get the right side of (7), we apply the GM–AM inequality and Theorem D; specifically
$$\left(\prod_{j=1}^{k} \langle A\mathbf{v}_j, \mathbf{v}_j \rangle\right)^{1/k} \leq \frac{1}{k}\sum_{j=1}^{k} \langle A\mathbf{v}_j, \mathbf{v}_j \rangle \leq \frac{1}{k}\sum_{j=1}^{k} \lambda_j \quad •$$

Theorem E gives almost immediately two well-known theorems on determinants.

## 63. Matrix Inequalities

**Theorem F (Hadamard's Determinant Theorem).** If $A$ is a real $n \times n$ matrix with entries $a_{ij}$, then

$$(\det A)^2 \leq \prod_{j=1}^{n} \left( \sum_{i=1}^{n} a_{ij}^2 \right) \tag{8}$$

If $A$ is nonnegative definite,

$$\det A \leq \prod_{j=1}^{n} a_{jj} \tag{9}$$

**Proof.** We consider (9) first. Recall that the determinant of any square matrix is equal to the product of its characteristic values (Problem B). If we let $\mathbf{e}_j$ be the standard unit vector and consider (7) with $k = n$, we get

$$\det A = \prod_{j=1}^{n} \lambda_j \leq \prod_{j=1}^{n} \langle A\mathbf{e}_j, \mathbf{e}_j \rangle = \prod_{j=1}^{n} a_{jj}$$

which is (9). Now if we apply (9) to the matrix $B = A^t A$, which is easily seen to be nonnegative definite, then

$$(\det A)^2 = \det B \leq \prod_{j=1}^{n} b_{jj} = \prod_{j=1}^{n} \left( \sum_{i=1}^{n} a_{ij}^2 \right)$$

which is (8). ●

**Theorem G (Minkowski's Determinant Theorem).** If $A$ and $B$ are real nonnegative definite $n \times n$ matrices, then

$$[\det(A + B)]^{1/n} \geq (\det A)^{1/n} + (\det B)^{1/n} \tag{10}$$

and

$$\det(A + B) \geq \det A + \det B \tag{11}$$

**Proof.** Let $\mathbf{v}_1, \ldots, \mathbf{v}_n$ be an orthonormal set of characteristic vectors for $(A + B)$. Then

$$[\det(A + B)]^{1/n} = \left[ \prod_{1}^{n} \langle (A + B) \mathbf{v}_j, \mathbf{v}_j \rangle \right]^{1/n}$$

$$= \left[ \prod_{1}^{n} (\langle A\mathbf{v}_j, \mathbf{v}_j \rangle + \langle B\mathbf{v}_j, \mathbf{v}_j \rangle) \right]^{1/n}$$

$$\geq \left[ \prod_{1}^{n} \langle A\mathbf{v}_j, \mathbf{v}_j \rangle \right]^{1/n} + \left[ \prod_{1}^{n} \langle B\mathbf{v}_j, \mathbf{v}_j \rangle \right]^{1/n}$$

$$\geq (\det A)^{1/n} + (\det B)^{1/n}$$

The next to last inequality follows from Theorem 61E, the last from Theorem E. Finally, (11) follows from (10) by raising both sides to the $n$th power. ●

The expressions occurring in Theorems D and E merit further exploration. Let the real symmetric $n \times n$ matrix $A$ have characteristic values $\lambda_1 \geq \lambda_2 \geq \cdots \geq \lambda_n$ and let

$$s_k(A) = \sum_{j=1}^{k} \lambda_{n-j+1}, \qquad S_k(A) = \sum_{j=1}^{k} \lambda_j$$

$$p_k(A) = \prod_{j=1}^{k} \lambda_{n-j+1}, \qquad P_k(A) = \prod_{j=1}^{k} \lambda_j \tag{12}$$

Note that

$$\lambda_n = s_1(A) = p_1(A), \qquad \lambda_1 = S_1(A) = P_1(A)$$

while

$$\text{trace } A = s_n(A) = S_n(A) \qquad \text{(Problem B)}$$

and

$$\det A = p_n(A) = P_n(A)$$

**Theorem H.** The functions $s_k$ and $S_k$ are, respectively, concave and convex on the class $\mathscr{S}_n$ of real symmetric $n \times n$ matrices. Thus if $A$ and $B$ belong to $\mathscr{S}_n$ and $\alpha > 0$, $\beta > 0$, with $\alpha + \beta = 1$, then

$$\alpha s_k(A) + \beta s_k(B) \leq s_k(\alpha A + \beta B)$$
$$\leq S_k(\alpha A + \beta B) \leq \alpha S_k(A) + \beta S_k(B)$$

**Proof.** Let $L_V(A) = \sum_1^k \langle A\mathbf{v}_j, \mathbf{v}_j \rangle$ where $V = \{\mathbf{v}_1, \ldots, \mathbf{v}_n\}$ is any orthonormal set in $\mathbf{R}^n$. From Theorem D (see also its proof where the case of equality is discussed),

$$-s_k(A) = \max_V [-L_V(A)], \qquad S_k(A) = \max_V L_V(A)$$

Since $L_V(A)$ is linear in $A$, both $-L_V$ and $L_V$ are convex on $\mathscr{S}_n$. Since the class of convex functions is closed under suprema (Problem 41B), $-s_k$ and $S_k$ are convex, which is what we wished to demonstrate. ●

We have shown, incidentally, that

$$\lambda_n = s_1(A) = \min_{\|\mathbf{v}\|=1} \langle A\mathbf{v}, \mathbf{v} \rangle; \qquad \lambda_1 = S_1(A) = \max_{\|\mathbf{v}\|=1} \langle A\mathbf{v}, \mathbf{v} \rangle \tag{13}$$

## 63. Matrix Inequalities

two striking results which are themselves special cases of Fischer's expressions for the characteristic values of a real symmetric matrix [Beckenbach and Bellman, 1965, p. 73].

**Theorem I.** The function $p_k$ defined in (12) is log-concave on the class of positive definite matrices. In fact, if $A$ and $B$ are nonnegative definite and $\alpha > 0, \beta > 0$ with $\alpha + \beta = 1$, then

$$[p_k(A)]^\alpha [p_k(B)]^\beta \leqslant p_k(\alpha A + \beta B)$$

and in particular

$$(\det A)^\alpha (\det B)^\beta \leqslant \det(\alpha A + \beta B)$$

**Proof.** We know from the proof of Theorem E that

$$p_k(A) = \min_V \prod_{j=1}^{k} \langle A\mathbf{v}_j, \mathbf{v}_j \rangle$$

where $V = \{\mathbf{v}_1, ..., \mathbf{v}_n\}$ is an orthonormal set in $\mathbf{R}^n$. Thus,

$$[p_k(A)]^\alpha [p_k(B)]^\beta = \left[\min \prod_1^k \langle A\mathbf{v}_j, \mathbf{v}_j \rangle\right]^\alpha \left[\min \prod_1^k \langle B\mathbf{v}_j, \mathbf{v}_j \rangle\right]^\beta$$

$$\leqslant \min \prod_1^k [\langle A\mathbf{v}_j, \mathbf{v}_j \rangle^\alpha \langle B\mathbf{v}_j, \mathbf{v}_j \rangle^\beta]$$

$$\leqslant \min \prod_1^k [\alpha \langle A\mathbf{v}_j, \mathbf{v}_j \rangle + \beta \langle B\mathbf{v}_j, \mathbf{v}_j \rangle]$$

$$= p_k(\alpha A + \beta B) \quad \bullet$$

### PROBLEMS AND REMARKS

**A.** Let $B$ be a nonsingular $n \times n$ matrix that can be partitioned as $[{}^P_Q]$ where $P$ and $Q$ are matrices with entries of 0's and 1's so distributed that no column of either $P$ or $Q$ has more than one entry of 1. Then $\det B = \pm 1$. [*Hint:* There is a column of $B$ with exactly one 1 (for if every column had two 1's, the rows of $Q$ could be added to the negative of the rows of $P$ to produce the zero vector). Evaluate $B$, expanding by minors of this column. Proceed by induction.]

**B.** The **trace** of an $n \times n$ real matrix $A = [a_{ij}]$ is defined to be the sum of the diagonal elements; $\text{tr}(A) = \sum_1^n a_{jj}$. Let $A$ be a symmetric matrix with characteristic values $\lambda_1, ..., \lambda_n$.

(1) Use Theorem D to show $\operatorname{tr}(A) = \sum_1^n \lambda_j$.
(2) $\sum_{i=1}^n \sum_{j=1}^n a_{ij}^2 = \sum_1^n \lambda_j^2$ [consider $\operatorname{tr}(A^2)$].
(3) If we order the $\lambda$'s so that $|\lambda_1| \geq |\lambda_2| \geq \cdots \geq |\lambda_n|$, then

$$\left(\sum_{j=1}^k |a_{jj}|^p\right)^{1/p} \leq \left(\sum_{j=1}^k |\lambda_j|^p\right)^{1/p}, \quad p \geq 1, \quad k \leq n$$

**C.** If $A$ is any $n \times n$ matrix with characteristic values $\lambda_1, \ldots, \lambda_n$,

$$\operatorname{tr}(A) = \sum_1^n \lambda_j; \quad \det(A) = \prod_1^n \lambda_j$$

**D.** Let $\mathscr{S}_n$ be the class of real symmetric $n \times n$ matrices. For $A \in \mathscr{S}_n$, let $\lambda_j = \lambda_j(A), j = 1, \ldots, n$, be the characteristic values of $A$ arranged so that $|\lambda_1| \geq |\lambda_2| \geq \cdots \geq |\lambda_n|$. For fixed $p \geq 1$, $k \leq n$, let $g(A) = (\sum_1^k |\lambda_j|^p)^{1/p}$.

(1) $g$ is convex on $\mathscr{S}_n$. [*Hint*: $g(A) = \sup(\sum_1^k |\langle A\mathbf{v}_j, \mathbf{v}_j\rangle|^p)^{1/p}$.
(2) $g$ is a norm for $\mathscr{S}_n$. In particular, $|\lambda_1|$ is a norm.
(3) $|\lambda_j(A)| = [\lambda_j(A^2)]^{1/2}$.
(4) $|\lambda_1(A)| = \sup_{\|\mathbf{v}\|=1} \|A\mathbf{v}\|$ and $|\lambda_n(A)| = \inf_{\|\mathbf{v}\|=1} \|A\mathbf{v}\|$.
This means that if $A$ is the matrix of a linear transformation $\mathbf{A}: \mathbf{R}^n \to \mathbf{R}^n$, then $\|\mathbf{A}\| = |\lambda_1(A)|$ (Section 22).
(5) If $B$ is any real $n \times n$ matrix and $A = B^t B$ has characteristic values (necessarily nonnegative) $\lambda_1 \geq \lambda_2 \geq \cdots \geq \lambda_n$, then

$$[\lambda_1(A)]^{1/2} = \sup_{\|\mathbf{v}\|=1} \|B\mathbf{v}\|, \quad [\lambda_n(A)]^{1/2} = \inf_{\|\mathbf{v}\|=1} \|B\mathbf{v}\|$$

**E.** If $A, B \in \mathscr{S}_n$ and $\lambda_1 \geq \lambda_2 \geq \cdots \geq \lambda_n$, then

$$\lambda_n(A) + \lambda_n(B) \leq \lambda_n(A+B) \leq \lambda_1(A+B) \leq \lambda_1(A) + \lambda_1(B)$$

**F.** Let $\mathscr{D}_n$ be the class of nonnegative definite $n \times n$ matrices.

(1) $g(A) = (\det A)^{1/n}$ is concave on $\mathscr{D}_n$.
(2) $A, B \in \mathscr{D}_n \Rightarrow \det(A+B) \geq 2^n (\det AB)^{1/2}$.
(3) $A$ symmetric and $B \in \mathscr{D}_n \Rightarrow \lambda_1(A) \leq \lambda_1(A+B)$ and $\lambda_n(A) \leq \lambda_n(A+B)$.

**G.** (**Oppenheim's inequality** [Beckenbach and Bellman, 1965, p. 71]). Let $A$, $B \in \mathscr{D}_n$ and let $p_k$ be defined as in (12). Then

$$[p_k(A+B)]^{1/k} \geq [p_k(A)]^{1/k} + [p_k(B)]^{1/k}$$

Consequently, $(p_k)^{1/k}$ is concave on $\mathscr{D}_n$.

*****H.** Let $W(A)$ be the set of all numbers of the form $\langle A\mathbf{x}, \mathbf{x}\rangle$ where $\|\mathbf{x}\| = 1$. If $A$ is symmetric, then $W(A) = [\lambda_n, \lambda_1]$ where $\lambda_1 \geq \lambda_2 \geq \cdots \geq \lambda_n$ are the characteristic values of $A$. More generally, if for symmetric $A$, $W_k(A)$ is the set of all vectors of the form $(\langle A\mathbf{x}_1, \mathbf{x}_1\rangle, \ldots, \langle A\mathbf{x}_k, \mathbf{x}_k\rangle)$ where $\{\mathbf{x}_1, \ldots, \mathbf{x}_n\}$ is an orthonormal set in $\mathbf{R}^n$, then $W_k(A)$ is a convex set in $\mathbf{R}^k$, the extreme points of which are of the form $(\lambda_{i_1}, \ldots, \lambda_{i_k})$ where $i_1, \ldots, i_n$ is a permutation of $1, \ldots, n$.

*****I.** (**Kantorovich's inequality** [Marcus and Minc, 1964, p. 117]). Let $f$ and $g$ be nonnegative and convex on $[a, b]$, $x_i \in [a, b]$, $\alpha_i > 0$, $\sum_1^n \alpha_i = 1$, and $c > 0$.

## 63. Matrix Inequalities

(1) $\left[\sum_1^n \alpha_i f(x_i)\right]^{1/2} \left[\sum_1^n \alpha_i g(x_i)\right]^{1/2} \leq \frac{1}{2} \max \left\{ cf(a) + \frac{1}{c} g(a),\, cf(b) + \frac{1}{c} g(b) \right\}$

(2) If in addition $f(x) g(x) \geq 1$ on $[a, b]$, then

$$1 \leq \left[\sum_1^n \alpha_i f(x_i)\right]^{1/2} \left[\sum_1^n \alpha_i g(x_i)\right]^{1/2}$$

(3) If $a > 0$,

$$1 \leq \sum_1^n \alpha_i x_i \sum_1^n \alpha_i x_i^{-1} \leq \tfrac{1}{4}[(a/b)^{1/2} + (b/a)^{1/2}]^2$$

[*Hint*: Let $f(x) = x$, $g(x) = 1/x$, $c = (ab)^{-1/2}$.]

(4) If $A$ is a real positive definite $n \times n$ matrix with characteristic values $\lambda_1 \geq \lambda_2 \geq \cdots \geq \lambda_n$ and $\mathbf{v}$ is a unit vector in $\mathbf{R}^n$, then

$$1 \leq \langle A\mathbf{v}, \mathbf{v}\rangle \langle A^{-1}\mathbf{v}, \mathbf{v}\rangle \leq \tfrac{1}{4}\left[\left(\frac{\lambda_n}{\lambda_1}\right)^{1/2} + \left(\frac{\lambda_1}{\lambda_n}\right)^{1/2}\right]^2$$

(5) From part (4) it follows that any positive definite $n \times n$ matrix $A$ satisfies $\det(A) \leq a_{11} a_{22} \cdots a_{nn}$. (*Hint*: Let $\mathbf{v}$ be the standard unit vector $\mathbf{e}_j$.)

*J. (**Bergstrom's inequality** [Beckenbach and Bellman, 1965, pp. 67–69]). Let $A$ be a real positive definite $n \times n$ matrix.

(1) $\langle A\mathbf{u}, \mathbf{u}\rangle \langle A^{-1}\mathbf{v}, \mathbf{v}\rangle \geq \langle \mathbf{u}, \mathbf{v}\rangle^2$ for all $\mathbf{u}, \mathbf{v} \in \mathbf{R}^n$. (*Hint*: Diagonalize and use the CBS inequality.)

(2) $\langle A^{-1}\mathbf{v}, \mathbf{v}\rangle = \min_{\mathbf{u}}(\langle A\mathbf{u}, \mathbf{u}\rangle / \langle \mathbf{u}, \mathbf{v}\rangle^2)$.

(3) If $A_i$ denotes the matrix $A$ with the $i$th row and the $i$th column deleted, and $\mathbf{u} = (u_1, \ldots, u_n)$, then

$$\frac{\det A_i}{\det A} = \min_{\mathbf{u}} \frac{\langle A\mathbf{u}, \mathbf{u}\rangle}{u_i^2} = \min_{u_i = 1} \langle A\mathbf{u}, \mathbf{u}\rangle$$

(*Hint*: Let $\mathbf{v}$ be the standard unit vector $\mathbf{e}_i$.)

(4) $$\frac{\det(A + B)}{\det(A_i + B_i)} \geq \frac{\det(A)}{\det(A_i)} + \frac{\det(B)}{\det(B_i)}$$

K. Let $U$ be a convex set in $\mathbf{R}^n$ such that for any $n \times n$ permutation matrix $P$, $\mathbf{x} \in U$ implies $P\mathbf{x} \in U$. Moreover, let $f: U \to \mathbf{R}$ be convex and symmetric $[f(P\mathbf{x}) = f(\mathbf{x})]$. If $\mathbf{x} = (x_1, \ldots, x_n) \in U$ and $\bar{x} = (1/n) \sum_1^n x_i$, then $f(\bar{x}, \ldots, \bar{x}) \leq f(\mathbf{x})$ (This is a special case of Schur-convexity discussed in Section 86).

L. If $A = (a_{ij})$ is a real $n \times n$ matrix with $|a_{ij}| \leq M$, then $|\det A| \leq M^n n^{n/2}$.

★ ★ ★ ★ ★

The best references for this section are the books by Marcus and Minc [1964] and Beckenbach and Bellman [1965], where many additional results and extended bibliographies may be found.

# VII

## Midconvex Functions

In most sciences one generation tears down what another has built, and and what one has established another undoes. In mathematics alone each generation builds a new story to the old structure.

HERMANN HANKEL

## 70. Introduction

The systematic study of convex functions was first undertaken by Jensen [1905, 1906]. He had in mind real-valued functions defined on an interval $I$ satisfying

$$f\left(\frac{x+y}{2}\right) \leq \tfrac{1}{2}[f(x) + f(y)] \tag{1}$$

for all $x$ and $y$ in $I$. We have called a function $f: I \to \mathbf{R}$ convex if and only if

$$f[\lambda x + (1-\lambda)y] \leq \lambda f(x) + (1-\lambda)f(y) \tag{2}$$

for all $x$ and $y$ in $I$, $\lambda \in (0, 1)$. Taking $\lambda = \tfrac{1}{2}$ shows that all functions satisfying (2) also satisfy (1), but the definitions are not equivalent since there are functions discontinuous on an open interval that satisfy (1), while all functions that satisfy (2) are continuous on open intervals (Theorem 11A).

Having used convex to describe functions that satisfy (2), we shall describe as **midconvex** those functions that satisfy (1). Both definitions can be applied to functions defined on a convex subset $U$ of a linear space $\mathbf{L}$, and it is in this context that we investigate the relationships between convex, midconvex, and related classes of real-valued functions.

After discussing the situations for functions having their domains in $\mathbf{L}$, we restrict attention in Section 72 to the special case $\mathbf{L} = \mathbf{R}$ that Jensen and other early writers had in mind. Following Jensen, there came a series of papers giving conditions under which midconvex functions would be continuous. These papers nicely illustrate the progression of mathematical ideas, and we trace this progress in its historical sequence. We conclude the section by noting several other definitions which have appeared in the literature, adding interest if not clarity to the concept of a convex function $f: I \to \mathbf{R}$.

## 71. Midconvex Functions on a Normed Linear Space

We have called a function $f: U \to \mathbf{R}$ defined on a convex set $U \subseteq \mathbf{L}$ **midconvex** if for any $\mathbf{x}, \mathbf{y} \in U$,

$$f\left(\frac{\mathbf{x}+\mathbf{y}}{2}\right) \leq \tfrac{1}{2}[f(\mathbf{x}) + f(\mathbf{y})]$$

As usual, saying $f: U \to \mathbf{R}$ is midconvex will mean the set $U$ is convex. An equivalent definition is sometimes given for midconvexity using $n$ points. Its statement depends on the notion of a rational convex combination of points. We say that $\sum_1^n \alpha_i \mathbf{x}_i$ is a **rational convex combination** of the points $\mathbf{x}_1, \ldots, \mathbf{x}_n$ if

$$\alpha_i \geq 0 \quad \text{for} \quad i = 1, \ldots, n \tag{1}$$

$$\sum_1^n \alpha_i = 1 \tag{2}$$

$$\alpha_i \text{ is rational} \quad \text{for} \quad i = 1, \ldots, n \tag{3}$$

**Theorem A.** $f$ is midconvex on the convex set $U \subseteq \mathbf{L}$ if and only if for any rational convex combination of points in $U$

$$f\left(\sum_1^n \alpha_i \mathbf{x}_i\right) \leq \sum_1^n \alpha_i f(\mathbf{x}_i) \tag{4}$$

**Proof.** It is obvious (take $\alpha_1 = \alpha_2 = \tfrac{1}{2}$) that any $f$ satisfying (4) is midconvex. The converse is much harder. We show first that a midconvex function satisfies (4) in the special case in which $\alpha_1 = \cdots = \alpha_n = 1/n$. When $n = 4$,

$$f\left[\tfrac{1}{2}\left(\tfrac{\mathbf{x}_1 + \mathbf{x}_2}{2}\right) + \tfrac{1}{2}\left(\tfrac{\mathbf{x}_3 + \mathbf{x}_4}{2}\right)\right] \leq \tfrac{1}{2}\left[f\left(\tfrac{\mathbf{x}_1 + \mathbf{x}_2}{2}\right) + f\left(\tfrac{\mathbf{x}_3 + \mathbf{x}_4}{2}\right)\right]$$
$$\leq \tfrac{1}{4}[f(\mathbf{x}_1) + f(\mathbf{x}_2) + f(\mathbf{x}_3) + f(\mathbf{x}_4)]$$

Similarly, for any $n = 2^k$,

$$f\left(\tfrac{\mathbf{x}_1 + \cdots + \mathbf{x}_n}{n}\right) \leq \tfrac{1}{n}[f(\mathbf{x}_1) + \cdots + f(\mathbf{x}_n)]$$

For $n = 3$,

$$f\left(\tfrac{\mathbf{x}_1 + \mathbf{x}_2 + \mathbf{x}_3}{3}\right)$$
$$= f\left[\tfrac{1}{2^2}\left(\mathbf{x}_1 + \mathbf{x}_2 + \mathbf{x}_3 + (2^2 - 3)\tfrac{\mathbf{x}_1 + \mathbf{x}_2 + \mathbf{x}_3}{3}\right)\right]$$
$$\leq \tfrac{1}{2^2}\left[f(\mathbf{x}_1) + f(\mathbf{x}_2) + f(\mathbf{x}_3) + (2^2 - 3)f\left(\tfrac{\mathbf{x}_1 + \mathbf{x}_2 + \mathbf{x}_3}{3}\right)\right]$$

## 71. Midconvex Functions on a Normed Linear Space

Thus,

$$\left(1 - \frac{2^2 - 3}{2^2}\right) f\left(\frac{x_1 + x_2 + x_3}{3}\right) \leq \frac{1}{2^2}[f(x_1) + f(x_2) + f(x_3)]$$

or

$$\frac{3}{2^2} f\left(\frac{x_1 + x_2 + x_3}{3}\right) \leq \frac{1}{2^2}[f(x_1) + f(x_2) + f(x_3)]$$

We now follow the same pattern for any $n \neq 2^k$. Choose $m$ so that $2^{m-1} < n < 2^m$. Then write

$$f\left(\frac{x_1 + \cdots + x_n}{n}\right) = f\left[\frac{1}{2^m}\left(x_1 + \cdots + x_n + (2^m - n)\frac{x_1 + \cdots + x_n}{n}\right)\right]$$

$$\leq \frac{1}{2^m}\left[f(x_1) + \cdots + f(x_n) + (2^m - n)f\left(\frac{x_1 + \cdots + x_n}{n}\right)\right]$$

We get

$$\left(1 - \frac{2^m - n}{2^m}\right) f\left(\frac{x_1 + \cdots + x_n}{n}\right) \leq \frac{1}{2^m}[f(x_1) + \cdots + f(x_n)]$$

from which algebraic simplification gives for any integer $n$

$$f\left(\frac{x_1 + \cdots + x_n}{n}\right) \leq \frac{1}{n}[f(x_1) + \cdots + f(x_n)]$$

Now given any set of $n$ positive rationals $\alpha_i$ such that $\sum_1^n \alpha_i = 1$, set $\alpha_i = u_i/d$ where $d$ is the least common denominator of the $\alpha_i$. It follows that $\sum_1^n u_i = d$, and we have

$$f\left(\frac{u_1}{d} x_1 + \cdots + \frac{u_n}{d} x_n\right) = f\left(\frac{1}{d}[\underbrace{(x_1 + \cdots + x_1)}_{u_1 \text{ terms}} + \cdots + \underbrace{(x_n + \cdots + x_n)}_{u_n \text{ terms}}]\right)$$

The desired inequality follows by appeal to the special case proved first. ●

Inequality (4) is called **Jensen's inequality**. It is instructive to examine the role played by each of the restrictions (1) $\alpha_i \geq 0$, (2) $\sum_1^n \alpha_i = 1$, (3) $\alpha_i$ rational, in the definition of a rational convex combination of points. If we remove some of the restrictions on the $\alpha_i$, thereby increasing the kinds of combinations of points $x_1, \ldots, x_n$ under consideration, it is to be expected that the class of functions still satisfying Jensen's inequality will be smaller.

**Theorem B.** Let $f: \mathbf{L} \to \mathbf{R}$. Then $f$ satisfies Jensen's inequality

(a) for all $\alpha_i$ $\Leftrightarrow$ $f$ is linear,
(b) for all $\alpha_i$ restricted by (2) $\Leftrightarrow$ $f$ is affine,
(c) for all $\alpha_i$ restricted by (1), (2) $\Leftrightarrow$ $f$ is convex,
(d) for all $\alpha_i$ restricted by (1), (2), (3) $\Leftrightarrow$ $f$ is midconvex.

**Proof.** Part (d) is just a restatement of Theorem A and (c) was cited in Section 40 as an alternate way to define convex functions. We have already (Theorem 22A) shown that $f$ is affine if and only if it satisfies Jensen's equality [that is (4) with = replacing $\leq$] for $\alpha_i$ restricted by (2), so surely an affine function satisfies (4) under the stated conditions. The proof of (b) will be completed if we show that a function $f$ satisfying Jensen's inequality for $\alpha_i$ restricted by (2) actually satisfies the corresponding equality. If $\sum_1^n \alpha_i = 1$, then at least one of the $\alpha_i$, say $\alpha_1$, is positive. Now

$$\mathbf{x}_1 = \frac{1}{\alpha_1}\left(\sum_1^n \alpha_i \mathbf{x}_i\right) + \frac{-\alpha_2}{\alpha_1}\mathbf{x}_2 + \cdots + \frac{-\alpha_n}{\alpha_1}\mathbf{x}_n$$

and since

$$\frac{1}{\alpha_1} + \frac{-\alpha_2}{\alpha_1} + \cdots + \frac{-\alpha_n}{\alpha_1} = \frac{\alpha_1}{\alpha_1} = 1$$

$$f(\mathbf{x}_1) \leq \frac{1}{\alpha_1} f\left(\sum_1^n \alpha_i \mathbf{x}_i\right) + \sum_2^n \frac{-\alpha_i}{\alpha_1} f(\mathbf{x}_i)$$

Multiplication by $\alpha_1$ and a bit of algebra gives

$$\sum_1^n \alpha_i f(\mathbf{x}_i) \leq f\left(\sum_1^n \alpha_i \mathbf{x}_i\right)$$

Since the opposite inequality holds by hypothesis, we have our desired equality.

Finally, to establish (a), it will be sufficient to show that if $f$ satisfies Jensen's inequality for all $\alpha_i$, then it is linear, the converse being obvious. Now if (4) holds for all $\alpha_i$, it certainly holds for $\alpha_i$ restricted by (2). Then according to part (b), $f$ is affine, which by definition means $f(\mathbf{x}) = g(\mathbf{x}) + b$ for some linear function $g$. It remains to show that $b = 0$. From $f(\mathbf{O}) = f(1 \cdot \mathbf{O} + 1 \cdot \mathbf{O}) \leq f(\mathbf{O}) + f(\mathbf{O})$, we learn that $0 \leq f(\mathbf{O})$; and $f(\mathbf{O}) = f(0 \cdot \mathbf{O}) \leq 0 \, f(\mathbf{O}) = 0$. We conclude that $b = f(\mathbf{O}) = 0$. ●

## 71. Midconvex Functions on a Normed Linear Space

Of the eight possible combinations of the restrictions (1), (2), (3), we have considered only four. The other possibilities also lead to interesting classes of functions. In fact, the requirement that $f$ satisfy Jensen's inequality for certain $\alpha$'s provides a unified way of identifying what have been historically some of the most studied classes of functions. We explore this in further detail in Problem F.

It is clear that a continuous midconvex function must be convex. For if $f$ satisfies Jensen's inequality for all rational $\alpha_i$ [satisfying (1) and (2)], the continuity of $f$ will force it to hold for irrational $\alpha_i$ as well. However, even when $\mathbf{L} = \mathbf{R}$, we can find examples of midconvex functions that are not continuous (Problem E). We conclude this section with a theorem that says midconvex functions, like convex functions, are continuous on an open set if they are bounded above on a neighborhood of just one point of the set. The difference is, of course, that unlike convex functions, a midconvex function with its domain in a finite-dimensional space need not be bounded from above in any neighborhood.

> **Theorem C.** Let $f$ be midconvex on an open set $U$ in a normed linear space $\mathbf{L}$. If $f$ is bounded above in a neighborhood of a single point $\mathbf{x}_0 \in U$, then $f$ is continuous and hence convex on $U$.

**Proof.** We first show that if a midconvex function $f$ is bounded above in a neighborhood of some point, then it is continuous at that point. Following the procedure discussed in Section 40, we may for convenience take the point to be $\mathbf{O}$, supposing $f$ to be bounded above on $N_r(\mathbf{O})$ by $B$ and normalized by $f(\mathbf{O}) = 0$. For a rational choice of $\varepsilon \in (0, 1)$ and an $\mathbf{x}$ such that $\|\mathbf{x}\| < \varepsilon r$, the expressions

$$\mathbf{x} = (1 - \varepsilon)\mathbf{O} + \varepsilon \left(\frac{\mathbf{x}}{\varepsilon}\right)$$

$$\mathbf{O} = \frac{\varepsilon}{1 + \varepsilon}\left(\frac{-\mathbf{x}}{\varepsilon}\right) + \frac{1}{1 + \varepsilon}\mathbf{x}$$

and the midconvexity of $f$ together with $f(\mathbf{O}) = 0$ enable us to write

$$f(\mathbf{x}) \leq \varepsilon f\left(\frac{\mathbf{x}}{\varepsilon}\right) \leq \varepsilon B$$

$$0 \leq \frac{\varepsilon}{1 + \varepsilon} B + \frac{1}{1 + \varepsilon} f(\mathbf{x})$$

Thus, $\|\mathbf{x}\| < r\varepsilon$ implies $|f(\mathbf{x})| \leqslant \varepsilon B$, showing $f$ to be continuous at $\mathbf{x} = \mathbf{O}$.

We now return to our theorem, taking $f$ be bounded from above by $B$ on $N_r(\mathbf{O})$. We will show $f$ to be bounded from above in a neighborhood of any other point $\mathbf{y} \neq \mathbf{O}$ in $U$, using a construction that parallels exactly the one used in the proof of Theorem 41A. Choose a rational $\rho > 1$ so that $\mathbf{z} = \rho\mathbf{y} \in U$ and let $\lambda = 1/\rho$. Then

$$M = \{\mathbf{v} \in \mathbf{L}: \ \mathbf{v} = (1 - \lambda)\mathbf{x} + \lambda\mathbf{z}, \mathbf{x} \in N\}$$

is a neighborhood of $\lambda\mathbf{z} = \mathbf{y}$ with radius $(1 - \lambda)r$ (Fig. 41.1). Moreover,

$$f(\mathbf{v}) \leqslant (1 - \lambda)f(\mathbf{x}) + \lambda f(\mathbf{z}) \leqslant B + f(\mathbf{z})$$

That is, $f$ is bounded above on $M$; and by the first remark of this proof, $f$ is continuous at $\mathbf{y}$. •

## PROBLEMS AND REMARKS

**A.** The first step in our proof of Theorem A was to prove

$$f\left(\frac{\mathbf{x}_1 + \cdots + \mathbf{x}_n}{n}\right) \leqslant \frac{1}{n}[f(\mathbf{x}_1) + \cdots + f(\mathbf{x}_n)] \qquad (*)$$

Artin [1964, p. 5] uses an unusual induction proof to obtain this result.

(1) If (*) holds for $n$, then it holds for $2n$.
(2) If (*) holds for $n + 1$, then it holds for $n$.
(3) Thus, (*) holds for all $n$.

**B.** Let $f$ be midconvex on an open convex set $U \subseteq \mathbf{L}$. If $\{(\mathbf{x}, z): \ \mathbf{x} \in U, z > f(\mathbf{x})\}$ is open in $\mathbf{L} \times \mathbf{R}$, then $f$ is continuous [Valentine, 1964, p. 130].

**C.** We say that $f: \mathbf{L} \to \mathbf{R}$ is a **gauge function** [**rational gauge function**] if

$$f(\mathbf{x} + \mathbf{y}) \leqslant f(\mathbf{x}) + f(\mathbf{y}); \qquad f(\alpha\mathbf{x}) = \alpha f(\mathbf{x})$$

for all $\mathbf{x}, \mathbf{y} \in \mathbf{L}$ and $\alpha \geqslant 0$ [$\alpha \geqslant 0$, $\alpha$ rational].

(1) $f$ is a gauge function $\Leftrightarrow f$ is a continuous rational gauge function.
(2) The gauge function of a convex set (defined in Problem 41E) is a gauge function in the sense above, but not conversely.
(3) It is clear from the definition that gauge functions are closely related to norms, (See Taylor [1958, pp. 134–137] in this regard.) The function $f: \mathbf{R}^2 \to \mathbf{R}$ defined by $f(r, s) = (r^2 + s^2)^{1/2} + r$ is a gauge but not a norm. The sum of a norm and a linear function is a gauge function.
(4) A gauge function is convex.

## 71. Midconvex Functions on a Normed Linear Space

**D.** We say $f: L \to R$ is an **additive function** if $f(\mathbf{x} + \mathbf{y}) = f(\mathbf{x}) + f(\mathbf{y})$ for all $\mathbf{x}, \mathbf{y} \in L$, and that it is a **Jensen function** if

$$f\left(\frac{\mathbf{x} + \mathbf{y}}{2}\right) = \tfrac{1}{2} f(\mathbf{x}) + \tfrac{1}{2} f(\mathbf{y})$$

The very extensive literature dealing with these classes of functions is summarized or otherwise indicated by Aczel [1966, pp. 43–48].

(1) $f: L \to R$ is additive $\Leftrightarrow f$ satisfies Jensen's equality for all rational $\alpha_i$.
(2) $f: L \to R$ is a Jensen function $\Leftrightarrow f$ satisfies Jensen's equality for rational $\alpha_i$ such that $\sum_1^n \alpha_i = 1 \Leftrightarrow f(\mathbf{x}) = g(\mathbf{x}) + b$ where $g$ is additive.
(3) A continuous additive function is linear.

**E.** Discontinuous additive functions (therefore discontinuous midconvex functions) exist. The construction of such functions depends on the notion of a so-called **Hamel basis** [Hamel, 1905] which is a basis $B$ for the linear space of real numbers over the scalar field of rational numbers. That is, $B$ is a subset of the real numbers with the properties

(i) if $b_1, \ldots, b_n \in B$ and $\alpha_1, \ldots, \alpha_n$ are rational, then $\alpha_1 b_1 + \cdots + \alpha_n b_n = 0 \Rightarrow \alpha_1 = \cdots = \alpha_n = 0$;
(ii) corresponding to any $x \in R$, there is a finite set of elements of $B$, say $b_1, \ldots, b_k$, and a set of rational numbers $\alpha_1, \ldots \alpha_k$ such that $x = \alpha_1 b_1 + \cdots + \alpha_k b_k$.

Such a basis exists as we might expect from Section 21; a proof may be based on Zorn's lemma [Taylor, 1958, pp. 172–173]. Given such a basis $B$, we may construct the desired additive function $f$ as follows. Assign values to $f$ on $B$ arbitrarily. For any other $x \in R$, expressed in the form guaranteed by (ii), define

$$f(x) = \alpha_1 f(b_1) + \cdots + \alpha_k f(b_k)$$

Then $f$ is additive and it will be discontinuous unless the values on $B$ are selected so that $f(b) = m_0 b$ for some $m_0 \in R$, $b \in B$.

**F.** Let $\Lambda = \{\boldsymbol{\lambda} = (\lambda_1, \ldots, \lambda_n): \lambda_i \in R, n = 1, 2, \ldots\}$, and denote with subscripts on $\Lambda$ the conditions (1), (2), (3) used in Theorem B. Thus,

$$\Lambda_{12} = \left\{\boldsymbol{\lambda} \in \Lambda: \boldsymbol{\lambda} \text{ satisfies (1) } \lambda_i \geq 0, (2) \sum_1^n \lambda_i = 1\right\}$$

(1) $f$ satisfies Jensen's inequality for all

$$\begin{aligned}
\boldsymbol{\lambda} \in \Lambda &\Leftrightarrow f \in L && \text{(linear)} \\
\boldsymbol{\lambda} \in \Lambda_1 &\Leftrightarrow f \in G && \text{(gauge)} \\
\boldsymbol{\lambda} \in \Lambda_2 &\Leftrightarrow f \in \text{Af} && \text{(affine)} \\
\boldsymbol{\lambda} \in \Lambda_3 &\Leftrightarrow f \in \text{Ad} && \text{(additive)} \\
\boldsymbol{\lambda} \in \Lambda_{12} &\Leftrightarrow f \in C && \text{(convex)} \\
\boldsymbol{\lambda} \in \Lambda_{13} &\Leftrightarrow f \in \text{RG} && \text{(rational gauge)} \\
\boldsymbol{\lambda} \in \Lambda_{23} &\Leftrightarrow f \in J && \text{(Jensen)} \\
\boldsymbol{\lambda} \in \Lambda_{123} &\Leftrightarrow f \in M && \text{(midconvex)}
\end{aligned}$$

(2) Let $V, W$ denote subsets of $\{1, 2, 3\}$, and let $F_V$ be the class of functions satisfying Jensen's inequality for all $\boldsymbol{\lambda} \in \Lambda_V$. Understanding no subscript to correspond to $\varnothing$, we have

(a) $V \subseteq W \Rightarrow F_V \subseteq F_W$,
(b) $V \subset W \Rightarrow F_V \subset F_W$,
(c) $F_V \cap F_W = F_{V \cap W}$.

(3) Parts (1) and (2) establish the validity of the dual Venn diagrams of Fig. 71.1.

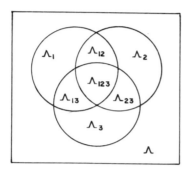
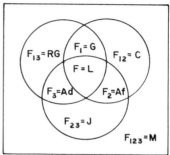

**Fig. 71.1**

*G. Let $f: U \to \mathbf{R}$ be defined on the open set $U \subseteq \mathbf{R}^n$. If for each $\mathbf{x} \in U$ and $\lambda > 0$ chosen so that $(\mathbf{x} + \lambda \mathbf{e}_i) \in U$, the difference $f(\mathbf{x} + \lambda \mathbf{e}_i) - f(\mathbf{x})$ has the same sign (or is zero), and if this is true for each standard basis vector $\mathbf{e}_i$, $i = 1,..., n$, we say $f$ is **partially monotone** on $U$. It is partially monotone at $\mathbf{x} \in U$ if there is a neighborhood of $\mathbf{x}$ on which it is partially monotone. If $f$ is midconvex on the open set $U \subseteq \mathbf{R}^n$, it is continuous if and only if $f$ is partially monotone at some $\mathbf{x} \in U$. This result, due to Bereanu [1969], has been generalized by Kuczma [1970a].

## 72. Midconvex Functions on R

We have seen (Problem 71E) that a midconvex function defined on $\mathbf{R}$ need not be continuous. The example given seems contrived, but there is a good reason for this. The thrust of the theorems in this section is that any discontinuous midconvex function must be a wild function indeed. It is our purpose to prove the principal theorems that lend substance to this assertion. Beyond this, we wish to illustrate the way in which mathematical knowledge grows by tracing the historical development of results related to these theorems.

In Jensen's pioneering studies [1905, 1906], he proved that if a midconvex function is bounded above on $(a, b)$, then $f$ is continuous there. By giving careful attention to both upper and lower bounds, Bernstein and Doetsch were able to prove [1915] the sharpened form of Jensen's result that we state as our first theorem.

## 72. Midconvex Functions on R

**Theorem A.** Let $f$ be midconvex on $(a, b)$. If $f$ is bounded from above in a neighborhood of a single point $x_0 \in (a, b)$, then $f$ is continuous on $(a, b)$.

**Proof.** This is a special case of Theorem 71C. ●

Theorem A has been proved by numerous methods; see Hardy, Littlewood, and Polya [1952, p. 92]. The work of Bernstein and Doetsch is outlined in Problems A through H. Their proof of Theorem A should be contrasted with the proof we have given to Theorem 71C. Our proof, due to Bourbaki [1953, p. 92], shows progress not only in its greater generality, but also in its simplicity. This is not to disparage either the work of Bernstein and Doetsch or the value of studying what they did. Indeed, it was just such study, indicated in Problems G and H, that led Ostrowski [1929a] to our next theorem.

**Theorem B.** Let $f$ be midconvex on $(a, b)$ and suppose there is a set $M \subseteq (a, b)$ having positive measure on which $f$ is bounded above. Then $f$ is continuous on $(a, b)$.

**Proof.** There is no loss of generality in assuming the upper bound of $f$ on $M$ to be 0. The set $M$ of positive measure can be covered by a set $J_n$ of nonoverlapping open intervals of $(a, b)$ such that

$$m(M) \leq \sum_1^\infty m(J_n) < \tfrac{4}{3} m(M)$$

Since the sets $M \cap J_n$ are all disjoint and measurable,

$$\sum_1^\infty m(M \cap J_n) = m \bigcup_1^\infty (M \cap J_n) = m(M),$$

so it is clear that there must be at least one $J_n$, call it $J$, such that

$$0 < m(M \cap J) \leq m(J) < \tfrac{4}{3} m(M \cap J)$$

Let the interval $J = (c, d)$ have midpoint $x_0$. Suppose $f$ is not bounded from above in any neighborhood of $x_0$. We may then choose $x_1$ such that

$$|x_1 - x_0| < \varepsilon = \frac{m(M \cap J)}{6} \quad \text{and} \quad f(x_1) > 1$$

Define
$$N = \{y: \ y = 2x_1 - x \text{ where } x \in (M \cap J)\}$$

Since $N$ is a reflection and a translation of $M \cap J$, it is clear that $m(N) = m(M \cap J) = 6\varepsilon$. For any $y \in N$,

$$|y - x_0| = |2x_1 - x - x_0| = |2(x_1 - x_0) + x_0 - x|$$
$$\leqslant 2|x_1 - x_0| + |x_0 - x| < 2\varepsilon + |x_0 - x|$$

so $N \subset (c - 2\varepsilon, d + 2\varepsilon)$, the containment being proper. Since $m(N) = 6\varepsilon$, $m(N \cap J) \geqslant 2\varepsilon$. Finally we note that $N \cap M = \varnothing$ since $f(x_1) \leqslant \frac{1}{2}[f(x) + f(y)] \leqslant \frac{1}{2}f(y)$; hence $f(y) \geqslant 2$. Putting all this together gives

$$m(J) \geqslant m[(M \cap J) \cup (N \cap J)] = m(M \cap J) + m(N \cap J)$$
$$\geqslant 6\varepsilon + 2\varepsilon = \tfrac{4}{3}m(M \cap J)$$

This contradicts the way in which $J$ was chosen, showing that $f$ must be bounded above on some neighborhood of $x_0$. Continuity of $f$ now follows from Theorem A. ●

The results of Jensen, Bernstein and Doetsch, and Ostrowski are the first entries to a sequence of theorems that have appeared over a 60-year period. All the theorems to which we refer draw conclusions about the continuity of midconvex functions known to be bounded on some subset of their domains.

*The midconvex function $f$ defined on $[a, b]$ will be continuous (hence convex) on $(a, b)$ if*

| | | |
|---|---|---|
| 1905 | $f$ is bounded above on $(a, b)$. | Jensen |
| 1915 | $f$ is bounded above on any nonempty open subset. | Bernstein and Doetsch |
| 1929 | $f$ is bounded above on a set $M$ with Lebesgue measure $m(M) > 0$. | Ostrowski [1929a] |
| 1956 | $f$ is bounded above on a set $M$ where $M + M = \{m_1 + m_2: \ m_i \in M\}$ has inner Lebesgue measure $m_*(M + M) > 0$. | Kurepa |
| 1957 | $f$ is bounded above on a set $M$ where for some $n$, $m_*(\sum_1^n M) > 0$. | Kemperman Marcus [1957a, 1959b] |
| 1959 | $f$ is bounded above on a set $M$ where $$m_*\left[\bigcup_{n=0}^{\infty} \frac{1}{2^n} \sum_{j=1}^{2^n} M\right] > 0$$ | Kuczma |

| 1964 | $f$ is bounded above on a second category Baire set. | Mehdi |
| 1970 | $f$ is bounded above on a set $T$ where $T$ is a set such that every additive function bounded above on $T$ is continuous. | Kuczma [1970b] |

Some mathematicians might be inclined to cite the later results as curious occupation with minutia rather than as examples of expanding mathematical knowledge. We too have expressed an opinion about the relative importance of the results by stating only two of the earliest ones as theorems. Nevertheless, we do have an indication of the way one idea suggests another, and as Ger showed [1969] with examples, the results of 1957 and 1959 (building on that of 1956) both represented improvements over what had previously been known.

We should also mention that these theorems gave rise to many results that do not fit neatly into our list. Mohr [1952], Csaszar [1958], and Marcus [1959a, b] have proved similar results for midconvex functions defined on $U \subseteq \mathbf{R}^n$ (Problem K). Deak [1962] developed a parallel set of theorems for functions he called $p$-convex, that is, for functions satisfying

$$f[px + (1-p)y] \leqslant pf(x) + (1-p)f(y)$$

for some $p \in (0, \tfrac{1}{2}]$ and all $x, y \in (a, b)$. Mehdi's work shifted attention from measure theoretic considerations to topological ones, making it possible to carry his work over to arbitrary topological spaces.

All the results mentioned so far establish continuity of $f$ from an analysis of a subset of $(a, b)$ on which $f$ is bounded from above. Our next theorem is of a different type. We show that measurability and the midconvexity of $f$ guarantee its continuity. This result was apparently first proved by Blumberg [1919], and independently by Sierpinski [1920].

**Theorem C.** If $f: (a, b) \to \mathbf{R}$ is measurable and midconvex, then $f$ is continuous on $(a, b)$.

**Proof.** Let us suppose that $f$ is not continuous, anticipating a contradiction. Choose $x_0 \in (a, b)$ and $c$ such that $(x_0 - 2c, x_0 + 2c) \subseteq (a, b)$, and let $B_n = \{x \in (a, b): f(x) > n\}$, a measurable set. For fixed $n$, select $u \in B_n \cap (x_0 - c, x_0 + c)$, something we can certainly do according to Theorem A. Now for any $\lambda \in [0, 1]$,

$$n < f(u) = f\left[\frac{u + \lambda c}{2} + \frac{u - \lambda c}{2}\right] \leqslant \tfrac{1}{2}[f(u + \lambda c) + f(u - \lambda c)]$$

It follows that either $f(u + \lambda c) > n$ or $f(u - \lambda c) > n$. Thus, either $u + \lambda c$ or $u - \lambda c$ is in $B_n$; equivalently, if $M_n = \{x: x = y - u, y \in B_n\}$, then either $\lambda c$ or $-\lambda c$ is in $M_n$ for all $\lambda \in [0, 1]$. This implies, as we shall show below, that $c \leqslant m(M_n) = m(B_n)$. Assuming this fact and noting that $B_1 \supseteq B_2 \supseteq \cdots$, we have by a standard theorem from measure theory [Natanson I, 1961, p. 70],

$$c \leqslant \lim_{n \to \infty} m(B_n) = m\left(\bigcap_1^\infty B_n\right)$$

and so $\bigcap_1^\infty B_n \neq \varnothing$. This means that there is a point $v \in (a, b)$ such that $f(v) > n$ for every $n$, an impossibility.

To complete the proof, we consider a measurable set $M$ for which $-\lambda c$ or $\lambda c$ is in $M$ for each $\lambda \in [0, 1]$. Set $A_1 = M \cap [-c, 0]$, $A_2 = M \cap [0, c]$. Then $-A_1 \cup A_2 = [0, c]$ and so

$$c = m[0, c] \leqslant m(-A_1) + m(A_2) = m(A_1) + m(A_2)$$
$$= m(A_1 \cup A_2) \leqslant m(M) \quad \bullet$$

We have used *midconvex* in this section to describe functions defined on an interval $I$ so as to satisfy

$$f\left(\frac{x+y}{2}\right) \leqslant \tfrac{1}{2}[f(x) + f(y)] \tag{1}$$

We have seen that such functions need not be continuous, but that when they are, they will be *convex*; that is for all $\alpha \in [0, 1]$,

$$f[\alpha x + (1 - \alpha)y] \leqslant \alpha f(x) + (1 - \alpha) f(y) \tag{2}$$

Many of the early writers and some more recent ones use *convex* to describe functions only known to satisfy (1). One must therefore check the definition being used when reading the literature, a practice made even more advisable by the appearance of several other definitions of convex that are used for functions defined on **R**.

Artin [1964, p. 1] says $f$ is convex on an interval $I$ if for all distinct triples $x_1, x_2, x_3 \in I$, the symmetric function

$$\psi(x_1, x_2, x_3) = \frac{(x_3 - x_2)f(x_1) + (x_1 - x_3)f(x_2) + (x_2 - x_1)f(x_3)}{(x_1 - x_2)(x_2 - x_3)(x_3 - x_1)}$$

remains nonnegative. We have seen that this is equivalent to our own definition (Problem 11I).

## 72. Midconvex Functions on R

Wright [1954], looking for functions satisfying a certain set of inequalities, calls a function convex if for each $x_2 > x_1$ and $\delta > 0$,

$$f(x_1 + \delta) - f(x_1) \leqslant f(x_2 + \delta) - f(x_2) \tag{3}$$

The relationship of Wright convexity (3) to the definitions described by (1) and (2) is taken up in Problem N.

### PROBLEMS AND REMARKS

Problems A through H outline the paper [Bernstein and Doetsch, 1915]. Assume throughout that $f$ is a midconvex function defined on $[a, b]$.

**A.** Call $r$ a rational division point of $[a, b]$ if $(b - r)/(b - a)$ is rational. For such a point, $f(r) \leqslant \max\{f(a), f(b)\}$. Suppose $\delta \in (0, (b - a)/2)$ and $Ra$ is the set of rational division points of $[a, b]$. Then $f$ is uniformly continuous on $D = [a + \delta, b - \delta] \cap Ra$, and there exists a function $g$ that is midconvex and continuous on $(a, b)$ such that for each $r \in Ra$, $g(r) = f(r)$. (*Hint*: Once the uniform continuity of $f$ is established on $D$, the existence of a continuous extension $g$ follows from the Tietze extension theorem [Buck, 1965, p. 86].)

**B.** If $f$ is not bounded above in $(a, b)$, it is not bounded above in any nonempty open subinterval of $(a, b)$.

**C.** If $f$ is not bounded below on $(a, b)$, it is not bounded below in any nonempty open subinterval of $(a, b)$.

**D.** If $f$ is bounded above in $(a, b)$, then $f$ is bounded below in $(a, b)$.

Suppose that $f$ is bounded from below on $(a, b)$. For $x \in (a, b)$ and $\delta > 0$ chosen so that $[x - \delta, x + \delta] \subset (a, b)$, there is a lower bound on the closed interval. Since any such lower bound is less than or equal to $f(x)$, there is a greatest lower bound that we denote by $m(x, \delta)$. Moreover,

$$\lim_{\delta \to 0} m(x, \delta) = m(x)$$

exists since $m(x, \delta)$ is monotone increasing with $\delta$ and bounded above by $f(x)$. Call $m(x)$ the **lower bounding function** for $f$ on $(a, b)$.

**E.** The lower bounding function is a continuous midconvex (that is, a convex) function on $(a, b)$.

**F.** Suppose $f$, bounded from below, fails to agree with its lower bounding function $m(x)$ in at least one point of $(a, b)$. Then $f$ is unbounded from above.

These assertions established, Bernstein and Doetsch proved Theorem A as follows. Suppose $f$ is bounded from above on some nonempty open subinterval of $(a, b)$. Then $f$ is bounded above on all of $(a, b)$ by Problem B, hence bounded below on $(a, b)$ by Problem D. It follows from Problem F that $f$ agrees with $m$ on $(a, b)$, known (Problem E) to be continuous.

They go on with two more observations important to later developments.

**G.** Let $f$ be discontinuous on $(a, b)$. If $f$ is bounded from below, then $V = \{(x, f(x)): x \in (a, b)\}$ is dense in the set $S = \{(x, y): x \in [a, b], y \geq m(x)\}$ where $m$ is the lower bounding function of $f$.

**H.** If in Problem G, $f$ is not bounded from below, then the set $V$ is dense in $S = \{(x, y): x \in [a, b]\}$.

Problems G and H fully vindicate our assertion that a discontinuous midconvex function is wild. Contemplating these same two results, Ostrowski observed that if $f$ was a discontinuous midconvex function that assumed the value $k_0$, and if $k_0 < k_1 < k_2$, then $f$ must take a value in $(k_1, k_2)$. Phrased another way, if $f$ takes on the value $k_0$, and if there exist values of $k_1$ and $k_2$, $k_0 < k_1 < k_2$, for which the set

$$M = \{x \in (a, b): k_1 < f(x) < k_2\}$$

is empty, then $f$ is continuous. This sets up his generalization [Ostrowski, 1929b] of these results.

**I.** Suppose $f$ is midconvex on $(a, b)$, assumes the value $k_0$, and that for some choice of $k_1$, $k_2$, $k_0 < k_1 < k_2$, the set $M$ described above has measure zero. Then $f$ is continuous.

**J.** A midconvex function may be bounded below, yet not bounded above and not continuous. We cannot, therefore, hope to prove an analog to Theorem B for functions bounded below, but we can go this far. If a midconvex function is bounded below on a set of positive measure, then it is bounded below on $(a, b)$ [Hukuhara, 1954].

**K.** Many of the results stated in this section can be proved for a midconvex function $f$ defined in the open set $U \subseteq \mathbf{R}^n$. Mohr [1952] proved an $n$-dimensional version of Theorem A, a result that now appears as a special case of our Theorem 71C.

(1) If $U \subseteq \mathbf{R}^n$ and $f: U \to \mathbf{R}$ is midconvex on $U$, then $f$ bounded above on a set $E \subseteq U$ with $m(E) > 0$ implies that $f$ is continuous.

(2) If $U \subseteq \mathbf{R}^n$ is bounded and $f: U \to \mathbf{R}$ is midconvex on $U$, then $f$ bounded below on a set $E \subseteq U$ with $m(E) > 0$ implies that $f$ is bounded below on $U$.

The proofs of these and similar theorems are given by Marcus [1959b].

**L.** We refer to the function of Theorem 62A defined by $g(t) = M_t(\mathbf{x}, \boldsymbol{\alpha})$. Hardy, Littlewood, and Polya [1952, p. 72] establish the convexity of $g$ by first showing that it is midconvex, and then appealing to the continuity of $M_t(\mathbf{x}, \boldsymbol{\alpha})$.

**M.** Theorem C can be proved for midconvex functions not assumed finite on all of $(a, b)$, requiring instead that they be finite almost everywhere on $(a, b)$ [Hirschman and Widder, 1955, p. 120].

**N.** Let $C$, $W$, and $M$ be the sets of convex, Wright-convex, and midconvex functions. Prove that $C \subset W \subset M$, each inclusion being proper. Whether or not the last inclusion is proper was posed as a question by Wright [1954] and answered by Kenyon [1956] and Klee [1956].

## REFERENCES

E. Artin (1964). "The Gamma Function." Holt, New York.

F. Bernstein and G. Doetsch (1915). Zur Theorie der konvexen Funktionen. *Math. Ann.* **76**, 514–526.

H. Blumberg (1919). On convex functions. *Trans. Amer. Math. Soc.* **20**, 40–44.
N. Bourbaki (1953). "Espaces Vectorial Topologiques." Hermann, Paris.
A. Csaszar (1958). Sur les ensembles et les fonctions convexes. *Mat. Lapok.* **9**, 273–282.
E. Deak (1962). Über konvexe und interne Funktionen, sowie eine gemeinsame Verallgemeinerung von beiden. *Ann. Univ. Sci. Budapest. Eötvös Sect. Math.* **5**, 109–154.
R. Ger (1969). Some remarks on convex functions. *Fund. Math.* **66**, 255–262.
G. H. Hardy, J. E. Littlewood, and G. Polya (1952). "Inequalities." Cambridge Univ. Press, London and New York.
I. I. Hirschman and D. V. Widder (1955). "The Convolution Transform." Princeton Univ. Press, Princeton, New Jersey.
M. Hukuhara (1954). Sur la fonction convexe. *Proc. Jap. Acad.* **30**, 683–685.
J. L. W. V. Jensen (1905). Om konvexe Funktioner og Uligheder mellem Middelvaerdier. *Nyt. Tidsskr. Math.* **16B**, 49–69.
J. L. W. V. Jensen (1906). Sur les fonctions convexes et les inegalités entre les valeurs moyennes. *Acta Math.* **30**, 175–193.
F. B. Jones (1942). Connected and disconnected plane sets and the functional equation $f(x) + f(y) = f(x + y)$. *Bull. Amer. Math. Soc.* **48**, 115–120.
J. H. B. Kemperman (1957). A general functional equation. *Trans. Amer. Math. Soc.* **86**, 28–56.
H. Kenyon (1956). Note on convex functions. *Amer. Math. Monthly* **63**, 107.
V. L. Klee (1956). Solution of a problem of E. M. Wright on convex functions. *Amer. Math. Monthly* **63**, 106–107.
M. E. Kuczma (1959). Note on convex functions. *Ann. Univ. Sci. Budapest. Eötvös Sect. Math.* **2**, 25–26.
M. E. Kuczma (1970a). Some remarks on convexity and monotonicity. *Rev. Roumaine Math. Pures Appl.* **15**, 1463–1469.
M. E. Kuczma (1970b). On discontinuous additive functions. *Fund. Math.* **66**, 384–392.
S. Kurepa (1956). Convex functions. *Glas. Mat. Fiz. Astron.* [2], **11**, 89–93.
S. Marcus (1957a). Critères de majoration pour les fonctions sous additives, convexes ou internes. *C.R. Acad. Sci. Ser. A-B* **244**, 2270–2272.
S. Marcus (1957b). Fonctions convexes et fonctions internes. *Bull. Sci. Math.* [2], **81**, 66–70.
S. Marcus (1957c). Sur un théorème de F. B. Jones. Sur un théorème de S. Kurepa. *Bull. Math. Soc. Sci. Math. Phys. R.P. Roumaine* (N.S.) [1], **49**, 433–434.
S. Marcus (1959b). Généralisation, aux fonctions de plusieurs variables, des théorèmes de Alexander Ostrowski et des Masuo Hukuhara concernant les fonctions convexes. *J. Math. Soc. Jap.* **11**, 171–176.
M. R. Mehdi (1964). Some remarks on convex functions. *J. London Math. Soc.* **39**, 321–326.
E. Mohr (1952). Beitrag zur theorie der konvexen Funktionen. *Math. Nachr.* **8**, 133–148.
A. Ostrowski (1929a). Über die Funktionalgleichung der Exponentialfunktionen und verwandte Funktionalgleichungen. *Jahresber. Deut. Math. Ver.* **38**, 54–62.
A. Ostrowski (1929b). Zur Theorie der konvexen Funktionen. *Comment. Math. Helv.* **1**, 157–159.
W. Sierpinski (1920). Sur les fonctions convexes mesurables. *Fund. Math.* **1**, 125–129.
E. M. Wright (1954). An inequality for convex functions. *Amer. Math. Monthly* **61**, 620–622.

# VIII

## Related Classes of Functions

> There is probably no other science which presents such different appearances to one who cultivates it and one who does not, as mathematics. To (the noncultivator) it is ancient, venerable, and complete; a body of dry, irrefutable, unambiguous reasoning. To the mathematician, on the other hand, his science is yet in the purple bloom of vigorous youth, everywhere stretching out after the "attainable but unattained," and full of the excitement of nascent thoughts.
>
> C. H. CHAPMAN

> Another characteristic of mathematical thought is that it can have no success where it cannot generalize.
>
> C. S. PIERCE

## 80. Introduction

We have come to the last chapter of our book. If we were to draw analogy with the erection of a building, then this final chapter ought to be the roof that ties the walls together and gives the structure a completed look. If our analogy were to a picture, then our conclusion should consist of those final deft strokes of the brush that give depth and beauty to the subject. But neither of these analogies is the right one. Our subject is too much alive and branches off in too many directions to be neatly wrapped up and finished off.

It is to the figure of a living growing tree that we turn for the proper analogy. Chapter I (Convex Functions on the Real Line) describes the seed, the germinating idea from which the rest has sprung. Chapters II and III (Normed Linear Spaces and Convex Sets) tell about the soil in which the mature tree grows. Chapter IV (Convex Functions on a Normed Linear Space) deals with the trunk, the basic core of the subject, and the support of all that follows. Chapters V, VI, and VII (Optimization, Inequalities, and Midconvex Functions) discuss branches which, having appeared at the earliest stages of life, are large and well developed. Now it is time to point out that there are other branches on our still growing tree.

In scanning these other branches, many of which reach over into other areas, we have chosen to direct attention to relationships with the rest of the tree and to note the direction of growth. It would take too much space to prove all that we shall notice in each of the branches; indeed some of them are themselves the subjects of small books. Thus the character of our last chapter is different from the rest of the book. With motivation and exposition of the main ideas as our goal, we give definitions, show the relation to convex functions, state results, and give references to the literature.

Quasiconvex functions form a branch that has exhibited recent spurts of growth. With the help of a survey article [Greenberg and Pierskalla, 1971] very much in the spirit of this chapter, we summarize facts about quasiconvex functions in a two-column chart in Section 81 that exhibits the parallel with convex functions.

Twice differentiable functions are convex if $f''(x) \geqslant 0$. Functions having derivatives of all orders are called completely convex if the even derivatives of all orders have prescribed signs. We take up this class of functions in Section 82.

Sections 83, 84, and 85 are related. In 83 we discuss functions convex of order $n$, a topic popularized by Popoviciu. An aesthetically pleasing development, embodying as it does the essential features of historic papers on convex functions, is pointed out in Section 84 where we treat the generalized convex functions introduced by Beckenbach. We then show in Section 85 how the topics of the previous two sections appear as special cases of a more abstract class of generalized convex functions.

Finally, in Section 86 we briefly mention still other classes of functions that have their genesis in the study of convex functions.

## 81. Quasiconvex Functions

Given a function $f: U \to \mathbf{R}$, $U \subseteq \mathbf{L}$, we have had several occasions to consider the **level sets**

$$L_\alpha = \{\mathbf{x} \in U: f(\mathbf{x}) \leqslant \alpha\}$$

For example, in our study of conjugate convex functions, our interest centered on whether $L_\alpha$ was closed for each real $\alpha$. We focus now on the simple observation that if $f$ is convex, then $L_\alpha$ is convex for each $\alpha$. De Finetti [1949] asked the obvious question: What about the converse? It led him to study the class of functions that are now called quasiconvex.

To set the stage for a more precise formulation of de Finetti's question, we follow Fenchel [1953] in making some remarks about level sets for an arbitrary (not necessarily convex) function $f: U \to \mathbf{R}$. For such a function, let $J$ be the smallest interval (open or closed, finite or infinite) containing the entire range of $f$. To exclude the trivial case of a constant function, we shall assume that $J$ has interior points. The following facts are evident.

$$\bigcup_{\alpha \in J} L_\alpha = U \tag{1}$$

If $\alpha, \beta \in J$ and $\alpha \leqslant \beta$, then $L_\alpha \subseteq L_\beta$ \qquad (2)

$$\bigcap_{\alpha > \beta} L_\alpha = L_\beta \quad \text{if} \quad \beta \in J \tag{3}$$

$$\bigcap_{\alpha \in J} L_\alpha = \varnothing \quad \text{if} \quad J \text{ does not contain a lower bound} \tag{4}$$

Figure 81.1 shows some of the level sets for a function $w = f(r, s)$ that has a surface of revolution as its graph.

So far, our steps are reversible; that is, given a family $\{L_\alpha\}$ of sets indexed by the real numbers of some interval $J$ and satisfying (1), (2),

## 81. Quasiconvex Functions

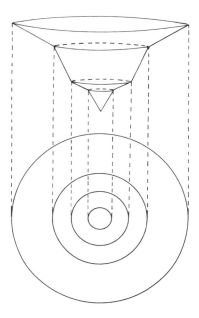

**Fig. 81.1**

(3), and (4), we may construct a function $f: U \to \mathbf{R}$ having the $L_\alpha$ as its level sets. We simply define $f$ by

$$f(\mathbf{x}) = \inf\{\alpha \in J: \mathbf{x} \in L_\alpha\}$$

and demonstrate (see Fenchel [1953, p. 116]) that

$$L_\alpha = \{\mathbf{x} \in U: f(\mathbf{x}) \leq \alpha\}$$

With $f$ now constructed from sets $\{L_\alpha\}$ satisfying (1) through (4), it is natural to wonder what additional knowledge about the sets $L_\alpha$ will enable us to draw conclusions about $f$. It is known, for example, that if the $L_\alpha$ are all closed, then $f$ will be lower semicontinuous [Natanson I, 1961, p. 152]. The question of de Finetti is of this type. If we add the assumption

$$L_\alpha \text{ is convex for each } \alpha \in J \tag{5}$$

what can we say about the corresponding function? It is plain from Fig. 81.1, where all the level sets of a nonconvex function are concentric circles, that we cannot conclude that $f$ is convex. Although both de Finetti [1949] and Fenchel [1953, 1956] have given further restrictions on the family $\{L_\alpha\}$ which together with (5) guarantee the convexity of $f$, our interest focuses on condition (5) taken without further restrictions.

If $U$ is a convex set in $\mathbf{L}$, we call $f: U \to \mathbf{R}$ **quasiconvex** if the level sets $L_\alpha$ are convex for all $\alpha \in \mathbf{R}$. What properties must such a function have?

**Theorem A.** A function $f$ defined on the convex set $U \subseteq \mathbf{L}$ is quasiconvex if and only if for all $\mathbf{x}, \mathbf{y} \in U$ and $\lambda \in [0, 1]$.

$$f[\lambda\mathbf{x} + (1 - \lambda)\mathbf{y}] \leqslant \max\{f(\mathbf{x}), f(\mathbf{y})\} \qquad (6)$$

**Proof.** We suppose first that the level sets are convex. For fixed $\mathbf{x}, \mathbf{y} \in U$, let $\alpha = \max\{f(\mathbf{x}), f(\mathbf{y})\}$. Then $\mathbf{x} \in L_\alpha$, $\mathbf{y} \in L_\alpha$, and since $L_\alpha$ is convex, $[\lambda\mathbf{x} + (1 - \lambda)\mathbf{y}] \in L_\alpha$. It follows that (6) holds.

On the other hand, suppose $L_\alpha$ is a level set of a function satisfying (6). If $\mathbf{x} \in L_\alpha$, $\mathbf{y} \in L_\alpha$, then of course $f(\mathbf{x}) \leqslant \alpha$, $f(\mathbf{y}) \leqslant \alpha$, and from (6), $f[\lambda\mathbf{x} + (1 - \lambda)\mathbf{y}] \leqslant \alpha$; the point $[\lambda\mathbf{x} + (1 - \lambda)\mathbf{y}]$ lies in $L_\alpha$, which means $L_\alpha$ is convex. ●

We have encountered quasiconvex functions before (Problem 42D) in the context of differentiable functions. The definition that appeared there is consistent with the present one, as we now show.

**Theorem B.** A function $f: U \to \mathbf{R}$ that is continuous and differentiable on the open convex set $U \subseteq \mathbf{L}$ is quasiconvex if and only if for each $\mathbf{x}, \mathbf{y} \in U$

$$f(\mathbf{y}) \leqslant f(\mathbf{x}) \quad \text{implies} \quad f'(\mathbf{x})(\mathbf{y} - \mathbf{x}) \leqslant 0 \qquad (7)$$

**Proof.** Suppose that $f$ is quasiconvex and that $f(\mathbf{y}) \leqslant f(\mathbf{x})$. By Theorem A, for $0 < \lambda < 1$,

$$f[(1 - \lambda)\mathbf{x} + \lambda\mathbf{y}] \leqslant \max\{f(\mathbf{x}), f(\mathbf{y})\} = f(\mathbf{x})$$

or

$$f[\mathbf{x} + \lambda(\mathbf{y} - \mathbf{x})] - f(\mathbf{x}) \leqslant 0$$

Multiplying by $1/\lambda$ and letting $\lambda \downarrow 0$ gives

$$f_+'(\mathbf{x}; \mathbf{y} - \mathbf{x}) \leqslant 0$$

provided this one-sided directional derivative exists. But $f$ is given to be differentiable, so

$$f'(\mathbf{x})(\mathbf{y} - \mathbf{x}) = f_+'(\mathbf{x}; \mathbf{y} - \mathbf{x}) \leqslant 0$$

Conversely, consider a differentiable function satisfying (7). For two arbitrary fixed points $\mathbf{x}_1, \mathbf{x}_2 \in U$ with the notation chosen so that $f(\mathbf{x}_1) \leqslant f(\mathbf{x}_2)$, define a differentiable function $\phi \colon [0, 1] \to \mathbf{R}$ by

$$\phi(\lambda) = f[(1-\lambda)\mathbf{x}_1 + \lambda\mathbf{x}_2] = f[\mathbf{x}_1 + \lambda(\mathbf{x}_2 - \mathbf{x}_1)]$$

Our problem is to show that $\phi(\lambda) \leqslant \phi(1)$ for all $\lambda \in (0, 1)$. Suppose on the contrary that $\phi(\lambda) > \phi(1)$ for some $\lambda$. Then there is also a $\lambda_0 \in (0, 1)$ for which $\phi(\lambda_0) > \phi(1)$ and $\phi'(\lambda_0) > 0$. By the chain rule (Theorem 23C).

$$\phi'(\lambda_0) = f'[\mathbf{x}_1 + \lambda_0(\mathbf{x}_2 - \mathbf{x}_1)](\mathbf{x}_2 - \mathbf{x}_1)$$
$$= f'(\mathbf{x}_0)(\mathbf{x}_2 - \mathbf{x}_1)$$

where $\mathbf{x}_0 = \mathbf{x}_1 + \lambda_0(\mathbf{x}_2 - \mathbf{x}_1)$. Now $f(\mathbf{x}_0) = \phi(\lambda_0) > \phi(1) = f(\mathbf{x}_2)$, so by (7)

$$f'(\mathbf{x}_0)(\mathbf{x}_2 - \mathbf{x}_0) \leqslant 0$$

Since $\mathbf{x}_2 - \mathbf{x}_0 = (1 - \lambda_0)(\mathbf{x}_2 - \mathbf{x}_1)$, we have

$$\phi'(\lambda_0) = f'(\mathbf{x}_0)(\mathbf{x}_2 - \mathbf{x}_1) = \frac{1}{1-\lambda_0} f'(\mathbf{x}_0)(\mathbf{x}_2 - \mathbf{x}_0) \leqslant 0$$

We have the desired contradiction. ●

Along with Theorem B, Ponstein [1967] gives several other characterizations of the class of quasiconvex functions. This class is considerably larger than the class of convex functions. It contains, in addition to the convex functions, all functions like that illustrated in Fig. 81.1, all monotone functions $f \colon I \to \mathbf{R}$, and many others. Yet most of the properties of convex functions have their counterparts for quasiconvex functions, an observation nicely illustrated in the accompanying table, taken with the permission of the authors from the previously mentioned survey article by Greenberg and Pierskalla [1971]. Here we suppose that $f \colon U \to \mathbf{R}$ where $U$ is a convex set in $\mathbf{R}^n$, and we use $E$ to denote the extreme points of $U$. Also, we write

$H(\mathbf{x})$ for $[f''(\mathbf{x})]$, the Hessian matrix,
$D(\mathbf{x})$ for

$$\begin{bmatrix} 0 & \nabla f(\mathbf{x}) \\ (\nabla f(\mathbf{x}))^t & H(\mathbf{x}) \end{bmatrix}$$

the bordered Hessian,
$|D_j(\mathbf{x})|$ for the $j$th principal minor of $D(\mathbf{x})$, $j = 0, \ldots, n$.
Note that $|D_0(\mathbf{x})| = 0$.

ANALOGOUS PROPERTIES

| Convex | Quasiconvex |
|---|---|
| 1a. $f$ is convex $\Leftrightarrow$ epi $f$ is a convex set | 1b. $f$ is quasiconvex $\Leftrightarrow L_\alpha$ is a convex set for all $\alpha \in \mathbf{R}$ |
| 2a. $f$ is continuous on $U^0$ | 2b. $f$ is continuous almost everywhere on $U^0$ |
| 3a. One-sided partial derivatives exist everywhere on $U^0$ | 3b. One-sided partial derivatives exist almost everywhere on $U^0$ |
| 4a. Suppose $f$ is twice continuously differentiable on $\mathbf{R}^n$. Then $f$ is convex on $\mathbf{R}^n \Leftrightarrow H(\mathbf{x})$ is nonnegative definite throughout $\mathbf{R}^n$ | 4b. Suppose $f$ is twice continuously differentiable on $\mathbf{R}^n$. If $f$ is quasiconvex on $\mathbf{R}_+^n$, then $\mid D_j(\mathbf{x}) \mid \leqslant 0$ for $j = 1,...,n$. If $\mid D_j(\mathbf{x}) \mid < 0$ for $j = 1,...,n$, then $f$ is quasiconvex on $\mathbf{R}_+^n$ |
| 5a. Suppose $f$ is continuously differentiable on $\mathbf{R}^n$. Then $f$ is convex on $\mathbf{R}^n \Leftrightarrow f(\mathbf{y}) - f(\mathbf{x}) \geqslant f'(\mathbf{x})(\mathbf{y} - \mathbf{x})$ | 5b. Suppose $f$ is continuously differentiable on $\mathbf{R}^n$. Then $f$ is quasiconvex on $\mathbf{R}^n \Leftrightarrow f(\mathbf{y}) \leqslant f(\mathbf{x}) \Rightarrow f'(\mathbf{x})(\mathbf{y} - \mathbf{x}) \leqslant 0$ |
| 6a. $\sup_{\mathbf{x} \in U} f(\mathbf{x}) = \sup_{\mathbf{x} \in E} f(\mathbf{x})$ if $U$ is compact | 6b. $\sup_{\mathbf{x} \in U} f(\mathbf{x}) = \sup_{\mathbf{x} \in E} f(\mathbf{x})$ if $U$ is compact |
| 7a. Every local minimum is a global minimum | 7b. Every local minimum is a global minimum or $f$ is constant in a neighborhood of the local minimum point |
| 8a. The set of global minimum points is convex | 8b. The set of global minimum points is convex |
| 9a. The Kuhn–Tucker constraint condition holds for $\{\mathbf{x}: \phi_j(\mathbf{x}) \leqslant 0, j = 1,...,m\}$ if there is an $\mathbf{x}$ such that $\phi_j(\mathbf{x}) < 0$ for all $j = 1,...,m$ | 9b. The Kuhn–Tucker constraint condition holds for $\{\mathbf{x}: \phi_j(\mathbf{x}) \leqslant 0, j = 1,...,m\}$ if $\nabla \phi_j(\mathbf{x}) \neq \mathbf{O}$ for $j$ such that $\phi_j(\mathbf{x}) = 0$, $j = 1,...,m$ |
| 10a. Let $U, V$ be compact subsets of $\mathbf{R}^m$ and $\mathbf{R}^n$ respectively, and let $K: U \times V \to \mathbf{R}$ be such that for each $\mathbf{y} \in V$, $K(\mathbf{x}, \cdot)$ is concave and for each $\mathbf{x} \in U$, $K(\cdot, \mathbf{y})$ is convex. Further, let $K$ be continuous on $U \times V$. Then $K$ has a saddle point $(\bar{\mathbf{x}}, \bar{\mathbf{y}}) \in U \times V$ | 10b. Let $U, V$ be compact subsets of $\mathbf{R}^m$ and $\mathbf{R}^n$ respectively, and let $K: U \times V \to \mathbf{R}$ be such that for each $\mathbf{y} \in V$, $K(\mathbf{x}, \cdot)$ is quasiconcave and upper semicontinuous, and for each $\mathbf{x} \in U$, $K(\cdot, \mathbf{y})$ is quasiconvex and lower semicontinuous. Then $K$ has a saddle point $(\bar{\mathbf{x}}, \bar{\mathbf{y}}) \in U \times V$ |
| 11a. $f(\lambda \mathbf{x}) \leqslant \lambda f(\mathbf{x})$ for $\lambda \in [0, 1]$ if $f(\mathbf{0}) \leqslant 0$ | 11b. $f(\lambda \mathbf{x}) \leqslant f(\mathbf{x})$ for $\lambda \in [0, 1]$ if $f(\mathbf{x}) \geqslant f(\mathbf{O})$ |
| 12a. $g(\lambda) = f(\lambda \mathbf{x})/\lambda$ is increasing for $\lambda > 0$ if $f(\mathbf{O}) = 0$ | 12b. $g(\lambda) = f(\lambda \mathbf{x})$ is increasing for $\lambda \geqslant 0$ if $f(\mathbf{x}) \geqslant f(\mathbf{O})$ |
| 13a. $g(\lambda) = f[\lambda \mathbf{x} + (1 - \lambda) \mathbf{y}]$ is convex over $\lambda \in [0, 1]$ for any $\mathbf{x}, \mathbf{y} \in U \Leftrightarrow f$ is convex | 13b. $g(\lambda) = f[\lambda \mathbf{x} + (1 - \lambda) \mathbf{y}]$ is quasiconvex over $\lambda \in [0, 1]$ for any $\mathbf{x}, \mathbf{y} \in U \Leftrightarrow f$ is quasiconvex |
| 14a. $g(\mathbf{x}) = \sup_{\alpha \in \Gamma} f_\alpha(\mathbf{x})$ is convex | 14b. $g(\mathbf{x}) = \sup_{\alpha \in \Gamma} f_\alpha(\mathbf{x})$ is quasiconvex |
| 15a. $g(\mathbf{x}) = F[f(\mathbf{x})]$ is convex if $F$ is convex and increasing. | 15b. $g(\mathbf{x}) = F[f(\mathbf{x})]$ is quasiconvex if $F$ is increasing |

We include in our list of references only those papers to which we have referred, since a complete bibliography is to be found in the paper by Greenberg and Pierskalla.

## REFERENCES

1949, B. de Finetti. Sulla stratificazoni convesse. *Ann. Mat. Pura Appl.* [4] 30, 173–183.
1953, W. Fenchel. Convex Cones, Sets, and Functions (mimeographed lecture notes). Princeton Univ. Press, Princeton, New Jersey.
1956, W. Fenchel. Über konvexe Funktionen mit vorgeschriebenen Niveaumannigfaltigkeiten. *Math. Z.* 63, 496–506.
1967, J. Ponstein. Seven kinds of convexity. *SIAM Rev.* 9, 115–119.
1971, H. J. Greenberg and W. P. Pierskalla. A review of quasi-convex functions. *Oper. Res.* 19, 1553–1570.

## 82. Completely Convex Functions

In 1914, Serge Bernstein initiated a series of interesting theorems about real-valued functions having derivatives of all orders on some interval $(a, b)$, that is, about functions in the class $C^\infty(a, b)$. Typical of his results is the following.

> **Theorem A.** Suppose that no derivative of $f$ changes sign in $(a, b)$. Suppose further that it is impossible to find more than $k$ (finite) successive derivatives of $f$ whose absolute values $|f^{(r)}(x)|$ all vary in the same direction [that is, all increase or all decrease in $(a, b)$]. Then $f$ can be analytically continued to a function defined on the complex plane, and this (uniquely determined) extension will be an entire function of exponential type.

A set of $n$ successive derivatives with absolute values all varying in the same way is called a block of derivatives of length $n$. Let $n_1, n_2, \ldots$ denote the lengths of the successive blocks into which the sequence $f(x), f'(x), f''(x), \ldots$ is thus decomposed. Theorem A draws a conclusion from the boundedness of the sequence $n_1, n_2, \ldots$. Further analysis of the sequence by Bernstein and others has led to more refined information about the character of the analytic extension. A survey of results of this type was given by Polya [1943].

In the course of these studies, Bernstein introduced the term

**absolutely monotonic** function, defining it to be one for which, on $(a, b)$,

$$f^{(k)}(x) \geqslant 0, \qquad k = 0, 1,...$$

He also studied functions satisfying

$$(-1)^k f^{(k)}(x) \geqslant 0, \qquad k = 0, 1,... \tag{1}$$

that he called **completely monotonic**. Such functions have been studied since then, and a good exposition of the results has been given by Widder [1941].

In the context of such results, Bernstein had proved that if all the even derivatives are at least 0 in $(a, b)$, then $f$ has an analytic continuation into the complex plane. Boas suggested to Widder that this might be proved by use of the Lidstone series. This seemed plausible because the Lidstone series, a generalization of the Taylor series, approximates a given function in the neighborhood of two points instead of one by using the even derivatives. Specifically, the series corresponding to $f$ is

$$f(1) \Lambda_0(x) + f(0) \Lambda_0(1 - x) + f''(1) \Lambda_1(x) + f''(0) \Lambda_1(1 - x) + \cdots \tag{2}$$

where $\Lambda_n(x)$ is a polynomial of degree $2n + 1$ defined by the relations

$$\Lambda_0(x) = x$$
$$\Lambda_n''(x) = \Lambda_{n-1}(x)$$
$$\Lambda_n(0) = \Lambda_n(1) = 0$$

It is known that the sum of a Lidstone series must be entire [Widder, 1941, pp. 177–179].

Investigation of this idea by Widder failed to yield the desired result (later obtained by Boas [1941b] using another method), but it did lead Widder to another theorem [Widder, 1940].

**Theorem B.** If $f \in C^\infty(a, b)$ satisfies

$$(-1)^k f^{(2k)}(x) \geqslant 0, \qquad k = 0, 1, 2,... \tag{3}$$

then $f$ can be analytically continued to an entire function of exponential type at most $\pi$.

This result drew immediate attention, and Boas suggested [Widder, 1942] that in analogy with the terminology of Bernstein's functions

## 82. Completely Convex Functions

satisfying (1), functions satisfying (3) should be called **completely convex**. Boas pointed out that since Widder's theorem places no requirements on the odd derivatives, it neither contains nor is contained in the results (see Theorem A) of Bernstein. Boas [1941a] gave another proof of Widder's theorem, and both he and Polya obtained generalizations of the work of Widder and Bernstein [Boas, 1941a; Polya, 1941]. By 1942, in joint papers [Boas and Polya, 1941, 1942], they had a theorem that contained both Theorem A and Theorem B as special cases.

Widder [1942] pursued the relationship between completely convex functions and functions that are represented by their Lidstone series (2). He noted that $\sinh x$ is not completely convex in any interval, but that it is represented by its Lidstone expansion. He showed that if $f$ has a Lidstone expansion with every term nonnegative on [0, 1], then $f$ would necessarily be completely convex. On the other hand, $\sin \pi x$ is completely convex, but has a Lidstone expansion that is identically 0. To get necessary and sufficient conditions for $f$ to be representable by its Lidstone series, he therefore introduced yet another notion of convexity. A function $f \in C^\infty[0, 1]$ is called a **minimal completely convex** function if $f(x)$ is completely convex and if $f(x) - \varepsilon \sin \pi x$ is not completely convex on [0, 1] for any choice of $\varepsilon > 0$. He then obtained the desired theorem.

**Theorem C.** A necessary and sufficient condition that $f(x)$ can be represented as an absolutely convergent Lidstone series is that it should be the difference of two minimal completely convex functions on [0, 1].

This theorem can be used to characterize functions of a complex variable having a Lidstone series representation, but Boas [1943] has given a neater result in this direction.

While on the topic of representing completely convex functions, we cite Witner's paper [1958] in which he gives an integral representation for all such functions. Not all functions representable by Witner's integral are completely convex, however, so it was left to Boas [1959] to give both necessary and sufficient conditions for an integral representation of completely convex functions.

In 1957, Protter generalized the notion of completely convex by rewriting (3) in the form

$$f^{(4k)}(x) \geq 0$$
$$f^{(4k+2)}(x) \leq 0 \qquad k = 0, 1, 2,\ldots \qquad (4)$$

for $x \in (a, b)$. He then defined **almost completely convex** functions by replacing the second inequality of (4) with

$$f^{(4k+2)}(a) + f^{(4k+2)}(b) \leqslant \frac{\pi^2}{(b-a)^2} [f^{(4k)}(a) + f^{(4k)}(b)]$$

Noting that this is a bigger class of functions ($e^{cx}$, $0 < c \leqslant \pi$, is almost completely convex but not completely convex on [0, 1]), Protter then proved an analog to Widder's theorem.

> **Theorem D.** If $f$ is almost completely convex on $(a, b)$, then $f$ can be analytically continued to an entire function of exponential type at most $\pi(b - a)$.

As seems to be the rule with any paper concerning completely convex functions, Protter's work was followed quickly by a paper from Boas [1958]. This time he gave a proof that was shorter and somewhat more general than the earlier paper.

Finally we note that our own survey of this material is no exception to the rule stated in the previous paragraph. Shortly after this was written, Boas [1971] published a similar expository account, and the interested reader is referred to this paper for a fuller account and a more extended bibliography.

## REFERENCES

1914, S. Bernstein, Sur la definition et les proprietes des fonctions analytiques d'une variable reelle. *Math. Ann.* **75**, 449–468.

1940, D. V. Widder, Functions whose even derivatives have a prescribed sign. *Proc. Nat Acad. Sci. U.S.* **26**, 657–659.

1941a, R. P. Boas, A note on functions of exponential type. *Bull. Amer. Math. Soc.* **47**, 750–754.

1941b, R. P. Boas, Functions with positive derivatives. *Duke Math. J.* **8**, 163–172.

1941, G. Polya, On functions whose derivatives do not vanish in a given interval. *Proc. Nat. Acad. Sci. U.S.* **27**, 216–217.

1941, D. V. Widder, "The Laplace Transform." Princeton Univ. Press, Princeton, New Jersey.

1941, R. P. Boas and G. Polya, Generalizations of completely convex functions. *Proc. Nat. Acad. Sci. U.S.* **27**, 323–235.

1942, R. P. Boas and G. Polya, Influence of the signs of the derivative of a function on its analytic character. *Duke Math. J.* **9**, 406–424.

1942, D. V. Widder, Completely convex functions and Lidstone series. *Trans. Amer. Math. Soc.* **51**, 387–398.

1943, R. P. Boas, Representation of functions by Lidstone series. *Duke Math. J.* **10**, 239–245.

## 83. Convex Functions of Higher Order

1943, G. Polya, On the zeroes of the derivatives of a function and its analytic character. *Bull. Amer. Math. Soc.* **49**, 178–191.
1952, S. Bernstein, Remarks on the theory of regularly monotonic functions. *Izv. Akad. Nauk SSSR Ser. Mat.* **16**, 3–16.
1957, M. H. Protter, A generalization of completely convex functions. *Duke Math. J.* **24**, 205–213.
1958, R. P. Boas, Almost completely convex functions. *Duke Math. J.* **25**, 193–195.
1958, A. Wintner, On cosine-like arches. *Amer. J. Math.* **80**, 125–130.
1959, R. P. Boas, Representations of completely convex functions. *Amer. J. Math.* **81**, 709–714.
1971, R. P. Boas, Signs of derivatives and analytic behavior. *Amer. Math. Monthly* **78**, 1085–1093.
1972, D. Leeming and A. Sharma, A generalization of the class of completely convex functions. *In* "Inequalities" (O. Shisha, ed.), Vol. III. Academic Press, New York.

## 83. Convex Functions of Higher Order

Suppose $f$ is defined on $[a, b]$ and that $x_0 < x_1 < \cdots < x_n$ are points in this interval. There is a uniquely determined polynomial $P(x)$ of degree at most $n$ that agrees with $f$ at each $x_i$. Using Newton's method of divided differences, we obtain this so-called Lagrange interpolating polynomial in the form

$$P(x) = a_0 + a_1(x - x_0) + \cdots + a_n(x - x_0)(x - x_1) \cdots (x - x_{n-1})$$

where the coefficients are given by

$$a_0 = f(x_0) = [x_0; f]$$

$$a_1 = \frac{f(x_0) - f(x_1)}{x_0 - x_1} = [x_0, x_1; f]$$

$$a_2 = \frac{[x_0, x_1; f] - [x_1, x_2; f]}{x_0 - x_2} = [x_0, x_1, x_2; f]$$

$$\vdots$$

$$a_n = \frac{[x_0, \ldots, x_{n-1}; f] - [x_1, \ldots, x_n; f]}{x_0 - x_n} = [x_0, \ldots, x_n; f]$$

The expression $[x_0, x_1; f]$ is called the **first divided difference**; $[x_0, x_1, x_2; f]$ is the **second divided difference**, or the **divided difference of order 2**; and so on.

We saw in Section 82 how completely convex functions arise from studying functions for which certain coefficients in the Taylor expansion never change sign in an interval. Following earlier work by Hopf [1926], Popoviciu [1934] initiated a study of functions for which certain of the coefficients in the interpolating polynomial $P(x)$ do not change sign, no matter how the $x_i$ are chosen. The results of many papers are summarized in his book [Popoviciu, 1944] where an extensive bibliography of early results may be found.

The principal definition necessary to this work may be motivated by first noting that if $[x_0, x_1; f] \geq 0$, then $f(x_0) \leq f(x_1)$. Thus, if $[x_0, x_1; f] \geq 0$ for all $x_0 < x_1$ in $[a, b]$, then $f$ must be increasing on $[a, b]$. Moving on to the second divided difference, we find after a bit of algebra that if $[x_0, x_1, x_2; f] \geq 0$, then

$$\frac{(x_1 - x_2)f(x_0) + (x_2 - x_0)f(x_1) + (x_0 - x_1)f(x_2)}{(x_0 - x_1)(x_1 - x_2)(x_0 - x_2)} \geq 0$$

It this holds for all $x_0 < x_1 < x_2$ chosen in $[a, b]$, then [Problem 11I) $f$ is convex on $[a, b]$.

With these observations in mind, we define an **n-convex function** to be a function $f: [a, b] \to \mathbf{R}$ such that for all choices of $x_0 < x_1 < \cdots < x_{n+1}$ in $[a, b]$, the divided difference $[x_0, \ldots, x_{n+1}; f] \geq 0$. Thus a 0-convex function is an increasing function, and a 1-convex function is an ordinary convex function.

The graph of a convex function $f$ (that is, a 1-convex function) lies below the chord (the graph of the first-degree polynomial) joining, for $x_1, x_2 \in [a, b]$, the points $(x_1, f(x_1))$ and $(x_2, f(x_2))$. It is natural to wonder if $n$-convex functions have a similar geometric description. They do; $f$ is $n$-convex if and only if for $a \leq x_1 \leq \cdots \leq x_{n+1} \leq b$, the graph of $f$ lies alternately above and below the graph of the unique Lagrange interpolating polynomial of degree at most $n$ passing through the points $(x_i, f(x_i))$ [Bullen, 1971a].

> **Theorem A.** Suppose for some $n \geq 1$ that $f: [a, b] \to \mathbf{R}$ is $n$-convex. Then $f^{(k)}(x)$, the derivative of order $k$, exists on $(a, b)$ for $k < n$ and is $(n - k)$-convex. In particular, $f^{(n-1)}(x)$ exists, is convex, and therefore has left and right derivatives on $(a, b)$.

Once again we have a theorem that generalizes Theorem 11B.

The study of divided differences is of course greatly simplified if the points $x_0, \ldots, x_n$ are equally spaced so that $x_j = x_0 + j\delta$ for some $\delta > 0$. In this case,

$$[x_0, \ldots, x_n; f] = \frac{1}{n!\, \delta^n} \sum_{j=0}^{n} (-1)^{n-j} \binom{n}{j} f(x_0 + j\delta)$$

## 83. Convex Functions of Higher Order

With this restriction to equally spaced points, to say that $f$ is 1-convex means

$$\frac{[f(x_0) - f(x_1)] - [f(x_1) - f(x_2)]}{2\delta^2} \geq 0$$

which is equivalent to

$$\frac{f(x_0) + f(x_2)}{2} \geq f\left(\frac{x_0 + x_2}{2}\right)$$

This is just the condition for midconvexity, a condition we know to be weaker than convexity. We therefore call a function **weakly n-convex** if for any set of equally spaced points $x_0, \ldots, x_{n+1}$ in $[a, b]$, we have $[x_0, \ldots, x_{n+1}; f] \geq 0$. For such functions, it will be necessary to impose further conditions in order to conclude that $f$ is continuous. One such condition is the subject matter of a paper by Ciesielski [1959] with which he won a prize in a student competition. His condition has a familiar ring, having already been sounded in Theorem 72B.

**Theorem B.** Let $f$ be weakly $n$-convex on $[a, b]$ for some $n \geq 1$. If $f$ is bounded on at least one subset $E$ of $(a, b)$ having positive measure, then $f$ is continuous on $(a, b)$.

When $n = 1$, then (Theorem 72B) it is enough to have $f$ bounded above, but for $n > 1$, we need $f$ bounded above and below [Kemperman, 1969]. In the spirit of the history of such theorems, already described in Section 72, this result has been generalized [Ger, 1972].

If we specify that $f$ is continuous, then the concepts of $n$-convex and weakly $n$-convex turn out to be equivalent. Building on this fact, Boas and Widder [1940] have given a simple proof of Theorem A.

Next we would like to generalize the material of Section 14 on differences of convex functions. Toward that end, let

$$a_0 = x_0 < x_1 < \cdots < x_m = b$$

be a partition of $[a, b]$, and let

$$\Delta_k^i = [x_i, x_{i+1}, \ldots, x_{i+k}; f]$$

Define the **$n$th total variation** $V_n(f) = V_n(f; [a, b])$ by

$$V_n(f) = \sup \sum_{i=1}^{m-n} |\Delta_n^i - \Delta_n^{i-1}|$$

where the supremum is taken over all partitions of $[a, b]$. If $V_n(f) < \infty$, we say that $f$ is of **bounded nth variation** on $[a, b]$. Now we can state a result of Hopf [1926] that should be compared with Theorem 14D.

**Theorem C.** Let $f: [a, b] \to \mathbf{R}$ have bounded $n$th variation on $[a, b]$ for some $n \geqslant 0$. Then $f$ may be represented as the difference $g - h$ of two $n$-convex functions each having bounded $n$th variation.

Divided differences and convex functions of higher order occasionally appear in the literature in connection with approximation by spline functions. They also occur, as we shall see in Section 85, as a special case of functions convex with respect to $n$-parameter families. We prepare the way for this by turning next to the study of functions convex with respect to 2-parameter families.

## REFERENCES

1926, E. Hopf, Über die Zusammenhänge zwischen gewissen höheren Differenzenquotienten reeller Funktionen einer reelen Variablen und deren Differenzierbarkeitseigenschaften. Thesis, Univ. of Berlin.

1934, T. Popoviciu, Sur quelques proprietés des fonctions d'une variable reele convexes d'ordre superieur. *Mathematica, Cluj* **8**, 1–85.

1940, R. P. Boas and D. V. Widder, Functions with positive differences. *Duke Math. J.* **7**, 496–503.

1944, T. Popoviciu, "Les Fonctions Convexes." Hermann, Paris.

1959, Z. Ciesielski, Some properties of convex functions of higher orders. *Ann. Polon. Math.* **7**, 1–7.

1971a, P. S. Bullen, A criterion for $n$-convexity. *Pac. J. Math.* **36**, 81–98.

1972, R. Ger, Convex functions of higher order in Euclidean spaces. *Ann. Polon. Math.* **25**, 293–302.

## 84. Generalized Convex Functions

A natural way to generalize the concept of a convex function was introduced by Beckenbach [1937]. His definition is motivated by consideration of the class of functions of a real variable defined by

$$\mathscr{F}^1 = \{F(x) = mx + b: \; m, b \in \mathbf{R}\}$$

This class has the following property (P) that is of critical importance to the generalization to follow.

## 84. Generalized Convex Functions

(P) Given any two points $x_1$ and $x_2$ in $(a, b)$ and any two real numbers $y_1$ and $y_2$, there is a unique member of $\mathscr{F}^1$ passing through $(x_1, y_1)$ and $(x_2, y_2)$.

We can now define both midconvex and convex functions in terms of the members of $\mathscr{F}^1$. Let $f$ be a real-valued function defined on $(a, b)$. Then corresponding to $x_1, x_2 \in (a, b)$, there is a uniquely determined function denoted by

$$F(x; x_1, x_2) = F(x) \tag{1}$$

which agrees with $f$ at $x_1$ and $x_2$. We observe that $f$ is midconvex if (Fig. 84.1)

$$\text{for each } x_1 < x_2, \quad f\left(\frac{x_1 + x_2}{2}\right) \leqslant F\left(\frac{x_1 + x_2}{2}\right) \tag{2}$$

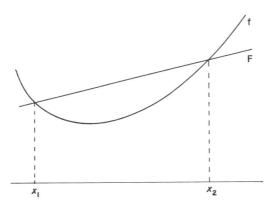

Fig. 84.1

and we say $f$ is convex if

$$\text{for each } x_1 < x_2, \quad f(x) \leqslant F(x) \quad \text{on } (x_1, x_2) \tag{3}$$

The way is now prepared for the desired generalization. We begin with a class $\mathscr{F}$ of functions continuous on $(a, b)$ and having property (P) above. Then for a fixed $f$, using the notation of (1), we say $f$ is **midconvex** or **convex with respect to** $\mathscr{F}$ (shortened to $\mathscr{F}$-**midconvex** or $\mathscr{F}$-**convex**) according as (2) or (3) is satisfied.

Besides the class $\mathscr{F}^1$ already mentioned, we list the following families as examples having property (P) on $(a, b)$.

$$\mathscr{F}^2 = \{x^2 + \alpha x + \beta: \; \alpha, \beta \in \mathbf{R}\}$$
$$\mathscr{F}^3 = \{\alpha \cos x + \beta \sin x: \; \alpha, \beta \in \mathbf{R}\}, \qquad b - a < \pi$$
$$\mathscr{F}^4 = \{(\alpha x + \beta)^3: \; \alpha, \beta \in \mathbf{R}\}$$
$$\mathscr{F}^5 = \{\alpha x + \beta + \phi(x): \; \alpha, \beta \in \mathbf{R}, \phi \text{ is continuous,}$$
$$\text{nowhere differentiable on } (a, b)\}$$

Consider now a function $f$ that is $\mathscr{F}$-convex for some family $\mathscr{F}$. How many of the properties of a convex function carry over to $f$? The theorems below give answers to this question, beginning with the subject of continuity for convex and midconvex functions. The first three theorems are due to Beckenbach and Bing [1945].

**Theorem A.** If $f$ is $\mathscr{F}$-convex on $(a, b)$, then $f$ is continuous on $(a, b)$.

This theorem clearly extends Theorem 11A to generalized convex functions, and our next two theorems similarly extend Theorems 72A and 72B.

**Theorem B.** If $f$ is $\mathscr{F}$-midconvex on $(a, b)$ and is bounded from above on some subinterval of $(a, b)$, then $f$ is continuous on $(a, b)$.

**Theorem C.** If $f$ is $\mathscr{F}$-midconvex and is bounded from above on a set $E$ of positive measure in $(a, b)$, then $f$ is continuous on $(a, b)$.

So far, the properties of generalized convex functions seem to parallel exactly those of convex functions. This happy situation does not extend, however, to properties related to differentiation. This is most easily seen by considering the family $\mathscr{F}^5$ above. Since any $F \in \mathscr{F}^5$ is convex with respect to $\mathscr{F}^5$, it is clear that a function may be convex with respect to a family while failing to have a derivative at even one point. The best results obtainable without putting further restrictions on the family $\mathscr{F}$ depend on extending the notion of support. A function $F \in \mathscr{F}$ is said to **support** $f$ defined on $(a, b)$ at $x_0 \in (a, b)$ if $F(x_0) = f(x_0)$ and $F(x) \leqslant f(x)$ for $x \in (a, b)$. With this notion we obtain a theorem that invites comparison with Theorem 12D.

## 84. Generalized Convex Functions

**Theorem D.** A function $f$ that is $\mathscr{F}$-convex has support at each $x_0 \in (a, b)$. If the support is not unique, there exist two extreme supports $F^+$ and $F^-$ in $\mathscr{F}$ such that

(a) $F^+(x) \leqslant F^-(x)$ on one side of $x_0$, and the inequality is reversed on the other side; they "cross" at $x_0$;
(b) on each side of $x_0$, any support to $f$ at $x_0$ falls between $F^+$ and $F^-$;
(c) every $F \in \mathscr{F}$ lying between $F^+$ and $F^-$ is a support function for $f$ at $x_0$.

The proof of Theorem D was given by M. M. Peixoto [1948a]. A later result [Green, 1953] comes closer to the kind of theorem one associates with convex functions (see Theorems 11C, 12E).

**Theorem E.** A function that is $\mathscr{F}$-convex has a unique support at all but a possibly countable set of points.

Theorem E says more than appears at first glance, since a function $f$ convex with respect to $\mathscr{F}$ may be differentiable at $x_0$ and still not have a unique support at $x_0$. Consider $f(x) = |x|^3$. It is convex with respect to $\mathscr{F}^4$ above, is differentiable at the origin, and has many supports at the origin.

To obtain results about differentiation instead of support, one has to impose further conditions on the family $\mathscr{F}$. Some conditions of this kind were suggested in the paper [1948a] by M. M. Peixoto. Others were suggested by Bonsall [1950]; in fact Bonsall thought he had conditions on $\mathscr{F}$ to guarantee the existence of second derivatives almost everywhere for functions convex with respect to his families $\mathscr{F}$. But he himself pointed out [1954] that his assertion was false. The results we include here are due to Green [1953] who imposed upon $\mathscr{F}$ some conditions that occurred to him while making application of generalized convex functions to a problem in geometry [Green, 1950]. He called a family $\mathscr{F}$ having property (P) an $\mathscr{F}\mathscr{S}$ family if

(i) each $F \in \mathscr{F}$ has a continuous derivative,
(ii) when $F_n \to F$, then $F_n' \to F'$ uniformly in every compact subinterval of $(a, b)$.

He then obtained a result parallel to Theorem 12E.

**Theorem F.** Let $f$ be $\mathscr{F}\mathscr{S}$-convex. If $f$ has unique support $F_0$ at $x_0$, then $f'(x_0)$ exists, and $f'(x_0) = F_0'(x_0)$.

Since $\mathscr{F}^4$ is an $\mathscr{F}\mathscr{S}$ family, the example of $f(x) = |x|^3$ used above precludes any possibility of proving the converse of this theorem. On the basis of Theorem E, however, we do have one obvious consequence of Theorem F that finally gives us a generalization of Theorem 11C.

**Theorem G.** A function $f$ that is $\mathscr{FS}$-convex has a derivative at all but a possibly countable set of points.

Green points out that his result neither implies nor is implied by the similar result of M. M. Peixoto since the conditions they add to the family $\mathscr{F}$ are different. In analogy with Theorem 11A, Green also shows the following.

**Theorem H.** A function $f$ that is $\mathscr{FS}$-convex satisfies a Lipschitz condition in every compact subinterval $J$ of $(a, b)$, and thus is absolutely continuous and has a derivative almost everywhere that is bounded in $J$.

There has also been attention given to convergence properties. In his original paper, Beckenbach [1937] considered two sequences of points $(x_n, y_n)$ and $(x_n', y_n')$ with $x_n$ and $x_n'$ in $(a, b)$. He used $F_n$ to designate the unique member of $\mathscr{F}$ determined by $(x_n, y_n)$ and $(x_n', y_n')$.

**Theorem I.** Suppose $(x_n, y_n) \to (x, y)$ and $(x_n', y_n') \to (x', y')$. Then if $F \in \mathscr{F}$ is determined by $(x, y)$ and $(x', y')$, it follows that $F_n \to F$ uniformly on every compact subinterval of $(a, b)$.

Green [1953] followed this with some results about the convex functions themselves, results to be compared with Theorem 13E and Problem 13C.

**Theorem J.** Let $f_n$ be a sequence of $\mathscr{F}$-convex functions on $(a, b)$ converging pointwise to $f$. Then $f$ is $\mathscr{F}$-convex and the convergence is uniform in every compact subinterval of $(a, b)$. And if $\mathscr{F}$ is an $\mathscr{FS}$ family, then with the exception of a countable set, $\lim_{n \to \infty} f_n'(x) = f'(x)$.

We have mentioned Bonsall's attempt to prove something about the existence of second derivatives of generalized convex functions. We know of no positive result in this direction, but there are results similar to those for convex functions known to be twice differentiable. Let us state the familiar properties in the following form. The family $\mathscr{F}^1$ is the set of continuous solutions to the differential equation $y''(x) = 0$.

## 84. Generalized Convex Functions

A twice differentiable function $y = f(x)$ is convex if and only if $y'' = f''(x) \geq 0$. The natural generalization of this observation relates generalized convex functions to the subject of differential inequalities. This relationship is responsible for much of the interest in generalized convex functions.

Consider the differential equation

$$y'' = G(x, y, y') \tag{4}$$

where $G$ is continuous on $S = (a, b) \times \mathbf{R} \times \mathbf{R}$. Let $\mathscr{F}$ be the family of (necessarily continuous) solutions to (4) on $(a, b)$. We further suppose that $G$ is such that members of $\mathscr{F}$ satisfy the boundary value conditions:

(B1) Corresponding to any $x_0 \in (a, b)$ and any real numbers $y_0$ and $y_0'$, there is a unique $y(x) \in \mathscr{F}$ such that $y(x_0) = y_0$, $y'(x_0) = y_0'$.

(B2) Corresponding to any $(x_0, y_0)$ and $(x_1, y_1)$ where $x_i \in (a, b)$, there is a unique $y(x) \in \mathscr{F}$ such that $y(x_0) = y_0$, $y(x_1) = y_1$.

Numerous things have been written about conditions on $G$ that guarantee (B1), (B2), or (B1) and (B2). For our purpose we merely quote the fundamental result due to M. M. Peixoto [1949].

**Theorem K.** Let the differential equation (4) be such that the family $\mathscr{F}$ of solutions has properties (B1) and (B2). Then a twice differentiable function $f$ is $\mathscr{F}$-convex if and only if

$$f''(x) \geq G(x, y, y')$$

If the inequality is strict, then $f$ is strictly $\mathscr{F}$-convex.

This theorem includes Theorem 12C as a special case. Reid [1959] characterizes the functions convex with respect to the solution space $\mathscr{F}$ of $y'' + p_1(x) y' + p_2(x) y = 0$. Hanai [1945] relates generalized convex functions to third-order differential inequalities. By means of these inequalities, generalized convex functions also come into the study of the behavior of analytic functions in angular domains [Hardy, Littlewood, and Polya 1952, p. 98; Valiron, 1932].

Many of the results mentioned in this section are proved in the book largely devoted to this topic by M. M. Peixoto [1948b]. We mention in conclusion that convex functions and then generalized convex functions are related to classical notions from differential geometry. Since we are not assuming familiarity with differential geometry here, we merely refer the interested reader to the papers of Beckenbach [1948a] and Clement [1953].

## REFERENCES

1932, G. Valiron, Fonctions convexes et fonctions entieres. *Bull. Soc. Math. France* **60**, 278–287.
1937, E. F. Beckenbach, Generalized convex functions. *Bull. Amer. Math. Soc.* **43**, 363–371.
1945, E. F. Beckenbach and R. H. Bing, On generalized convex functions. *Trans. Amer. Math. Soc.* **58**, 220–230.
1945, S. Hanai, On the inequalities $y''' \geq G(x, y, y', y'')$. *Proc. Jap. Acad.* **21**, 378–381.
1948a, E. F. Beckenbach, Properties of surfaces of negative curvature. *Amer. Math. Monthly* **55**, 285–301.
1948a, M. M. Peixoto, On the existence of derivatives of generalized convex functions. *Summa Brasil. Math.* **2**, No. 3, 35–42.
1948b, M. M. Peixoto, Convexity of curves. *Notas Mat.* **No. 6**, Livraria Boffoni, Rio de Janeiro.
1949, M. M. Peixoto, Generalized convex functions and second order differential inequalities. *Bull. Amer. Math. Soc.* **55**, 563–572.
1950, E. F. Bonsall, The characterization of generalized convex functions. *Quart. J. Math. Oxford Ser.* **1**, 100–111.
1950, J. W. Green, Sets subtending a constant angle on a circle. *Duke Math. J.* **17**, 263–267.
1953, E. F. Bonsall, The characterization of generalized convex functions. *Quart J. Math. Oxford Ser.* **4**, 253.
1953, P. A. Clement, Generalized convexity and surfaces of negative curvature. *Pac. J. Math.* **3**, 333–368.
1953, J. W. Green, Support, convergence, and differentiability properties of generalized convex functions. *Proc. Amer. Math. Soc.* **4**, 391–396.
1959, W. Reid, Variational aspects of generalized convex functions. *Pac. J. Math.* **9**, 571–581.

## 85. More about Generalized Convex Functions

New life was breathed into the topic of generalized convex functions when Tornheim [1950], apparently following the lead of Popoviciu [1944], suggested using $n$ points instead of just two to determine the functions of the family $\mathscr{F}$. An **n-parameter family** $\mathscr{F}_n$ is a collection of functions continuous on $(a, b)$ having the property (P).

(P)  Given any $n$ distinct points $x_1, ..., x_n$ in $(a, b)$ and any $n$ real numbers $y_1, ..., y_n$, there is a unique member $F \in \mathscr{F}_n$ satisfying $F(x_i) = y_i$, $i = 1, ..., n$.

Let $f: (a, b) \to \mathbf{R}$. Given a set of $n$ distinct points $x_1 < \cdots < x_n$ in $(a, b)$, we use $F(x; x_1, ..., x_n)$ [or just $F(x)$ where confusion is unlikely] to denote the unique member of $\mathscr{F}_n$ which agrees with $f$ at $x_1, ..., x_n$. Choosing from among several equivalent definitions, we follow Kemper-

## 85. More about Generalized Convex Functions

man, [1969] in calling $f$ an $\mathscr{F}_n$-**convex function** if for every choice of $n+1$ points $x_i$ satisfying $a < x_1 < \cdots < x_n < x_{n+1} < b$,

$$f(x_{n+1}) \geqslant F(x_{n+1}\,;\,x_1,...,x_n) = F(x_{n+1})$$

that is, if $f$ majorizes $F$ on the interval $(x_n, b)$. In turn, we say $f$ is **weakly** $\mathscr{F}_n$-**convex** if for each choice of $\delta > 0$ and $x_1$ such that $a < x_1 < b - n\delta$, the $F$ agreeing with $f$ at $x_1$, $x_2 = x_1 + \delta,..., x_n = x_1 + (n-1)\delta$ satisfies $f(x_1 + n\delta) \geqslant F(x_1 + n\delta)$.

It is obvious that the families $\mathscr{F}$ described in Section 84 are $\mathscr{F}_2$ families in the terminology of the present section. It is less obvious that a function $f$ is $\mathscr{F}$-convex if and only if it is $\mathscr{F}_2$-convex. To see this, consider an $\mathscr{F}_2$-convex function on an interval $(x_1, x_2) \subseteq (a, b)$. The function $F(x; x_1, x_2)$, to be denoted by $F_{12}(x)$ in this discussion, is determined, and according to the definition of $\mathscr{F}_2$-convexity, $F_{12}(x) \leqslant f(x)$ for $x > x_2$. Suppose there is an $x_3 \in (x_1, x_2)$ such that $f(x_3) > F_{12}(x_3)$. Letting $F_{13} \in \mathscr{F}_2$ be the function agreeing with $f$ at $x_1$ and $x_3$, it follows from $x_2 > x_3$ that $F_{13}(x_2) \leqslant f(x_2) = F_{12}(x_2)$, while $F_{13}(x_3) = f(x_3) > F_{12}(x_3)$. Since $F_{12}$ and $F_{13}$ are continuous, there must be an $x_4 \in (x_3, x_2]$ such that $F_{12}(x_4) = F_{13}(x_4)$. This, however, contradicts the uniqueness of the function in $\mathscr{F}_2$ passing through $(x_1, F_{12}(x_1))$ and $(x_4, F_{12}(x_4))$. We conclude that for all $x \in (x_1, x_2)$, $f(x) \leqslant F_{12}(x)$, which says $f$ is $\mathscr{F}$-convex as defined in Section 84.

One sees by similar arguments that an $\mathscr{F}$-convex function is $\mathscr{F}_2$-convex and that weakly $\mathscr{F}$-convex functions are exactly the weak $\mathscr{F}_2$-convex functions. Thus the study of $\mathscr{F}_n$-convex functions does include as a special case the generalized convex functions of Section 84.

Even more has been accomplished. Consider the case where $\mathscr{F}_n$ is chosen to be the set of polynomials of degree at most $n - 1$. Given $f : (a, b) \to \mathbf{R}$ and $n$ points $a < x_1 < \cdots < x_n < b$, the member $F \in \mathscr{F}_n$ agreeing with $f$ at each $x_i$, $i = 1,..., n$, is the uniquely determined interpolating polynomial encountered in Section 83; that is,

$$F(x; x_1,..., x_n) = a_0 + a_1(x - x_1) + \cdots + a_{n-1}(x - x_1)(x - x_2) \cdots (x - x_{n-1})$$

where

$$a_k = [x_1,..., x_{k+1}\,;f]$$

Thus for $x_{n+1} \in (x_n, b)$

$$\begin{aligned}
f(x_{n+1}) &- F(x_{n+1}\,;\,x_1,..., x_n) \\
&= F(x_{n+1}\,;\,x_1,..., x_n, x_{n+1}) - F(x_{n+1}\,;\,x_1,..., x_n) \\
&= a_n(x_{n+1} - x_1)(x_{n+1} - x_2) \cdots (x_{n+1} - x_n) \\
&= [x_1,..., x_{n+1}\,;f](x_{n+1} - x_1)(x_{n+1} - x_2) \cdots (x_{n+1} - x_n)
\end{aligned}$$

We conclude that $f$ is $\mathscr{F}_n$-convex (that is, convex with respect to the family of polynomials of degree at most $n-1$) if and only if $[x_1,\ldots,x_{n+1};f] \geqslant 0$. But this is just what we called $(n-1)$-convex in Section 83. In the same way a function is weakly $\mathscr{F}_n$-convex if and only if it is weakly $(n-1)$-convex. As we suggested in Section 83, higher order convexity now appears as a special case of $\mathscr{F}_n$-convexity.

By giving attention to the number of times the graph of an $\mathscr{F}_n$-convex function can intersect the graph of a function in $\mathscr{F}_n$, it is possible [Kemperman, 1969] to characterize $\mathscr{F}_n$-convex functions as follows.

**Theorem A.** A function $f:(a,b) \to \mathbf{R}$ is $\mathscr{F}_n$-convex if and only if for every choice of $x_i$ satisfying $a < x_1 < \cdots < x_n < b$,

$$(-1)^{n+i-1}[f(x) - F(x)] \geqslant 0 \quad \text{on} \quad (x_{i-1}, x_i), \qquad i = 2,\ldots,n$$

where as usual $F$ is the member of $\mathscr{F}_n$ agreeing with $f$ at $x_1,\ldots,x_n$.

The theorem of Bullen [1971a] mentioned in Section 83 as a geometric description of $n$-convexity is, of course, a special case of Theorem A. Mathsen [1972] uses Theorem A as the definition of $\mathscr{F}_n$-convexity. Tornheim derives Theorem A, having defined $f$ to be $n$-convex if it intersects no member of $\mathscr{F}_n$ more than $n$ times, multiplicities not being counted, but he required that $f$ be continuous. Apparently it was Moldovan [1959] who first noted that $\mathscr{F}_n$-convexity implies the continuity of $f$. The proof of this fact follows easily [Mathsen, 1972] from Theorem A.

**Theorem B.** If $f:(a,b) \to \mathbf{R}$ is $\mathscr{F}_n$-convex, then $f$ is continuous.

As usual, a weakly $\mathscr{F}_n$-convex function is $\mathscr{F}_n$-convex if and only if it is continuous [Kemperman, 1969].

We have come to expect that whatever the form of the generalization, functions called convex on $(a,b)$ will be continuous, but for functions weakly (or midpoint) convex, some kind of boundedness condition is essential to establishing continuity. So it is in the case of weakly $\mathscr{F}_n$-convex functions. Owing to the inclusion of both generalized convexity and $n$th-order convexity as special cases of $\mathscr{F}_n$-convexity, the relevant theorem appears as a generalization of both Theorem 83B and Theorem 84C.

## 85. More about Generalized Convex Functions

**Theorem C.** If $f$ is weakly $\mathscr{F}_n$-convex and is bounded on a measurable subset $E$ of positive measure in $(a, b)$, then $f$ is continuous on $(a, b)$.

While it is enough to have $f$ bounded only from above in the case of an $\mathscr{F}_2$ family, such a result could not be expected in the general case because of the remarks following Theorem 83B. However, we do have an unexpected result, proved early in the study of $n$-parameter families [Tornheim, 1950] which makes it appear that things in fact become more orderly when $n > 2$.

**Theorem D.** Suppose we have an $\mathscr{F}_n$ family in which all the members are differentiable. If $n \geq 3$, then an $\mathscr{F}_n$-convex function is differentiable.

This result is surprising because of the difficulty surrounding questions of differentiability for the case $n = 2$. On the other hand, when $n \geq 3$ and $\mathscr{F}_n$ is chosen so that $\mathscr{F}_n$-convexity corresponds to $(n - 1)$-convexity, it is natural to compare Theorem D with Theorem 83A. This suggests that an $\mathscr{F}_n$-convex function might even be more than once differentiable. Hartman [1958] obtained a result of this kind using a slightly weaker definition for the $k$th derivative at $x_0$ based on representing $f$ in a neighborhood of $x_0$ by a $k$th-degree Taylor polynomial (defined for $k = 2$ in Problem 44H). In the same direction, conditions on members of $\mathscr{F}_n$ have been given [Kemperman, 1969] which will ensure that an $\mathscr{F}_n$-convex function will have a $k$th derivative at all but a countable set of points.

Special attention has been given in the literature (see particularly [Karlin and Studden, 1966, Chapter 11]) to the case in which $\mathscr{F}_n$ is a **linear family**. By definition, this is an $\mathscr{F}_n$ family in which there are $n$ functions $u_1, \ldots, u_n$ continuous on $(a, b)$ such that any $F \in \mathscr{F}_n$ may be expressed in the form

$$F = c_1 u_1 + \cdots + c_n u_n$$

The conditions on the $\mathscr{F}_n$ family mean that given any $n$ points $x_1, \ldots, x_n$, and any real numbers $y_1, \ldots, y_n$, we must be able to find unique $c_i$'s satisfying

$$\begin{bmatrix} u_1(x_1) & \cdots & u_n(x_1) \\ \vdots & & \vdots \\ u_1(x_n) & \cdots & u_n(x_n) \end{bmatrix} \begin{bmatrix} c_1 \\ \vdots \\ c_n \end{bmatrix} = \begin{bmatrix} y_1 \\ \vdots \\ y_n \end{bmatrix}$$

Thus the condition that $\mathscr{F}_n$ always have a unique $F$ with $F(x_i) = y_i$, $i = 1, \ldots, n$, means that

$$D = \det[u_j(x_i)] \neq 0$$

whenever the $x_i$'s are distinct. Since this determinant is a continuous function, it must everywhere be the same sign, and we shall assume the basis functions chosen so that the determinant is positive whenever $x_1 < \cdots < x_n$. Such a set of functions $u_1(x), \ldots, u_n(x)$ is called a **Tchebycheff system**. Then the following theorem characterizes functions that are $\mathscr{F}_n$-convex.

**Theorem E.** A function $f: (a, b) \to \mathbf{R}$ is convex with respect to the linear family $\mathscr{F}_n$ if and only if for $a < x_1 < \cdots < x_n < x < b$,

$$\det \begin{bmatrix} u_1(x_1) & \cdots & u_n(x_1) & f(x_1) \\ \vdots & & \vdots & \vdots \\ u_1(x_n) & \cdots & u_n(x_n) & f(x_n) \\ u_1(x) & \cdots & u_n(x) & f(x) \end{bmatrix} \geq 0$$

To see that this is so, we expand the determinant above according to the last row, obtaining (after a bit of algebra) $[f(x) - F(x)]D$. Consequently the determinant is nonnegative if and only if $f(x) \geq F(x)$ on $(x_n, b)$. This is equivalent to our original definition of $\mathscr{F}_n$-convexity.

For illustrative purposes, let $n = 2$, $u_1(x) = 1$, and $u_2(x) = x$. Then for $x_1 < x_2$, we have

$$\det \begin{bmatrix} 1 & x_1 \\ 1 & x_2 \end{bmatrix} > 0$$

and Theorem E says that $f$ is convex with respect to the linear family $\mathscr{F}_2$ if and only if

$$\det \begin{bmatrix} 1 & x_1 & f(x_1) \\ 1 & x_2 & f(x_2) \\ 1 & x & f(x) \end{bmatrix} \geq 0$$

This is a restatement of Problem 12P.

Among many results known for the special case of a linear family $\mathscr{F}_n$, we cite only one. It generalizes the classic result that a twice differentiable function is convex if and only if $f''(x) \geq 0$.

**Theorem F.** Let $\mathscr{F}_n$ be a linear family in which all members are $n$-times differentiable. Then an $n$-times differentiable function $f$ is $\mathscr{F}_n$-convex if and only if for all $x \in (a, b)$,

$$\det \begin{bmatrix} u_1(x) & u_2(x) & \cdots & u_n(x) & f(x) \\ u_1'(x) & u_2'(x) & \cdots & u_n'(x) & f'(x) \\ \vdots & \vdots & & \vdots & \vdots \\ u_1^{(n)}(x) & u_2^{(n)}(x) & \cdots & u_n^{(n)}(x) & f^{(n)}(x) \end{bmatrix} \geq 0$$

## 85. More about Generalized Convex Functions

Much of the interest in $n$-parameter families stems from their connection with approximation theory. This was evident in the paper by Tornheim [1950] in which he proved the following useful convergence theorem.

**Theorem G.** Let $\{x_{ij}\}_{j=1}^{\infty}$, $i = 1,\ldots, n$, be sequences such that $a < x_{1j} < \cdots < x_{nj} < b$ for each $j$, and such that they respectively converge to $x_1 < \cdots < x_n$. Similarly let $\{y_{ij}\}_{j=1}^{\infty}$, $i = 1,\ldots, n$, be sequences converging to $y_1,\ldots, y_n$. If, for each $j$, $F_j \in \mathscr{F}_n$ is the member satisfying $F_j(x_{ij}) = y_{ij}$, and if $F \in \mathscr{F}_n$ is the member such that $F(x_i) = y_i$, $i = 1,\ldots, n$, then the sequence $\{F_j\}$ converges, uniformly on every compact subinterval of $(a, b)$, to $F$.

Using this theorem, Tornheim is able to give answers to the two questions associated with the approximation problem (Section 56). Our linear space is the set $C[a, b]$ of functions continuous on $[a, b]$, and the norm is the familiar $\|f - g\| = \max |f(x) - g(x)|$; $\mathscr{F}_n$ is of course a subset of $C[a, b]$.

**Theorem H.** For any function $f \in C[a, b]$, there is a unique best approximation in $\mathscr{F}_n$.

Other papers developing the relationship of $n$-parameter families to approximation theory are by Motzkin [1949], Curtis [1959], and Hartman [1967]. Especially to be noted is the book by Karlin and Studden [1966, Chapter 11] in which linear $\mathscr{F}_n$ families are related to approximation by spline functions.

We saw in Section 84 that there was a natural application of generalized convexity to differential inequalities. Such applications exist for $\mathscr{F}_n$-convex functions as well, but a still further generalization seems best suited to this study. Let $\lambda(n) = (\lambda_1,\ldots, \lambda_k)$ where $\lambda_i$, $i = 1,\ldots, k$, are positive integers satisfying $\lambda_1 + \cdots + \lambda_k = n$. We say that $\lambda(n)$ is an ordered partition of $n$, and the set of all such partitions is designated by $P(n)$.

For $\lambda(n) \in P(n)$, we define a **$\lambda(n)$ family** $\mathscr{F}_{\lambda(n)}$ to be a collection of functions on $(a, b)$ having the following properties.

(P1) Each function is $r$-times differentiable on $(a, b)$, $r$ being the maximum of $\lambda_i - 1$, $i = 1,\ldots, k$.
(P2) For every choice of $k$ points $x_1 < \cdots < x_k$ in $(a, b)$ and every set $y_i^{(j)}$ of $n$ real numbers, there is a unique $F \in \mathscr{F}_{\lambda(n)}$ satisfying $F^{(j)}(x_i) = y_i^{(j)}$, $i = 1,\ldots, k$; $j = 0,\ldots, \lambda_i - 1$.

Then a function $f$, $r$-times differentiable, is said to be $\mathscr{F}_{\lambda(n)}$-**convex** [or simply $\lambda(n)$-convex] if for every choice of $k$ points $a < x_1 < \cdots < x_k < b$, the unique $F \in \mathscr{F}_\lambda(n)$ satisfying

$$F^{(j)}(x_i) = f^{(j)}(x_i), \quad i = 1,\ldots,k; \quad j = 0,\ldots, \lambda_i - 1$$

also satisfies

$$(-1)^{M(i)}[f(x) - F(x)] \geqslant 0 \quad \text{for} \quad x \in (x_{i-1}, x_i), \quad i = 2,\ldots,k$$

where $M(i) = \lambda_1 + \cdots + \lambda_{i-1} + n$.

If $\lambda(n) = (1,\ldots,1)$, then $\mathscr{F}_{\lambda(n)}$ is simply an $\mathscr{F}_n$ family of the type we have been discussing and the definition of $\lambda(n)$-convexity reduces to Theorem A. If $\lambda(n) = (n)$ so that all the conditions are satisfied at one point, we say that initial-valued problems are uniquely solvable in $\mathscr{F}_{\lambda(n)}$. If $\mathscr{F}$ is a $\lambda(n)$ family for all $\lambda(n) \in P(n)$, $\mathscr{F}$ is called an **unrestricted n-parameter family** on $(a, b)$. These definitions at least indicate the content of a number of papers appearing in the literature, and enable us to understand the statement of the one theorem we have chosen for illustrative purposes.

**Theorem I.** Let $\mathscr{F}$ be a family of functions in $C^{(n-1)}(a, b)$. Then $\mathscr{F}$ is an unrestricted $n$-parameter family on $(a, b)$ if and only if $\mathscr{F}$ is an $\mathscr{F}_n$ family and initial value problems are uniquely solvable in $\mathscr{F}$.

This theorem, first proved by Hartman [1958], has been proved more simply [Opial, 1967] in the case where $\mathscr{F}$ is the solution set for an $n$th-order homogeneous linear differential equation with summable coefficients. Mathsen [1969] shows that the theorem fails when the interval is closed.

Most of the papers relating to $\lambda(n)$-convexity are oriented toward differential equations and differential inequalities.

Finally we mention that Hartman [1967] has extended the notion of $n$-parameter families $\mathscr{F}_n$ to functions defined on some region in $\mathbf{R}^n$.

## REFERENCES

1944, T. Popoviciu, "Les Fonctions Convexes." Hermann, Paris.
1949, T. S. Motzkin, Approximation by curves of a unisolvent family. *Bull. Amer. Math. Soc.* **55**, 789–793.
1950, L. Tornheim, On $n$-parameter families of functions and associated convex functions. *Trans. Amer. Math. Soc.* **69**, 457–467.

1958, P. Hartman, Unrestricted $n$-parameter families. *Rend. Circ. Mat Palermo* [2] **7**, 123–142.
1959, P. C. Curtis, Jr., $N$-parameter families and best approximation. *Pac. J. Math.* **9**, 1013–1027.
1959, E. Moldovan, Sur une généralisation des fonctions convexes. *Mathematica (Cluj)* [2] **1**, 49–80.
1966, S. Karlin and W. J. Studden, "Tchebycheff Systems: With Applications in Analysis and Statistics." Wiley (Interscience), New York.
1967, P. Hartman, Interpolating families and generalized convex functions. *Duke Math. J.* **34**, 511–518.
1967, Z. Opial, On a theorem of O. Arama. *J. Differential Equations* **3**, 88–91.
1969, J. H. B. Kemperman, On the regularity of generalized convex functions. *Trans. Amer. Math. Soc.* **135**, 69–93.
1969, R. M. Mathsen, $\lambda(n)$-parameter families. *Can. Math. Bull.* **12**, 185–191.
1970, I. V. Čebaevskaja, Generalized convexity and Jensen's Inequality. *Izv. Vysš. Učebn. Zaved. Mathematika* **3** (**94**), 91–95.
1972, R. M. Mathsen, $\lambda(n)$-convex functions. *Rocky Mt J. Math.* **2**, 31–43.

## 86. Other Related Topics

Thus far in Chapter VIII we have examined a number of topics one is certain to come across in any review of the literature relating to convex functions. To press the analogy with a tree made at the beginning of the chapter, it could be said that each of these topics is a branch that has had a significant development of its own. We begin this section by taking note of another well-developed subject, subharmonic functions, which, though it grows out of roots in complex variable theory, exhibits many properties we recognize as characteristic of convex functions. We follow this with a description of a second topic growing out of complex function theory, not so much because it resembles our main interest, but because it is referred to in the literature as the class of (complex) convex functions. Finally, we discuss several offshoots more surely stemming from convex functions, but having a rather sparse development.

### SUBHARMONIC FUNCTIONS

Although we have insisted that a linear function of a single real variable must be of the form $f(x) = mx$, it is common in elementary courses and in certain areas of applied mathematics to call a function linear if it is described by $l(x) = mx + b$. We adopt this usage here so as to be consistent with the literature relating to our present subject.

Suppose $f: I \to \mathbf{R}$ is twice differentiable on an open interval $I$.

Then $f$ is linear if and only if $d^2f/dx^2 = 0$, and it is convex if and only if $d^2f/dx^2 \geq 0$. It also turns out that $f$ is convex on the interval $I$ if and only if, whenever a linear function $l(x) \geq f(x)$ on the boundary of a subinterval $[a, b] \subseteq I$, $l(x) \geq f(x)$ throughout $[a, b]$.

Now consider $f: U \to \mathbf{R}$ where $f(r, s)$ is twice differentiable on an open set $U \subseteq \mathbf{R}^2$. Then $f$ is **harmonic** if $\partial^2 f/\partial r^2 + \partial^2 f/\partial s^2 = 0$, and it is **subharmonic** if $\partial^2 f/\partial r^2 + \partial^2 f/\partial s^2 \geq 0$. Riesz [1925] first drew attention to the class of functions subharmonic on $U \subseteq \mathbf{R}^2$ by characterizing such functions as follows. The function $f: U \to [-\infty, \infty)$ is subharmonic on the open set $U$ if and only if

(1) $f$ is not identically $-\infty$ on $U$
(2) $f$ is upper semicontinuous on $U$
(3) whenever a function $h$ is harmonic on an open set $D$, continuous on $\bar{D} \subseteq U$, and such that $h(x) \geq f(x)$ on the boundary of $D$, then $h(x) \geq f(x)$ throughout $D$.

The class of subharmonic functions, exhibiting many properties of the class of convex functions, is interesting in itself. It has also attracted attention because of the connection of this class with many applications; for example, the holomorphic function $f(\mathbf{x}) = h_1(r, s) + ih_2(r, s)$ gives rise to a subharmonic function $|f(\mathbf{x})|$, a relationship that has been greatly exploited.

The fullest account of subharmonic functions is given by Rado [1949]. Selecting from a wealth of results, we mention the following as illustrative of the way in which properties of subharmonic functions parallel those of convex functions.

Let $f, g$ be subharmonic in an open set $U \subseteq \mathbf{R}^2$.

A. If $h$ is a harmonic function majorizing $f$ on some open set $D$ as in (3) above, then either $h(\mathbf{x}) > f(\mathbf{x})$ on $D$, or else $h = f$ on $D$.  Problem 11A(3)

B. If $f$ assumes a global maximum in $U$, then $f$ is constant.  Theorem 51C

C. $f + g$ and $f \vee g$ are subharmonic; $f \cdot g$ is not.  Theorem 13A

D. The uniform limit of subharmonic functions is subharmonic.  Theorem 13E

The same kind of relationship exists between functions convex on an open set $U \subseteq \mathbf{R}^n$ and functions plurisubharmonic on $\mathbf{C}^n$ where $\mathbf{C}^n$ is the $n$-dimensional linear space of complex $n$-tuples. The convexity of $f$ on $U$ may be defined by requiring that for $f: U \to \mathbf{R}$, the restriction of $f$ to

$$U \cap \{\mathbf{x}: \mathbf{x} = \mathbf{x}_0 + t\mathbf{b}\} \tag{4}$$

## 86. Other Related Topics

is, for any choice of $\mathbf{x}_0$, $\mathbf{b} \in \mathbf{R}^n$, a convex function of the real variable $t$. In analogy, $f: U \to [-\infty, \infty)$, upper semicontinuous on $U$, is **plurisubharmonic** if the restriction of $f$ to (4) is, for any choice of $\mathbf{x}_0$, $\mathbf{b} \in \mathbf{C}^n$, a subharmonic function of the complex variable $t$.

Once again it is possible to list properties that hold for convex and plurisubharmonic functions. Bremermann [1956] carries out such a program in some detail, concluding with the comment that "This list could be continued rather indefinitely. By substituting *linear* for *harmonic*, *straight line* for *analytic plane*, **x** [a real variable] for **z** [a complex variable], etc., we obtain corresponding theorems for plurisubharmonic and convex functions."

We limit our references below to those mentioned in our summary. The book by Rado gives an extensive bibliography for subharmonic functions, and this is updated by Beckenbach and Bellman [1965, p. 160]. A similar list for plurisubharmonic functions is given by Bremermann [1956].

## REFERENCES

1925, F. Riesz, Über subharmonische Funktionen und ihre Rolle in der Funktionentheorie und in der Potentialtheorie. *Acta Litt. Sci. Szeged* **2**, 87–100.
1949, T. Rado, "Subharmonic Functions." Chelsea, Bronx, New York
1956, H. J. Bremermann, Complex convexity. *Trans. Amer. Math. Soc.* **82**, 17–51.
1965, E. F. Beckenbach and R. Bellman, "Inequalities," 2nd rev. printing, Springer-Verlag, Berlin and New York.

## COMPLEX CONVEXITY

In studying the mapping properties of complex functions of a convex variable, considerable attention is given to a class of functions that are called convex. Properties of functions in this class are not particularly related to any class we have studied, and the methods of study are entirely different, drawing as they do on complex function theory. However, literature relating to this class of functions commonly appears with a title mentioning convex functions, so we give here a summary aiming to help the reader recognize such papers when he encounters them.

A univalent (that is one-to-one) complex function $f(z)$ that is holomorphic on the open unit disk $D$ may without loss of generality be assumed to map 0 into 0 and satisfy $f'(0) = 1$. The Taylor expansion of a function so normalized is of the form

$$f(z) = z + a_2 z^2 + a_3 z^3 + \cdots$$

If the image of the open unit disk $D$ under such a function is convex, then the function $f$ is naturally enough called **convex**. It can be proved that $f$ is convex if and only if

$$1 + \operatorname{Re} \frac{zf''(z)}{f'(z)} > 0$$

Closely related to convex functions are starlike functions. The function $f$ is called **starlike** if the image of $D$ under $f$ is star shaped with respect to the origin, that is, if for any point $w$ in the image set, the straight line segment joining the origin and $w$ is also in the set. The function $f$ is starlike if and only if

$$\operatorname{Re} \frac{zf'(z)}{f(z)} > 0$$

on $D$.

A function $f$ holomorphic in $D$ is said to be **close to convex** in $D$ if there exists a convex function $\phi$ such that

$$\operatorname{Re} \frac{f'(z)}{\phi'(z)} > 0$$

for all $z \in D$.

The characterizations of convex and starlike functions, together with other basic results, are proved in Sansone and Gerretsen [1969, pp. 185–216]. Close to convex functions were first defined and discussed by Kaplan [1952], and have been studied in a number of subsequent papers.

## REFERENCES

1952, K. W. Kaplan, Close to convex schlicht functions. *Mich. Math. J.* 1, 169–185.
1969, G. Sansone and J. Gerretsen, "Lectures on the Theory of Functions of a Complex Variable," Vol. II. Wolters, Noordhoff, Groningen.

## APPROXIMATELY CONVEX FUNCTIONS

Solutions to the linear functional equation $f(x + y) = f(x) + f(y)$ have been shown [Hyers, 1941] to be stable in the following sense. If $f$ is $\varepsilon$-approximately linear, that is, if

$$|f(x + y) - f(x) - f(y)| \leqslant \varepsilon$$

then there is an actual solution $g$ to the linear functional equation such that $|g(x) - f(x)| \leqslant \varepsilon$. This led to defining and studying approximately convex functions. Let $U$ be a convex set in $\mathbf{R}^n$. Then $f: U \to \mathbf{R}$ is **$\varepsilon$-approximately convex** if for all $\mathbf{x}, \mathbf{y} \in U$ and $\lambda \in (0, 1)$,

$$f[\lambda \mathbf{x} + (1 - \lambda)\mathbf{y}] \leqslant \varepsilon + \lambda f(\mathbf{x}) + (1 - \lambda) f(\mathbf{y})$$

Hyers and Ulam [1952] showed that if $f$ is $\varepsilon$-approximately convex in an open convex set $U \subseteq \mathbf{R}^n$, then there is a convex function $g: U \to \mathbf{R}$ such that $|g(\mathbf{x}) - f(\mathbf{x})| \leqslant k_n \varepsilon$ where $k_n = (n^2 + 3n)/(4n + 4)$. The same result is discussed by Green [1952a].

Things were then taken one step further [Green, 1952b]. Call $f: U \to \mathbf{R}$ **$\varepsilon$-approximately subharmonic** if (a) it is upper semicontinuous, and (b) if $h$ is harmonic in an open set $D$ interior to $U$, continuous on the boundary of $D$, and majorizes $f$ on this boundary, then $f(\mathbf{x}) \leqslant \varepsilon + h(\mathbf{x})$. Then if $f$ is $\varepsilon$-approximately subharmonic on $U$, there exists a function $g: U \to \mathbf{R}$ subharmonic on $U$ such that $g(\mathbf{x}) \leqslant f(\mathbf{x}) \leqslant \varepsilon + g(\mathbf{x})$.

## REFERENCES

1941, D. H. Hyers, On the stability of the linear functional equation. *Proc. Nat. Acad. Sci. U.S.* **27**, 222–224.
1952, D. H. Hyers and S. M. Ulam, Approximately convex functions. *Proc. Amer. Math. Soc.* **3**, 821–828.
1952a, J. W. Green, Approximately convex functions. *Duke Math. J.* **19**, 499–504.
1952b, J. W. Green, Approximately subharmonic functions. *Proc. Amer. Math. Soc.* **3**, 829–833.

## ALMOST CONVEX FUNCTIONS

Interested in the study of midconvex functions, Kuczma has defined almost convex as follows. Let $I$ be an open interval of the real line. The function $f: I \to \mathbf{R}$ is **almost convex** if

$$f\left(\frac{x+y}{2}\right) \leqslant \tfrac{1}{2}[f(x) + f(y)]$$

holds for $(x, y) \in I \times I$ except on a set $M \subseteq I \times I$ of planar Lebesgue measure zero. He then proves that an almost convex function is equal almost everywhere to a convex function.

## REFERENCE

1970c, M. E. Kuczma, Almost convex functions. *Colloq. Math.* 21, 279–284.

## SCHUR-CONVEX FUNCTIONS

Let $I$ be an interval of the real line $\mathbf{R}$ and $I^n = I \times I \times \cdots \times I$ ($n$ factors). Recognizing Schur [1923] who introduced the idea, call $f: I^n \to \mathbf{R}$ ($n > 1$) **Schur-convex** if

$$f(S\mathbf{x}) \leqslant f(\mathbf{x}) \tag{5}$$

holds for all $\mathbf{x} \in I^n$ and every doubly stochastic matrix $S$. (See Section 63 for a discussion of doubly stochastic matrices, permutation matrices, and their relation to convex functions.) A permutation matrix $P$ is doubly stochastic as is its inverse $P^{-1}$. Thus if $f$ is Schur-convex,

$$f(P\mathbf{x}) \leqslant f(\mathbf{x}) = f(P^{-1}P\mathbf{x}) \leqslant f(P\mathbf{x})$$

so $f(P\mathbf{x}) = f(\mathbf{x})$. This is the definition of a **symmetric function** and so we have proved that every Schur-convex function is symmetric. Call $f: I^n \to \mathbf{R}$ **strictly Schur-convex** if (5) holds and is for each $\mathbf{x}$ a strict inequality except when $S\mathbf{x} = P\mathbf{x}$ for some permutation matrix $P$. For example, $f(x_1, x_2) = x_1^2 + x_2^2$ is strictly Schur-convex on $\mathbf{R}^2$ while $g(x_1, x_2) = x_1 + x_2$ is merely Schur-convex.

A Schur-convex function need not be convex [consider $f(x_1, x_2) = |x_2 - x_1|^{1/2}$ on $\mathbf{R}^2$] and a convex function need not be Schur-convex [consider $f(x_1, x_2) = x_1 + x_2^2$ on $\mathbf{R}^2$]. However, we do have the following result.

**Theorem A.** A convex function $f: I^n \to \mathbf{R}$ is Schur-convex if and only if it is symmetric.

***Proof.*** We have already shown that a Schur-convex function is symmetric. Suppose therefore that $f$ is convex and symmetric on $I^n$. Being convex, we know from Theorem 63B that for each fixed $\mathbf{x} \in I^n$, there is a permutation matrix $P_x$ such that $f(S\mathbf{x}) \leqslant f(P_x\mathbf{x})$ for all doubly stochastic matrices $S$. But $f$ is also symmetric, so $f(P_x\mathbf{x}) = f(\mathbf{x})$. We conclude that $f(S\mathbf{x}) \leqslant f(\mathbf{x})$ as desired. ●

In particular we note that if $\phi: I \to \mathbf{R}$ is convex, then $f(\mathbf{x}) = \phi(x_1) + \phi(x_2) + \cdots + \phi(x_n)$ is Schur-convex on $I^n$. This allows us to prove the following theorem which has a surprising number of applications [Beckenbach and Bellman, 1965, p. 30].

## 86. Other Related Topics

**Theorem B.** Suppose that we have $2n$ numbers $\{x_k, y_k\}$, $k = 1, 2, ..., n$, satisfying

(a) $x_1 \geq x_2 \geq \cdots \geq x_n, y_1 \geq y_2 \geq \cdots \geq y_n$,
(b) $y_1 + \cdots + y_k \leq x_1 + \cdots + x_k$ $(k = 1, ..., n-1)$,
(c) $y_1 + \cdots + y_n = x_1 + \cdots + x_n$.

Then if $\phi: [x_n, x_1] \to \mathbf{R}$ is convex,

$$\phi(y_1) + \cdots + \phi(y_n) \leq \phi(x_1) + \cdots + \phi(x_n)$$

The proof consists in showing that $\mathbf{y} = S\mathbf{x}$ for some doubly stochastic matrix $S$ [Ostrowski, 1952, p. 257; Hardy, Littlewood, and Polya 1952, p. 49], and then using the Schur-convexity of $\phi(x_1) + \cdots + \phi(x_n)$.

Just as ordinary convexity can be characterized by a monotonicity property of the derivative, so also can Schur-convexity as was noted by Schur [1923] and developed by Ostrowski [1952].

**Theorem C.** Let $f(x) = f(x_1, ..., x_n)$ be symmetric and have continuous partial derivatives on $I^n$ where $I$ is an open interval. Then $f: I^n \to \mathbf{R}$ is Schur-convex if and only if

$$(x_2 - x_1)\left(\frac{\partial f}{\partial x_2} - \frac{\partial f}{\partial x_1}\right) \geq 0 \qquad (6)$$

on $I^n$. It is strictly Schur-convex if (6) is a strict inequality for $x_1 \neq x_2$.

For a proof of this theorem and many related results and applications we refer the reader to Ostrowski [1952].

### REFERENCES

1923, I. Schur, Über eine Klasse von Mittelbildungen mit Anwendungen auf die Determinantentheorie. *Sitzungsber. Berlin. Math. Ges.* **22**, 9–20.
1952, A. Ostrowski, Sur quelques applications des fonctions convexes et concave au sens de I. Schur. *J. Math. Pures Appl.* **31**, 253–292.
1965, E. F. Beckenbach and R. Bellman, "Inequalities," 2nd rev. printing, pp. 30–32. Springer-Verlag, Berlin and New York.

### MATRIX-CONVEX FUNCTIONS

Let $\mathbf{S}_n$ denote the set of all real $n \times n$ symmetric matrices and introduce a (partial) order relation by saying $A \leq B$ if $B - A$ is nonnegative definite. If $A \in S_n$ has $\{\lambda_1, ..., \lambda_m\}$ as its **spectrum** (set of distinct characteristic values), then by the spectral theorem [Halmos, 1958, p. 156], $A = \sum_1^m \lambda_i E_i$ where the $E_i$ are uniquely determined orthogonal idempotent matrices. Now if $p(x) = a_0 + a_1 x + \cdots + a_k x^k$ is an ordinary polynomial, it follows [Halmos, 1958, p. 165] that

$$p(A) = a_0 + a_1 A + \cdots + a_k A^k = \sum_1^m p(\lambda_i) E_i$$

This suggests a definition for $f(A)$ where $f$ is any real-valued function with domain containing the spectrum of $A$. We simply write $f(A) = \sum_1^m f(\lambda_i) E_i$. Viewed this way, $f: I \to \mathbf{R}$ becomes a matrix-valued function defined on a set of symmetric matrices.

With these preliminaries understood, we can define the two main ideas to be discussed. Call $f: I \to \mathbf{R}$ **matrix-monotone increasing of order** $n$ if $A \leqslant B$ implies $f(A) \leqslant f(B)$ and call it **matrix-convex of order** $n$ if $\lambda \in (0, 1)$ implies

$$f[\lambda A + (1 - \lambda)B] \leqslant \lambda f(A) + (1 - \lambda) f(B)$$

In each case the implication is to hold for all $A, B \in \mathbf{S}_n$ whose spectra lie in $I$. These definitions give, for $n = 1$, ordinary monotonicity and convexity. These definitions were introduced by Löwner [1934] who is also responsible for the original statements of our first three theorems.

**Theorem A.** Let $f : (a, b) \to \mathbf{R}$ be matrix-monotone increasing of order $n > 1$. Then $f$ is $2n - 3$ continuously differentable and its $(2n - 3)$th derivative is an ordinary convex function.

**Theorem B.** $f: (a, b) \to \mathbf{R}$ is matrix-monotone increasing of all orders if and only if it is analytic in $(a, b)$, can be analytically continued into the whole upper half-plane, and there represents an analytic function whose imaginary part is nonnegative.

From this latter theorem it follows, for example, that $f(x) = x^\alpha$ ($0 < \alpha \leqslant 1$) and $g(x) = \log x$ are matrix-monotone increasing of all orders on $(0, \infty)$. C. Davis [1963, p. 199] gives a direct proof for $f(x) = \sqrt{x}$.

Before stating the next theorem, we recall (Section 83) that the symbol $[x_0, \ldots, x_n; f]$ denotes the divided difference of order $n$. It is well defined so long as $x_0, \ldots, x_n$ are distinct points, and if $f$ has a continuous derivative

## 86. Other Related Topics

of order $n$, the definition may be extended by continuity to allow for equal arguments [C. Davis, 1963, p. 190]. In particular,

$$[x_0, x_0; f] = \lim_{x_1 \to x_0} [x_0, x_1; f] = \lim_{x_1 \to x_0} \frac{f(x_1) - f(x_0)}{x_1 - x_0} = f'(x_0)$$

and more generally for the $n$th divided difference

$$[x_0, \ldots, x_0; f] = \frac{f^{(n)}(x_0)}{n!}$$

**Theorem C.** (See C. Davis [1963, p. 191].) A continuously differentiable function $f: (a, b) \to \mathbf{R}$ is matrix-monotone increasing of order $n$ if and only if for any choice of $x_1, \ldots, x_n \in (a, b)$, the $n \times n$ matrix $a_{ij} = [x_i, x_j; f]$ is nonnegative definite.

We now state two characterizations of matrix-convexity. The first is due to Krauss [1936], the second to Bendat and Sherman [1955].

**Theorem D.** A twice continuously differentiable function $f: (a, b) \to \mathbf{R}$ is matrix-convex of order $n$ if and only if for any choice of $x_1, \ldots, x_n \in (a, b)$, the $n \times n$ matrix $b_{ij} = [x_1, x_i, x_j; f]$ is nonnegative definite.

**Theorem E.** A twice continuously differentiable function $f: (a, b) \to \mathbf{R}$ is matrix-convex of order $n$ if and only if $g(x) = [x_0, x; f]$ is matrix-monotone increasing of order $n$ for each fixed $x_0 \in (a, b)$.

In spite of these two theorems it is not easy to give specific examples of matrix-convex functions. One example is $f(x) = x^2$ (more generally $ax^2 + bx + c$, $a \geqslant 0$). Davis [1963] shows this directly, and also demonstrates that $f(x) = |x|$ is not matrix-convex of any order $n > 1$ on an interval containing the origin.

The subject of matrix-convexity has been generalized to functions of several variables by Davis [1963] where there is an interesting connection with Schur-convexity.

## REFERENCES

1934, K. Löwner, Über monotone Matrixfunktionen. *Math. Z.* **38**, 177–216.
1936, F. Krauss, Über konvexe Matrixfunktionen. *Math. Z.* **41**, 18–42.

1955, J. Bendat and S. Sherman, Monotone and convex operator functons. *Trans. Amer. Math. Soc.* **79**, 58–71.

1963, C. Davis, Notions generalizing convexity for functions defined on spaces of matrices. In "Convexity." *Proc. Symp. Pure Math. Amer. Math. Soc.*, **7**, (V. L. Klee, ed.), pp. 187–201. Providence Rhode Island.

## AND STILL MORE

Having said in the Introduction that certain decisions had to be made about where to start this book, we here acknowledge that similar somewhat arbitrary decisions had to be made about where to stop. Our discussion of related classes of functions could continue almost indefinitely. We might write about pseudoconvex functions (Problem 42D), $p$-convex functions (end of Section 72), Wright-convex functions (Problem 72N), strongly convex functions (Appendix, Project P), uniformly convex functions (Appendix, Project Q), $\mathscr{G}$-convex functions (Appendix, Project R), and many others not mentioned explicitly in our book. All that we have written is about convexity, but we have not written all that can or is likely to be written about the topic. This is as it must be when one writes of a subject having the innate vigor and virility of convexity with its ability to penetrate new areas and sprout new branches.

# *Appendix*

> But if logic is the hygiene of the mathematician, it is not his source of food; the great problems furnish the daily bread on which he thrives.
>
> A. WEIL

> ... we can conceive of science as living, if besides solved problems, it has unsolved ones, and if the answers to earlier questions lead to new questions and stimulate new research.
>
> HEINRICH TIETZE

## Introduction

What does a working mathematician do? He solves problems—practical problems and abstract problems, old ones and new ones, but especially hard problems. To live with a problem day and night for weeks or months, to look in every cranny and travel every blind alley, finally to discover a long circuitous route to a solution—this is a mathematician's life. Then having found a solution, to explore every facet of it, to put it in its most elegant setting, and to show it to the world in clear striking beauty—this is the goal toward which he works.

Unfortunately, few undergraduate mathematicians have had a significant experience of this kind. The pressure to graduate in four years while learning something about a variety of subjects and the desire to master a number of topics within mathematics seem to preclude devoting large blocks of time to a single hard problem. However, more and more colleges are recognizing the importance of a deep creative independent study experience and are structuring their curricula to foster it. Some require a senior thesis, others a more modest independent study project. Many have set up senior seminars where students investigate and present material that is not part of regular courses. This book was conceived originally as a source book for such a senior seminar. It seems appropriate that it should end with a number of projects and problems for individual or group study.

We have described first of all 26 projects that we think merit serious investigation and could profitably be carried out by most readers of this book. The proposals are specific and rest on ideas developed earlier, but they are intended to be taken as open-ended suggestions leading the student to explorations constrained only by his time and imagination. For the very ambitious student wishing to confront genuine research problems, we have also posed a few famous unsolved problems more or less related to the subject matter of our book. The list is short, but we have indicated a number of references where many other unsolved problems may be found.

## Independent Study Projects

**A** (Measures of convexity). For the class of convex functions $f\colon [a, b] \to \mathbf{R}$, a measure of convexity $\mu$ might be expected to satisfy some or all of the following properties.

(1) $\mu(f) \geqslant 0$ and $\mu(f) = 0 \Leftrightarrow f$ is affine.
(2) $\mu(f + g) = \mu(f)$ if $g$ is affine.

# Independent Study Projects

(3) $\mu(f+g) \leq \mu(f) + \mu(g)$.
(4) $\mu(\alpha f) = \alpha \mu(f)$ if $\alpha \geq 0$.
(5) $a < c < b \Rightarrow \mu_a^c(f) + \mu_c^b(f) \leq \mu_a^b(f)$.
(6) $f_n \to f \Rightarrow \mu(f_n) \to \mu(f)$.

Try to think of several candidates for $\mu$ and then study their properties and interrelationships. Three possibilities one might consider are

$$\mu_1(f) = f_-'(b) - f_+'(a) \quad \text{(Section 14)}$$

$$\mu_2(f) = \max_{0 \leq \lambda \leq 1} [\lambda f(a) + (1-\lambda) f(b) - f(\lambda a + (1-\lambda)b)]$$

$$\mu_3(f) = \int_a^b [f(x) - A(x)] \, dx$$

where $A$ is a support function for $f$ at $(a+b)/2$ (Problem 12M).

**B** (Second derivatives). Pull together and expand all the results that relate convexity to second derivatives of various kinds. Such a study might be limited to functions $f: (a, b) \to \mathbf{R}$, beginning with material in Section 12 (including Problem 12N); or it might be carried out for functions with their domain in $\mathbf{R}^n$ or $\mathbf{L}$, thereby addressing some of the questions raised in Problem 44I. Some notions of the second derivative that should be considered are

(1) second Schwarz derivative [Natanson II, 1961, p. 137],
(2) other modified second derivatives [Busemann, 1958, pp. 8–10],
(3) second distribution derivatives [Halperin, 1952],
(4) second-degree Taylor approximation (Problem 44H).

**C** (Convex sequences). Call a sequence $\{a_n\}$, $n = 0, 1, \ldots$, **convex** if $\Delta^2 a_n = a_n - 2a_{n+1} - a_{n+2} \geq 0$ for all $n$. It is clear that if $f: [0, \infty) \to \mathbf{R}$ is a convex function, then $a_n = f(n)$ is a convex sequence. Investigate the theory of convex sequences. In particular, show that if $\{a_n\}$ is convex and bounded, then $\sum_0^\infty (n+1) \Delta^2 a_n < \infty$. Call a sequence **quasiconvex** if $\sum_0^\infty (n+1) |\Delta^2 a_n| < \infty$. Study the properties of such sequences. Basic references are Zygmund [1968] and Bary [1964].

**D** (A Banach algebra). Let $BCN[a, b]$ be the subspace of $BC[a, b]$ (defined in Section 14) consisting of functions $f$ for which $f(a) = f_+'(a) = 0$. It is a nonseparable Banach space under the norm $\|f\| = K_a^b(f)$ and becomes a commutative Banach algebra if multiplication is defined by

$$(f * g)(t) = \int_a^t f(t - s - a) \, d\hat{g}(s)$$

where $\hat{g}(s) = g_-'(s)$ for $s \in (a, b]$ and $\hat{g}(a) = 0$. Prove these facts and investigate the corresponding structures. Can you introduce an order relation so that $BCN[a, b]$ is also a vector lattice? A helpful reference for convolution products is Widder [1941], for Banach algebras, Lorch [1962] and Rudin [1966].

**E** (Generalized bounded convexity). Let $u$ be strictly increasing on $[a, b]$ and let

$$\Box f_j = \frac{f(x_j) - f(x_{j-1})}{u(x_j) - u(x_{j-1})}$$

Try to develop all the results of Section 14 in this context. Related references are Russell [1970] and Huggins [1972].

**F** (Complex bounded convexity). For $f : [a, b] \to \mathbf{R}^n$, define

$$\Box f_j = \frac{f(x_j) - f(x_{j-1})}{x_j - x_{j-1}}$$

$$K_a^b(f) = \sup \sum_1^{n-1} \| \Box f_{j+1} - \Box f_j \|$$

where $\| \; \|$ is the Euclidean norm. Generalize the results of Section 14, paying special attention to the case $\mathbf{R}^2$ which corresponds to complex-valued functions of a real variable. Relate $K_a^b(f)$ to the notion of total curvature from differential geometry.

**G** (French railroad space). In the plane, let the distance between **x** and **y** be the length of the shortest path between **x** and **y** via the French railroad (a system of straight lines emanating from Paris). More precisely, let $\text{dist}(\mathbf{x}, \mathbf{y}) = \| x - y \|$ if **x** and **y** lie on the same line through **O** and $\text{dist}(\mathbf{x}, \mathbf{y}) = \| \mathbf{x} \| + \| \mathbf{y} \|$ otherwise. Investigate the topological properties of this space (is addition continuous?) and study ordinary convex sets in the space. Introduce a new notion of convexity by calling a set $F$-convex if with each pair of its points it also contains the shortest (railroad) path connecting them. Develop a theory of $F$-convex sets. The French railroad space is mentioned by Hewitt and Stromberg [1965, p. 60].

**H** (Function classes determined by inequalities). Consider functions $f: \mathbf{L} \to \mathbf{R}$ where **L** is a linear space. We say a subset $S$ of $\mathbf{R} \times \mathbf{R}$ is **determining** for a class $\mathscr{F}(S)$ of functions if $f \in \mathscr{F}(S) \Leftrightarrow f(\alpha \mathbf{x} + \beta \mathbf{y}) \leq \alpha f(\mathbf{x}) + \beta f(\mathbf{y})$ for all $(\alpha, \beta) \in S$. Each of the eight classes of Problem 71F as well as many others of considerable importance arise in this way (the subadditive functions are another example; $S = \{(1, 1)\}$). A class may have many determining sets [$\{(1, 1)\}$ and $\{(1, 1), (2, 2)\}$ determine the same class], but there is always a unique maximal determining set. Find it for each of the classes referred to above. Can there be disjoint determining classes for the same class of functions? Are there minimal determining classes? If so, are they unique? Does the linear class have a countable determining set? Investigate the classes and continuity properties of classes of functions determined by a countable determining set. Generalize the whole idea by replacing the basic inequality with $f(\alpha x + \beta y) \leq p(\alpha) f(x) + p(\beta) f(y)$ for appropriate choices of $p$ (see Quintas and Supnick [1963] for some ideas).

**I** (Generalized ellipses [Hammer, 1967]). Let a curve be traced by holding a pencil against a loop of string of length $M$ which goes around a convex set $U$ of circumference $m < M$. An ellipse results if $U$ is a line segment, a circle if $U$ is a single point. What happens if $U$ is a triangle? Investigate the area, circumference, and optical properties. Consider other sets $U$.

**J** (Counterexamples in convexity). Interest in such books as "Counterexamples in Analysis" [Gelbaum and Olmsted, 1964] and "Counterexamples in Topology" [Steen and Seebach, 1970] underlines the usefulness of a collection of examples which illustrate possible pathologies. Organize in a systematic way a list of convex functions (or convex sets, or both) that exhibit the unusual, the unexpected in convexity theory. As a start, consider the examples in Problems 12F, 12G, 22B, 41J, 41K, 43B, 44D, and 44I.

**K** (M. Riesz's convexity theorem). Let $\mathbf{x} = (x_1, \ldots, x_m)$ and $\mathbf{y} = (y_1, \ldots, y_n)$ be

# Independent Study Projects

vectors with complex entries and let $(a_{ij})$ be a nonzero $m \times n$ complex matrix. Introduce norms $M_\alpha(\mathbf{x})$ and $N_\beta(\mathbf{y})$

$$M_\alpha(\mathbf{x}) = \begin{cases} \left(\sum_1^m |x_i|^{1/\alpha}\right)^\alpha & \text{if } \alpha > 0 \\ \max_i |x_i| & \text{if } \alpha = 0 \end{cases}$$

$$N_\beta(\mathbf{y}) = \begin{cases} \left(\sum_1^n |y_i|^{1/\beta}\right)^\beta & \text{if } \beta > 0 \\ \max_i |y_i| & \text{if } \beta = 0 \end{cases}$$

and let

$$M(\alpha, \beta) = \sup \left| \sum_{i=1}^m \sum_{j=1}^n a_{ij} x_i y_j \right|$$

where the supremum is taken over $\mathbf{x}$ and $\mathbf{y}$ for which $M_\alpha(\mathbf{x}) \leq 1$ and $N_\beta(\mathbf{y}) \leq 1$. Then $\log M(\alpha, \beta)$ is convex on $\mathbf{R}_+ \times \mathbf{R}_+$. Prove this fact and develop its many consequences [Hardy, Littlewood, and Polya, 1952, Chapter 7; Taylor, 1958, pp. 221–224; Thorin, 1948; Salem, 1949].

**L** (Hadamard's three circles theorem). Let $f(z)$ be a complex-valued function of a complex variable, holomorphic on the whole plane, and let $M(r)$ be the maximum of $|f(z)|$ on the circle $|z| = r$. Then $g(t) = \log M(e^t)$ is convex on $\mathbf{R}$. Prove this theorem and investigate its consequences. Titchmarsh [1939, pp. 172–173] is a good reference, but there is other literature to consult on this subject.

**M** (Convex functions associated with convex sets). In the literature we find many convex functions associated with a convex set or a pair of convex sets. Organize and study these functions. Some examples are

(1) support function (Problem 41E),
(2) gauge function (Problem 41E),
(3) distance function (Problem 41E),
(4) lower boundary function [Rockafellar, 1970a, p. 33],
(5) inner distance function $-\log n_0(x)$ [Bremermann, 1956],
(6) various extremum functions [Forsythe, 1970],
(7) Brunn–Minkowski function [Grünbaum, 1967, p. 338],
(8) some volume and surface area functions [Valentine, 1964, p. 185].

**N** (Infinite convex combinations). Call $\sum_1^\infty \lambda_i x_i$ an infinite convex combination of points $x_1, x_2, \ldots$ if the series converges and $\lambda_i \geq 0$, $\sum_1^\infty \lambda_i = 1$. Show that $U \subseteq \mathbf{R}^n$ is convex if and only if $U$ is closed under such combinations, but that this result may be false in infinite-dimensional spaces [Rubin and Wesler, 1958]. Does the inequality

$$f\left(\sum_1^\infty \lambda_i x_i\right) \leq \sum_1^\infty \lambda_i f(x_i) \tag{1}$$

similarly characterize convex functions on $\mathbf{R}^n$? What about infinite-dimensional spaces? Let $\Lambda$ be the set of real infinite sequences, and consider the classes of functions that satisfy (1) for various subsets of $\Lambda$. Study these classes from the point of view of Section 71. Generalize the whole idea by considering integrals in place of sums.

**O** (Convexity at a point). Let $U$ be a fixed, not necessarily convex set in a linear space **L**. Call $f: U \to \mathbf{R}$ **convex at a point** $\mathbf{x} \in U$ if for each representation $\mathbf{x} = \lambda \mathbf{y} + (1 - \lambda)\mathbf{z}$ with $\mathbf{y}, \mathbf{z} \in U$ and $\lambda \in [0, 1]$, we have $f(\mathbf{x}) \leq \lambda f(\mathbf{y}) + (1 - \lambda) f(\mathbf{z})$. In turn, say that $f$ is **convex on a subset** $V$ of $U$ if $f$ is convex at each point of $V$. Note that if $U$ is a convex set, then to say $f$ is convex on $U$ is equivalent to our usual notion of a convex function. On the other hand, if $U$ is the set of nonnegative integers and $f(n) = a_n$, then to say that $f$ is convex on $U$ is equivalent to saying that $\{a_n\}$ is a convex sequence (Project C). Investigate this new notion of a convex function. Some questions you might consider are suggested.

(1) Is being convex at a point equivalent to having support at a point?
(2) Does convexity at a point imply that Jensen's inequality holds at the point?
(3) For what kinds of subsets $V$ of $U = \mathbf{R}$ is it possible to find a function $f: \mathbf{R} \to \mathbf{R}$ that is convex on $V$?
(4) Suppose $f$ is continuous on $U$ and convex on $V \subseteq U$. What kinds of conditions on $V$ (denseness, full measure, etc.) ensure that $f$ is convex on $U$?
(5) Can you develop a differentiation theory that will characterize this new convexity theory?

**P** (Strong convexity). Call $f: (a, b) \to \mathbf{R}$ **strongly convex** if there is an $\alpha > 0$ such that

$$f\left(\frac{x + y}{2}\right) \leq \tfrac{1}{2}f(x) + \tfrac{1}{2}f(y) - \tfrac{1}{4}\alpha(x - y)^2$$

for all $x, y \in (a, b)$. Show that this is equivalent to each of the following conditions.

(1) $f[\lambda x + (1 - \lambda)y] \leq \lambda f(x) + (1 - \lambda) f(y) - \lambda(1 - \lambda)\alpha(x - y)^2$, $\lambda \in [0, 1]$.
(2) For each $x_0 \in (a, b)$, there is a linear function $T$ such that $f(x) \geq f(x_0) + T(x - x_0) + \alpha(x - x_0)^2$
(3) For differentiable $f$, $f(x) \geq f(x_0) + f'(x_0)(x - x_0) + \alpha(x - x_0)^2$
(4) For differentiable $f$, $[f'(x) - f'(y)] [x - y] \geq 2\alpha(x - y)^2$.
(5) For twice differentiable $f$, $f''(x) \geq 2\alpha$.

If $f$ is strongly convex on $(a, b)$, then $f$ is bounded below, $\{x: f(x) < \lambda\}$ is bounded for all $\lambda$, and $f$ has a unique minimum on any closed subinterval of $(a, b)$. What else is true? Generalize the whole idea to the setting of a normed linear space. See Polzak [1966] and Lyubich and Maistrovskii [1970].

**Q** (Uniform convexity). Let $U$ be an open convex set in a normed linear space **L**. Call $f: U \to \mathbf{R}$ **uniformly convex** if there is an increasing function $\delta: [0, \infty) \to [0, \infty)$ with $\delta(0) = 0$ and $\delta(t) > 0$ for $t > 0$ such that

$$f\left(\frac{x + y}{2}\right) \leq \tfrac{1}{2}f(x) + \tfrac{1}{2}f(y) - \delta(\| x - y \|)$$

for all $x, y \in U$. Call $f$ **uniformly quasiconvex** if

$$f\left(\frac{x + y}{2}\right) \leq \max\{f(x), f(y)\} - \delta(\| x - y \|)$$

**Independent Study Projects**                                                                 269

for all $x, y \in U$. Develop the theory of such functions. See Project $P$ [Polzak, 1966; Levitan and Polzak, 1966; Lyubich and Maistrovskii, 1970].

**R** (Convex functions in terms of suprema). We begin with the following observation. A function $f: (a, b) \to \mathbf{R}$ is convex if and only if it is the supremum of some family of affine functions (see Theorems 12D and 13D). Now let $\mathscr{G}$ be an arbitrary family of functions $g: I \to \mathbf{R}$. Call $f: I \to \mathbf{R}$ $\mathscr{G}$-**convex** if $f$ has the representation

$$f(x) = \sup_{g \in G} g(x)$$

for some subfamily $G$ of $\mathscr{G}$. Investigate this notion for various families $\mathscr{G}$ ($\mathscr{G}$ = affine functions on $[a, b]$, $\mathscr{G}$ = quadratic polynomials on $[a, b]$, $\mathscr{G}$ = Jensen functions on $[a, b]$, and so on). What properties of ordinary convex functions hold for $\mathscr{G}$-convex functions? A recent paper related to this idea is that by Kutateladze and Rubinov [1971].

**S** (Approximation of convex functions). Collect and organize results that relate to the approximation of convex functions. You might begin by proving that if $f: I \to \mathbf{R}$ is convex and $[a, b]$ is a closed subinterval of $I^0$, then for any positive integer $k$, there exists a sequence of $k$-times differentiable convex functions $\{g_i\}$ such that $g_i(x) \geqslant g_{i+1}(x) \geqslant f(x)$ and $g_i(x) \to f(x)$ on $[a, b]$ (see Bremermann [1956, p. 38]). Here is another problem that has received wide attention. For convex $f: [a, b] \to \mathbf{R}$, let $E_n(f) = \|f - f_n\|$ where $f_n$ is a polynomial (or a trigonometric polynomial, or a rational function) of degree at most $n$ which is closest to $f$. It turns out to be of interest [Shisha, 1965; Bulanov, 1970, 1971] to investigate the rate at which $E_n(f)$ goes to 0 with $n$.

**T** (Extreme points). There are dozens of papers in the mathematical literature addressed to the problem of identifying the extreme points of a given convex set, the problem for doubly stochastic matrices discussed in Section 63 being an example that has itself been the subject of several papers. Develop a catalog of known results. While the *Mathematical Reviews* will be your basic resource, you will also want to look at reference books on functional analysis [Köthe, 1969, pp. 333–337], matrix theory, and convexity.

**U** (Epigraphs and level sets). A function is convex if and only if its epigraph is a convex set. It is quasiconvex if and only if its level sets are convex. Investigate the properties of the epigraphs and level sets for the various kinds of convexity described in Chapter VIII.

**V** (Generalized convex functions). The literature associated with Section 85 is scattered, sometimes sketchy, and based on several different looking definitions of $\mathscr{F}_n$-convexity. Give a careful development of this material, reconciling definitions and proving the basic theorems.

**W** (Convex solutions of functional equations). In Problem 13H we pointed out that the gamma function is characterized as the unique solution to $f(x + 1) - xf(x) = 0$, $f(1) = 1$, that is log-convex on $(0, \infty)$. The problem of finding convex solutions to various functional equations has attracted considerable attention, as is illustrated by the following papers.

(1)   $1/f(x + 1) - xf(x) = 0$       [John, 1939],
(2)   $1/f(x + a) - x^p f(x) = 0$     [Thielman, 1941],
(3)   $f(x + 1) - f(x) = \log x$      [Leipnik and Oberg, 1967].

Investigate and collect results of this type. The book by Aczel [1966] has an extensive bibliography on functional equations.

**X** (Differentiability and continuity). Suppose $f: U \to \mathbf{M}$ is defined on an open set $U \subseteq \mathbf{L}$, and that $f'(\mathbf{x}_0)$ exists at $\mathbf{x}_0 \in U$. Write an expository paper relating differentiability at $\mathbf{x}_0$ to the continuity of $f$, and to the continuity of $f'$ when $f$ is differentiable in a neighborhood of $\mathbf{x}_0$. You may wish to include the following.

(1) Give examples in which $f'(\mathbf{x}_0)$ is not continuous. Problem 23E suggests an example in which $f$ is linear. Can you find an example in which $f$ is not affine?

(2) $f'(\mathbf{x}_0)$ is continuous $\Leftrightarrow f$ is continuous at $\mathbf{x}_0$.

(3) If $\mathbf{L}$ is finite dimensional, then $f$ is continuous at $\mathbf{x}_0$. In particular, if $\mathbf{L} = \mathbf{R}$, then the existence of $f'(x_0)$ implies the continuity of $f$ at $x_0$.

(4) Suppose $f$ is continuous on $V = U \setminus \{\mathbf{x}_0\}$, and that $f'(\mathbf{x}_0)$ exists. Must $f'(\mathbf{x}_0)$ then be continuous? If so, can this result be extended to other proper subsets $V \subseteq U$?

(5) Show by example that $f$ may be continuous and differentiable on $U$ while $f'$: $U \to \mathscr{L}(\mathbf{L}, \mathbf{M})$ is discontinuous.

(6) If $\mathbf{L}$ and $\mathbf{M}$ are finite dimensional, then for $f$ continuous and differentiable throughout $U$ we have for $\mathbf{x} \in U$,

$$f(\mathbf{x} + \mathbf{h}) = f(\mathbf{x}) + f'(\mathbf{x})\mathbf{h} + \|\mathbf{h}\| \varepsilon(\mathbf{x}, \mathbf{h})$$

and $f': U \to \mathscr{L}(\mathbf{L}, \mathbf{M})$ is continuous if and only if $\lim_{\|\mathbf{h}\| \to 0} \varepsilon(\mathbf{x}, \mathbf{h}) = 0$ uniformly on any compact subset of $U$.

(7) Give necessary and sufficient conditions, similar to those of part (6), for the continuity of $f'$ when $\mathbf{L}$ and $\mathbf{M}$ are not finite dimensional [Dieudonné, 1960, p. 159].

**Y** (Geometry of the unit sphere). We have seen (Problem 44E) that the differential properties of the norm are related to geometrical properties of the unit sphere in a normed linear space $\mathbf{L}$. Intimately related to the same topic are geometrical properties of the unit sphere in the dual space $\mathbf{L}^* = \mathscr{L}(\mathbf{L}, \mathbf{R})$.

(1) Among the geometrical properties of the unit sphere, one comes across the notions of *smooth, rotund, uniformly convex, uniformly smooth*, and *weak uniform convexity*. Find definitions of these concepts in the literature, explore the relationships between them, and give examples exhibiting the various possible properties [Day, 1955, 1962; Cudia, 1963].

(2) Find and prove the relations between concepts mentioned in part (1) and various notions of the differentiability of the norm function. Problem 44E provides a start.

(3) A normed linear space $\mathbf{L}$ is uniformly smooth [uniformly convex] $\Leftrightarrow$ the dual space $\mathbf{L}^*$ (strong topology) is uniformly convex [uniformly smooth] [Köthe, 1969]. Find other such results.

(4) Since equivalent norms (see Section 21) are not necessarily differentiable or nondifferentiable together, there is interest in the possibility of renorming spaces so that the new (equivalent) norm will exhibit desirable differentiability properties. As a typical result we state the following: A separable Banach space admits an equivalent norm of class $C' \Leftrightarrow$ its dual space is separable [Restrepo, 1965].

Besides the papers mentioned above, we also refer the reader to Cudia [1964], Smulian [1940], and Sundaresan [1967].

**Z** (Derivatives and matrices). Let $f: U \to \mathbf{R}^n$ be twice differentiable on the open set $U \subseteq \mathbf{R}^n$. Then $f'(\mathbf{x})$ is a linear transformation represented with respect to the standard basis by an $n \times n$ matrix $[f'(\mathbf{x})]$.

(1) If for each $\mathbf{x} \in U$, $[f'(\mathbf{x})]$ is skew symmetric, then $f$ is affine.

(2) If each $\mathbf{x} \in U$, $[f'(\mathbf{x})]$ is orthogonal, then $f$ is a Euclidean motion.

Find other results of this sort. Then consider $f: U \to \mathbf{R}$, again twice differentiable on $U \subseteq \mathbf{R}^n$. Now $f''(\mathbf{x})$ is an $n \times n$ matrix, and we may try to classify $f$ according to properties of the matrix $[f''(\mathbf{x})]$. For example, if $[f''(\mathbf{x})]$ is nonnegative definite, we know that $f$ is convex. What if $[f''(x)]$ is skew symmetric? orthogonal? and so forth? Some questions of this sort are considered by Roberts [1969].

## Unsolved Problems

**A** (Four-color problem [Ore, 1967; Klee, 1971; Saaty, 1972]). Can a convex polyhedral set in $\mathbf{R}^3$ be colored in four colors so that neighboring faces never receive the same color?

**B** (Sphere packing [Rogers, 1964; Klee, 1971]). What arrangement of congruent spherical balls in $\mathbf{R}^3$ provides the densest packing?

**C** (An area problem [Yaglom and Boltyanskii, 1961, p. 18]). Find a closed convex set (known to exist) of least area that will cover every plane figure of diameter 1. A circle of radius $1/\sqrt{3}$ will cover as will a regular hexagon inscribed in such a circle.

**D** (Borsuk's covering problem [Hadwiger *et al.*, 1964, pp. 15, 46; Grünbaum, 1963]). Can every convex set of diameter 1 in $\mathbf{R}^n$ be covered by $n + 1$ sets of diameter less than 1 (unsolved for $n > 3$)?

**E** (Equichordal points [Klee, 1969b]). Can a plane closed bounded convex set with nonempty interior have two equichordal points? An equichordal point is one through which all chords have equal length.

**F** (Hadamard matrices [Brenner and Cummings, 1972]). Hadamard showed that a real $n \times n$ matrix $A = (a_{ij})$ with $|a_{ij}| \leq 1$ satisfies $|\det A| \leq n^{n/2}$ (Problem 63L) and that equality can conceivably occur only if $n = 1, 2$, or a multiple of 4. For each such $n$, is there a matrix for which equality occurs?

**G** (Van der Waerden's conjecture [Marcus and Minc, 1964, pp. 129–132]). It is conjectured that the permanent $p(S)$ of an $n \times n$ doubly stochastic matrix $S$ satisfies

$$p(S) > \frac{n!}{n^n}$$

with equality if and only if $S$ has all its entries equal to $1/n$. The conjecture is unresolved, though partial results are known.

**H** (Nearest point problem in Hilbert space [Klee, 1971, p. 628]). A closed subset of $\mathbf{R}^n$ is convex if and only if each point of the space admits a unique nearest point in the set (Problem 55N). Is the same true in Hilbert space? See Valentine [1964, pp. 94–98], Klee, [1961, 1967], and Brøndsted [1965, 1966a].

**I** (Level sets [Fenchel, 1953, p. 115 ff.]). What "nice" conditions on a nested family of convex sets will ensure that it is the family of level sets of a convex function?

Hosts of other unsolved problems may be found in the following references.

Aequationes Math. (1968–). Problem section.
Amer. Math. Monthly (1969–). Research problem section.

Elemente der Math. (1954–). Ungelöste Probleme
W. Fenchel, ed. (1967). *Proc. Colloq. Convexity*, Mat. Inst., Copenhagen, 1965.
B. Grünbaum, (1967). "Convex Polytopes." Wiley, New York.
H. Hadwiger, H. Debrunner, and V. Klee (1964). "Combinatorial Geometry in the Plane." Holt, New York.
V. Klee, ed. (1963). Convexity. *Proc. Symp. Pure Math.* 7. Amer. Math. Soc., Providence, Rhode Island.
Z. Melzak (1965). Problems connected with convexity. *Can. Math. Bull.* 8, 565–573.
H. Meschkowski (1966). "Unsolved and Unsolvable Problems in Geometry." Ungar, New York.
F. Valentine (1964). "Convex Sets," pp. 163–194. McGraw-Hill, New York.

# Bibliography

J. Aczel
1966 "Lectures on Functional Equations and their Applications." Academic Press, New York.

J. C. Aggeri
1966 Les fonctions convexes continues et le théorème de Krein-Milman. *C. R. Acad. Sci. Ser. A-B* **262**, 229–232.

A. D. Alexandroff
1939 Almost everywhere existence of the second differential of a convex function and some properties of convex surfaces connected with it. *Uch. Zap. Leningrad. Gos. Univ. Ann. Ser. Math.* **6**, 3–35 (Russian).
1949 On surfaces represented as the difference of convex functions. *Izv. Akad. Nauk Kaz. SSR* 60, *Ser. Mat. Mekh.* **3**, 3–20 (Russian).
1950 Surfaces represented by the differences of convex functions. *Dokl. Akad. Nauk SSSR* [N.S.] **72**, 613–616 (Russian).

B. J. Anderson
1968 An inequality for convex functions. *Nord. Mat. Tidsskr.* **6**, 25–26.

R. D. Anderson and V. L. Klee
1952 Convex functions and upper semi-continuous collections. *Duke Math. J.* **19**, 349–357.

M. G. Arsove
1953 Functions representable as the difference of subharmonic functions. *Trans. Amer. Math. Soc.* **75**, 327–365.

E. Artin
1964 "The Gamma Function." Holt, New York.

G. Ascoli
1932 Sugli spazi lineari metrici e le loro varieta lineari. *Ann. Math. Pura Appl.* [4] **10**, 33–81.

E. Asplund
1968 Fréchet differentiability of convex functions. *Acta Math.* **121**, 31–47.

E. Asplund and R. T. Rockafellar
1969 Gradients of convex functions. *Trans. Amer. Math. Soc.* **139**, 443–467.

S. Banach
1932 "Théorie des Opérations Linéaires." Warsaw.

N. K. Bary
1964 "A Treatise on Trigonometric Series." Macmillan, New York.

E. M. L. Beale, ed.
1968 Applications of Mathematical Programming Techniques. NATO Sci. Affairs Committee, Academic Press, New York.

E. F. Beckenbach
1937 Generalized convex functions. *Bull. Amer. Math. Soc.* **43**, 363–371.
1948a Properties of surfaces of negative curvature. *Amer. Math. Monthly* **55**, 285–301.
1948b Convex functions. *Bull. Amer. Math. Soc.* **54**, 439–460.
1953 Convexity (unpublished).

E. F. Beckenbach and R. Bellman
1965 "Inequalities," 2nd rev. printing. Springer-Verlag, Berlin and New York.

E. F. Beckenbach and R. H. Bing
1945 On generalized convex functions. *Trans. Amer. Math. Soc.* **58**, 220–230.

J. Bendat and S. Sherman
1955 Monotone and convex operator functions. *Trans. Amer. Math. Soc.* **79**, 58–71.

B. Bereanu
1969 Partial monotonicity and J-convexity. *Rev. Roumaine Math. Pures Appl.* **14**, 1085–1087.

C. Berge and A. Ghouila-Houri
1965 "Programming, Games and Transportation Networks." Wiley, New York.

B. Bernstein and R. A. Toupin
1962 Some properties of the Hessian matrix of a strictly convex function. *J. Reine Angew. Math.* **210**, 65–72.

F. Bernstein and G. Doetsch
1915 Zur Theorie der konvexen Funktionen. *Math. Ann.* **76**, 514–526.

S. Bernstein
1914 Sur la definition et les proprietes des fonctions analytiques d'une variable reelle. *Math. Ann.* **75**, 449–468.
1952 Remarks on the theory of regularly monotonic functions. *Izv. Akad. Nauk SSSR Ser. Mat.* **16**, 3–16 (Russian).

A. S. Besicovitch and R. O. Davies
1965 Two problems on convex functions. *Math. Gaz.* **49**, 66–69.

G. Birkoff
1946 Tres observaciones sobre el lineal. *Univ. Nac. Tucumán Rev. Ser. A* **5**, 147–150.

Z. Birnbaum and W. Orlicz
1931 Über die Verallgemeinerung des Bergriffes der zueinander konjugierten Potenzen. *Studia Math.* **3**, 1–67.

# Bibliography

D. BLACKWELL AND M. A. GIRSHICK
1954 "Theory of Games and Statistical Decisions." Wiley, New York.

H. BLUMBERG
1919 On convex functions. *Trans. Amer. Math. Soc.* **20**, 40–44.

R. P. BOAS
1941a A note on functions of exponential type. *Bull. Amer. Math. Soc.* **47**, 750–754.
1941b Functions with positive derivatives. *Duke Math. J.* **8**, 163–172.
1943 Representation of functions by Lidstone series. *Duke Math. J.* **10**, 239–245.
1958 Almost completely convex functions. *Duke Math. J.* **25**, 193–195.
1959 Representations of completely convex functions. *Amer. J. Math.* **81**, 709–714.
1971 Signs of derivatives and analytic behavior. *Amer. Math. Monthly* **78**, 1085–1093.

R. P. BOAS AND G. POLYA
1941 Generalizations of completely convex functions. *Proc. Nat. Acad. Sci. U.S.* **27**, 323–325.
1942 Influence of the signs of the derivatives of a function on its analytic character. *Duke Math. J.* **9**, 406–424.

R. P. BOAS AND D. V. WIDDER
1940 Functions with positive differences. *Duke Math. J.* **7**, 496–503.

T. BONNESEN AND W. FENCHEL
1934 "Theorie der Konvexen Körper." Springer-Verlag, Berlin and New York (reprint, Chelsea, Bronx, New York, 1948).

F. F. BONSALL
1950 The characterization of generalized convex functions. *Quart. J. Math. Oxford Ser.* **1**, 100–111.
1953 The characterization of generalized convex functions. *Quart. J. Math. Oxford Ser.* **4**, 253.
1954 *Math. Rev.* **15**, p. 16.

E. BOREL
1921 La théorie du jeu et les équations intégrales à noyau symétrique. *C.R. Acad. Sci.* **173**, 1304–1308.
1927 Sur les sytèmes des formes linéaires à determinant symétrique gauche et la théorie générale du jeu. *C. R. Acad. Sci.* **184**, 52–54.

N. BOURBAKI
1953 "Espaces Vectorial Topologiques." Hermann, Paris.

J. BRACKEN AND G. P. MCCORMICK
1968 "Selected Applications of Nonlinear Programming." Wiley, New York.

H. J. BREMERMANN
1956 Complex convexity. *Trans. Amer. Math. Soc.* **82**, 17–51.

J. BRENNER AND L. CUMMINGS
1972 The Hadamard maximum determinant problem. *Amer. Math. Monthly* **79**, 626–630.

W. A. BROCK AND R. G. THOMPSON
1966 Convex solutions of implicit relations. *Math. Mag.* **39**, 208–211.

A. BRØNDSTED
1964 Conjugate convex functions in topological vector spaces. *Kgl. Dan. Vidensk. Selsk. Mat. Fys. Medd.* **34**, 1–26.
1965 Convex sets and Chebyshev sets. *Math. Scand.* **17**, 5–16.

1966a Convex sets and Chebyshev sets II. *Math. Scand.* **18**, 5–15.
1966b Milman's theorem for convex functions. *Math. Scand.* **19**, 5–10.

A. BRØNDSTED AND R. T. ROCKAFELLAR
1965 On the subdifferentiability of convex functions. *Proc. Amer. Math. Soc.* **16**, 605–611.

A. M. BRUCKNER AND E. OSTROW
1962 Some function classes related to the class of convex functions. *Pac. J. Math.* **12**, 1203–1215.

H. BRUNN
1889 Über Curven ohne Wendepunkte, München.

R. C. BUCK
1959 Linear spaces and approximation theory. *In* "On Numerical Analysis" (R. E. Langer, ed.), pp. 11–24. Univ. of Wisconsin Press, Madison.
1965 "Advanced Calculus," 2nd ed. McGraw-Hill, New York.

A. P. BULANOV
1970 The best rational approximation of convex functions and functions of bounded variation. *Sov. Math. Dokl.* **11**, 5–7.
1971 Rational approximation to convex functions with given modulus of continuity. *Math. USSR-Sb.* **13**, 473–490.

P. S. BULLEN
1971a A criterion for $n$-convexity. *Pac. J. Math.* **36**, 81–98.
1971b Rado's inequality. *Aequationes Math.* **6**, 149–156.

H. BUSEMANN
1958 "Convex Surfaces." Wiley (Interscience), New York.

H. BUSEMANN AND W. FELLER
1936 Krümmungseigenschaften konvexer Flächen. *Acta Math.* **66**, 1–47.

I. V. ČEBAEVSKAJA
1970 Generalized convexity and Jensen's inequality, *Izv. Vysš. Ucebn. Zaved. Mathematika* **3** (94), 91–95 (Russian).

A. CHARNES, W. W. COOPER, AND A. HENDERSON
1953 "An Introduction to Linear Programming." Wiley, New York.

T. W. CHAUNDY AND C. J. A. EVELYN
1967 Some inequalities. *J. London Math. Soc.* **42**, 110–124.

E. W. CHENEY
1966 "Introduction to Approximation Theory." McGraw-Hill, New York.

Z. CIESIELSKI
1959 Some properties of convex functions of higher orders. *Ann. Polon. Math.* **7**, 1–7.

P. A. CLEMENT
1953 Generalized convexity and surfaces of negative curvature. *Pac. J. Math.* **3**, 333–368.

L. COOPER AND D. STEINBERG
1970 "Introduction to Methods of Optimization." Saunders, Philadelphia, Pennsylvania.

R. M. CROWNOVER AND C. L. SIMMONS
1970 Functions of bounded convexity on rectifiable arcs. *Monatsh. Math.* **74**, 389–397.

A. CSASZAR
1958 On convex sets and functions. *Mat. Lapok.* **9**, 273–282 (Hungarian).

# Bibliography 277

D. F. CUDIA
1963 Rotundy. *In* "Convexity" (V. L. Klee, ed.), *Symp. Pure Math.* **7**, 73–98, Amer. Math. Soc., Providence, Rhode Island.
1964 The geometry Banach spaces. Smoothness. *Trans. Amer. Math. Soc.* **110**, 284–314.

P. C. CURTIS, JR.
1959 N-parameter families and best approximation. *Pac. J. Math.* **9**, 1013–1027.

G. B. DANTZIG
1951 Maximization of a linear function of variables subject to linear inequalities. *In* "Activity Analysis of Production and Allocation" (T. C. Koopmans, ed.) Cowles Comm. Monogr. No. 13, pp. 339–347. Wiley, New York.
1963 "Linear Programming and Extensions." Princeton Univ. Press, Princeton, New Jersey.

L. DANZER, B. GRÜNBAUM, AND V. L. KLEE
1963 Helly's theorem and its relatives. *In* "Convexity" (V. L. Klee, ed.) *Symp. Pure Math.* **7**, 101–180, Amer. Math. Soc., Providence, Rhode Island.

C. DAVIS
1963 Notions generalizing convexity for functions defined on spaces of matrices. *In* "Convexity" (V. L. Klee, ed.) *Symp. Pure Math.* **7**, 187–201, Amer. Math. Soc., Providence, Rhode Island.

M. M. DAY
1955 Strict convexity and smoothness of normed spaces. *Trans. Amer. Math. Soc.* **78**, 516–528.
1962 "Normed Linear Spaces." Springer-Verlag, Berlin and New York.

D. E. DAYKIN AND C. J. ELIEZER
1969 Elementary proofs of basic inequalities. *Amer. Math. Monthly* **76**, 543-546.

E. DEAK
1962 Über konvexe und interne Funktionen, sowie gemeinsame Verallgemeinerung von beiden. *Ann. Univ. Sci. Budapest. Eötvös. Sect. Math.* **5**, 109–154.

B. DE FINETTI
1949 Sulle stratificazioni convesse. *Ann. Mat. Pura Appl.* [4] **30**, 173–183.

C. DE LA VALLÉE POUSSIN
1908 Sur la convergence des formules d'interpolation entre ordonnées equidistantes. *Bull. Acad. Sci. Belg.* 319–410.

J. DIEUDONNÉ
1960 "Foundations of Modern Analysis." Academic Press, New York.

L. L. DINES
1938 On convexity. *Amer. Math. Monthly* **45**, 199–209.

R. DORFMAN, P. A. SAMUELSON, AND R. M. SOLOW
1958 "Linear Programming and Economic Analysis." McGraw-Hill, New York.

M. DRESHER
1961 "Games of Strategy." Prentice-Hall, Englewood Cliffs, New Jersey.

R. J. DUFFIN, E. L. PETERSON, AND C. ZENER
1967 "Geometric Programming-Theory and Application." Wiley, New York.

N. DUNFORD AND J. T. SCHWARTZ
1958 "Linear Operators." Vol. I, Wiley (Interscience), New York.

H. G. EGGLESTON
1958 "Convexity." Cambridge Univ. Press, London and New York.

C. J. Eliezer and B. Mond
1972   Generalizations of the Cauchy-Schwarz and Hölder inequalities. *In* "Inequalities" (O. Shisha, ed.), Vol. III, pp. 97–101. Academic Press, New York.

K. Fan
1953   Minimax theorems. *Proc. Nat. Acad. Sci. U.S.* **39**, 42–47.

J. Farkas
1901   Über die Theorie der einfachen Ungleichungen. *J. Reine Angew. Math.* **124**, 1–27.

W. Fenchel
1949   On conjugate convex functions. *Canad. J. Math.* **1**, 73–77.
1953   "Convex Cones, Sets and Functions" (mimeographed lecture notes). Princeton Univ. Press, Princeton, New Jersey.
1956   Über konvexe Funktionen mit vorgeschriebenen Niveaumannigfaltigkeiten. *Math. Z.* **63**, 496–506.
1967   *Proc. Colloq. Convexity* (W. Fenchel, ed.), Mat. Inst., Copenhagen, 1965.

L. R. Ford and D. R. Fulkerson
1956   Maximal flow through a network. *Canad. J. Math.* **8**, 399–404.
1962   "Flows in Networks." Princeton Univ. Press, Princeton, New Jersey.

G. Forsythe
1970   The max and min of a positive definite quadratic polynomial on a sphere are convex functions of the radius. *SIAM J. Appl. Math.* **19**, 551.

B. Friedman
1940   A note on convex functions. *Bull. Amer. Math. Soc.* **46**, 473–474.

D. Gale
1960   "The Theory of Economic Models." McGraw-Hill, New York.

D. Gale, V. L. Klee, and R. T. Rockafellar
1951   Linear programming and the theory of games. *In* "Activity Analysis of Production and Allocation" (T. C. Koopmans, *ed.*). Wiley, New York.

D. Gale, H. W. Kuhn, and A. W. Tucker
1968   Convex functions on convex polytopes. *Proc. Amer. Math. Soc.* **19**, 867–873.

B. Gelbaum and J. M. H. Olmsted
1964   "Counter Examples in Analysis." Holden-Day, San Francisco, California.

R. Ger
1969   Some remarks on convex functions. *Fund. Math.* **66**, 255–262.
1972   Convex functions of higher order in Euclidean spaces. *Ann. Polon. Math.* **25**, 293–302.

J. W. Green
1950   Sets subtending a constant angle on a circle. *Duke Math. J.* **17**, 263–267.
1952a  Approximately convex functions. *Duke Math. J.* **19**, 499–504.
1952b  Approximately subharmonic functions. *Proc. Amer. Math. Soc.* **3**, 829–833.
1953   Support, convergence, and differentiability properties of generalized convex functions. *Proc. Amer. Math. Soc.* **4**, 391–396.
1954   Recent applications of convex functions. *Amer. Math. Monthly* **61**, 449–454.

H. J. Greenberg and W. P. Pierskalla
1971   A review of quasi-convex functions. *J. Operations Res.* **19**, 1553–1570.

B. Grünbaum
1963   Borsuk's problem and related questions. *In* "Convexity" (V. L. Klee, ed.) *Symp. Pure Math.* **7**, 271–284, Amer. Math. Soc., Providence, Rhode Island.
1967   "Convex Polytopes." Wiley, New York.

**Bibliography**

I. Ja. Guberman
1970 On uniform convergence of convex functions in a closed domain. *Amer. Math. Soc. Transl.* [2] **80**, 41–55.

J. Hadamard
1893 Étude sur les propriétés des fonctions entières et en particulier d'une fonction considérée par Riemann. *J. Math. Pures Appl.* **58**, 171–215.

G. Hadley
1962 "Linear Programming." Addison-Wesley, Reading, Massachusetts.
1964 "Nonlinear and Dynamic Programming." Addison-Wesley, Reading, Massachusetts.

H. Hadwiger, H. Debrunner, and V. L. Klee
1964 "Combinational Geometry in the Plane." Holt, New York.

P. Halmos
1958 "Finite Dimensional Vector Spaces," 2nd ed., Van Nostrand-Reinhold, New York.

I. Halperin
1952 "Introduction to the Theory of Distributions," based on lectures by Laurent Schwartz, Univ. of Toronto Press, Toronto.

G. Hamel
1905 Eine Basis aller Zahlen und die unstetigen Lösungen der Funktionalgleichung $f(x + y) = f(x) + f(y)$. *Math. Ann.* **60**, 459–462.

P. Hammer
1967 Problem. *In* Proc. Colloq. Convexity (W. Fenchel, ed.) p. 316, Mat. Inst., Copenhagen.

S. Hanai
1945 On the inequalities $y''' \geqslant G(x, y, y', y'')$. *Proc. Jap. Acad.* **21**, 378–381.

G. H. Hardy, J. E. Littlewood, and G. Polya
1952 "Inequalities," 2nd ed., Cambridge Univ. Press, London and New York.

P. Hartman
1958 Unrestricted $n$-parameter families. *Rend. Circ. Mat. Palermo* [2] **7**, 123–142.
1959 On functions representable as a difference of convex functions. *Pac. J. Math.* **9**, 707–713.
1967 Interpolating families and generalized convex functions. *Duke Math. J.* **34**, 511–518.
1972 Convex functions and mean value inequalities. *Duke Math. J.* **39**, 351–360.

E. Hewitt and K. Stromberg
1965 "Real and Abstract Analysis." Springer-Verlag, Berlin and New York.

W. M. Hirsch and A. J. Hoffman
1961 Extreme varieties, concave functions, and the fixed charge problem. *Commun. Pure Appl. Math.* **14**, 355–369.

R. A. Hirschfield
1958 On a minimax theorem of K. Fan. *Nederl. Akad. Wetensch. Indag. Math.* **20**, 470–474.

I. I. Hirschman and D. V. Widder
1955 "The Convolution Transform." Princeton Univ. Press, Princeton, New Jersey.

O. Hölder
1889 Über einen Mittelwertsatz. *Nachr. Ges. Wiss. Goettingen*, 38–47.

E. Hopf
1926 Über die Zusammenhänge zwischen gewissen höheren Differenzen-quotienten reeller Funktionen einer reellen Variablen und deren Differenzierbarkeitseigenschaften. Thesis, Univ. of Berlin, Berlin.

F. Huggins
1972 A generalization of a theorem of F. Riesz. *Pac. J. Math.* **39**, 695–701.

M. Hukuhara
1954 Sur la fonction convexe. *Proc. Jap. Acad.* **30**, 683–685.

D. H. Hyers
1941 On the stability of the linear functional equation. *Proc. Nat. Acad. Sci. U.S.* **27**, 222–224.

D. H. Hyers and S. M. Ulam
1952 Approximately convex functions. *Proc. Amer. Math. Soc.* **3**, 821–828.

A. Ioffe and V. Tikhomirov
1968 Duality of convex functions and extremum problems, *Russ. Math. Surveys* **23**, 53–124.

D. Jackson
1930 "The Theory of Approximations." Amer. Math. Soc. Colloq. Publ., Amer. Math. Soc., Providence, Rhode Island.

J. L. W. V. Jensen
1905 Om konvexe Funktioner og Uligheder mellem Middelvaerdier, *Nyt Tidsskr. Math.* **16B**, 49–69.
1906 Sur les fonctions convexes et les inegalités entre les valeurs moyennes. *Acta Math.* **30**, 175–193.

F. John
1939 Special solutions of certain difference equations. *Acta Math.* **71**, 175–189.

F. B. Jones
1942 Connected and disconnected plane sets and the functional equation $f(x) + f(y) = f(x + y)$, *Bull. Amer. Math. Soc.* **48**, 115–120.

K. W. Kaplan
1952 Close to convex schlicht functions. *Mich. Math. J.* **1**, 169–185.

S. Karlin
1959 "Mathematical Methods and Theory in Games, Programming and Economics," Vol. 1. Addison-Wesley, Reading, Massachusetts.

S. Karlin and W. J. Studden
1966 "Tchebycheff Systems: With Applications in Analysis and Statistics." Wiley (Interscience), New York.

N. D. Kazarinoff
1961 "Analytic Inequalities." Holt, New York.

J. L. Kelley
1955 "General Topology." Van Nostrand-Reinhold, New York.

J. L. Kelley and I. Namioka
1963 "Linear Topological Spaces." Van Nostrand-Reinhold, New York.

J. H. B. Kemperman
1957 A general functional equation. *Trans. Amer. Math. Soc.* **86**, 28–56.
1969 On the regularity of generalized convex functions. *Trans. Amer. Math. Soc.* **135**, 69–93.

# Bibliography

H. KENYON
1956  Note on convex functions. *Amer. Math. Monthly* **63**, 107.

V. L. KLEE
1956  Solution of a problem of E. M. Wright on convex functions. *Amer. Math. Monthly* **63**, 106–107.
1961  Convexity of Chebyshev sets. *Math. Ann.* **142**, 292–304.
1963  "Convexity." (V. L. Klee, ed.), *Proc. Symp. Pure Math.* **7**. Amer. Math. Soc., Providence, Rhode Island.
1964  Convex polytopes and linear programming. *Proc. IBM Sci. Comput. Symp., Combinatorial Problems*, pp. 123–158.
1967  Remarks on nearest point, in normed linear spaces. *In* Proc. Colloq. Convexity (W. Fenchel, ed.) pp. 168–176, Mat. Inst., Copenhagen.
1969a Separation and support properties of convex sets—a survey. *In* "Control Theory and the Calculus of Variations" (A. V. Balakrishnan, ed.), pp. 235–303. Academic Press, New York.
1969b Can a plane convex body have two equichordal points. *Amer. Math. Monthly* **76**, 54–55.
1971  What is a convex set. *Amer. Math. Monthly* **78**, 616–631.

V. L. KLEE AND M. MARTIN
1971  Semicontinuity of the face-function of a convex set. *Comment. Math. Helv.* **46**, 1–12.

G. KÖTHE
1969  "Topological Vector Spaces." Springer-Verlag, Berlin and New York.

M. A. KRASNOSEL'SKIĬ AND Y. B. RUTICKIĬ
1961  "Convex Functions and Orlicz Spaces." Noordhoff, Groningen.

F. KRAUSS
1936  Über konvexe Matrixfunktionen. *Math. Z.* **41**, 18–42.

B. KREKÓ
1968  "Linear Programming." Amer. Elsevier, New York.

J. B. KRUSKAL
1969  Two convex counterexamples: A discontinuous envelope function and a non-differentiable nearest point mapping. *Proc. Amer. Math. Soc.* **23**, 697–703.

M. E. KUCZMA
1959  Note on convex functions. *Ann. Univ. Sci. Budapest. Eötvös Sect. Math.* **2**, 25–26.
1970a Some remarks on convexity and monotonicity. *Rev. Roumaine Math. Pures Appl.* **15**, 1463–1469.
1970b On discontinuous additive functions. *Fund. Math.* **66**, 384–392.
1970c Almost convex functions. *Colloq. Math.* **21**, 279–284.

H. W. KUHN AND A. W. TUCKER
1951  Nonlinear programming. *In* Proc. of Second Sympo. Math. Statist. Probability, (J. Neyman, ed.), pp. 481–492.

S. KUREPA
1956  Convex functions. *Glas. Mat. Fiz. Astron.* [2] **11**, 89–93.

S. KUTATELADZE AND A. RUBINOV
1971  Some classes of H-convex functions and sets. *Sov. Math. Dokl.* **12**, 665–668.

E. M. LANDIS
1951  On functions representable as the difference of two convex functions. *Dokl. Akad. Nauk SSSR* (N.S.) **80**, 9–11.

D. Leeming and A. Sharma
1972 A generalization of the class of completely convex functions. *In* "Inequalities, (O. Shisha, ed.), Vol. III. Academic Press, New York.

R. Leipnik and R. Oberg
1967 Subvex functions and Bohr's uniqueness theorem. *Amer. Math. Monthly* **74**, 1093–1094.

E. Levitan and B. Polzak
1966 Convergence of minimizing sequences in conditional extremum problems. *Sov. Math. Dokl.* **7**, 764–767.

E. R. Lorch
1951 Differentiable inequalities and the theory of convex bodies. *Trans. Amer. Math. Soc.* **71**, 243–266.
1962 "Spectral Theory." Oxford Univ. Press, London and New York.

G. Lorentz
1966 "Approximation of Functions." Holt, New York.

K. Löwner
1934 Über monotone Matrixfunktionen. *Math. Z.* **38**, 177–216.

Yu. I. Lyubich and G. D. Maistrovskii
1970 The general theory of relaxation processes for convex functionals. *Russ. Math. Surveys* **25**, 57–117.

O. L. Mangasarian
1969 "Nonlinear Programming." McGraw-Hill, New York.

M. Marcus and H. Minc
1964 "A Survey of Matrix Theory and Matrix Inequalities." Allyn & Bacon, Rockleigh, New Jersey.

S. Marcus
1957a Critères de majoration pour les fonctions sous additives, convexes ou internes. *C.R. Acad. Sci. Sér. A-B* **244**, 2270–2272.
1957b Fonctions convexes et fonctions internes. *Bull. Sci. Math.* [2] **81**, 66–70.
1957c Sur un théorème de F. B. Jones. Sur un théorème de S. Kurepa. *Bull. Math. Soc. Sci, Math. Phys. R.P. Roumaine (N.S.)* [1], **49**, 433–434.
1959a Sur un théorème de G. Szekeres, concernant les fonctions monotone et convexes. *Canad. J. Math.* **11**, 521–526.
1959b Generalisation, aux fonctions de plusieurs variables, des theoremes de A. Ostrowski et de M. Hukuhara concernant les fonctions convexes. *J. Math. Sci. Japan* **11**, 171–176.

R. M. Mathsen
1969 $\lambda(n)$-parameter families. *Can. Math. Bull.* **12**, 185–191.
1972 $\lambda(n)$-convex functions. *Rocky Mt. J. Math.* **2**, 31–43.

S. Mazur
1933 Über konvexe Mengen in linearen normierten Raümen. *Studia Math.* **4**, 70–84.

M. R. Mehdi
1964 Some remarks on convex functions. *J. London Math. Soc.* **39**, 321–326.

Z. Melzak
1965 Problems connected with convexity. *Canad. Math. Bull.* **8**, 565–573.

H. Meschkowski
1966 "Unsolved and Unsolvable Problems in Geometry." Ungar, New York.

# Bibliography

M. J. MILES
1969  An extremum property of convex functions. *Amer. Math. Monthly* **76**, 921–922.

G. J. MINTY
1962  Monotone (nonlinear) operators in Hilbert space. *Duke Math. J.* **29**, 341–346.
1964  On the monotonicity of the gradient of a convex function. *Pac. J. Math.* **14**, 243–247.

D. S. MITRINOVIĆ
1970  "Analytic Inequalities." Springer-Verlag, Berlin and New York.

E. MOHR
1952  Beitrag zur Theorie der konvexen Funktionen. *Math. Nachr.* **8**, 133–148.

E. MOLDOVAN
1959  Sur une généralisation des fonctions convexes. *Mathematica (Cluj)* [2] **1**, 49–80.

J.-J. MOREAU
1962  Fonctions convexes en dualité, multigraph. *Sémin. Math. Fac. Sci. Univ. Montpellier.*

T. S. MOTZKIN
1949  Approximation by curves of a unisolvent family. *Bull. Amer. Math. Soc.* **55**, 780–793.

M. A. NAIMARK
1970  "Normed Rings," 2nd Engl. ed. Walters-Noordhoff, Groningen.

M. Z. NASHED
1966  Some remarks on variations and differentials. Slaught Mem. Paper 11. *Amer. Math. Monthly* **73**, 63–78.

I. P. NATANSON
1961  "Theory of Functions of a Real Variable," Vol. I rev. ed. and Vol. II, Ungar, New York.

F. NEVANLINNA AND R. NEVANLINNA
1959  "Absolute Analysis." Springer-Verlag, Berlin and New York.

T. NISHIURA AND F. SCHNITZER
1965  Monotone functions and convex functions. *Mich. Math. J.* **12**, 481–483.
1972  Moments of convex and monotone functions, *Monatshefte für Mathematik* **76** 135–137.

Z. OPIAL
1967  On a theorem of O. Arama. *J. Differential Equations* **3**, 88–91.

O. ORE
1967  "The Four Color Problem." Academic Press, New York.

A. OSTROWSKI
1929a  Über die Funktionalgleichung der Exponentialfunktionen und verwandte Funktionalgleichungen. *Jahresber. Deut. Math. Ver.* **38**, 54–62.
1929b  Zur Theorie der konvexen Funktionen. *Comment. Math. Helv.* **1**, 157–159.
1952  Sur quelques applications des fonctions convexes et concave au sens de I. Schur. *J. Math. Pures Appl.* **31**, 253–292.

G. OWEN
1968  "Game Theory." Saunders, Philadelphia, Pennsylvania.

M. M. PEIXOTO
1948a On the existence of derivatives of generalized convex functions. *Summa Brasil. Math.* **2**, No. 3, 35–42.
1948b "Convexity of Curves." *Notas Mat.*, No. 6, Livraria Boffoni, Rio de Janeiro.
1949 Generalized convex functions and second order differential inequalities. *Bull. Amer. Math. Soc.* **55**, 563–572.

R. R. PHELPS
1960 A representation theorem for bounded convex sets. *Proc. Amer. Math. Soc.* **11**, 976–983.

G. POLYA
1941 On functions whose derivatives do not vanish in a given interval. *Proc. Nat. Acad. Sci. U.S.* **27**, 216–217.
1943 On the zeros of the derivatives of a function and its analytic character. *Bull. Amer. Math. Soc.* **49**, 178–191.

B. POLZAK
1966 Existence theorems and convergence of minimizing sequences in extremum problems with restrictions, *Sov. Math. Dokl.* **7**, 72–75.

J. PONSTEIN
1967 Seven kinds of convexity. *SIAM Rev.* **9**, 115–119.

T. POPOVICIU
1934 Sur quelques proprietes des fonctions d'une variable reele convexes d'order superieur. *Mathematica (Cluj)* **8**, 1–85.
1944 "Les Fonctions Convexes." Hermann, Paris.

M. H. PROTTER
1957 A generalization of completely convex functions. *Duke Math. J.* **24**, 205–213.

L. QUINTAS AND F. SUPNICK
1963 Semi-homogeneous functions. *Proc. Amer. Math. Soc.* **14**, 620–625.

H. RADEMACHER
1919 Differenzierbarkeit von Funktionen mehrerer Variabeln und über die Transformation der Dappelintegrale. *Math. Ann.* **79**, 340–359.

T. RADO
1935 On convex functions. *Trans. Amer. Math. Soc.* **37**, 266–285.
1949 "Subharmonic Functions." Chelsea, Bronx, New York.

J. F. RANDOLPH
1968 "Basic Real and Abstract Analysis." Academic Press, New York.

W. REID
1959 Variational aspects of generalized convex functions. *Pac. J. Math.* **9**, 571–581.

G. RESTREPO
1965 Differentiable norms. *Bol. Soc. Mat. Mexicana* [2] **10**, 47–55.

J. R. RICE
1964 "The Approximation of Functions," Vol. I. Addison-Wesley, Reading, Massachusetts.
1969 "The Approximation of Functions," Vol. II. Addison-Wesley, Reading, Massachusetts.

F. RIESZ
1911 Sur certain systems singuliers d'equations integrales. *Ann. Ecole Norm.* **28**, 33–62.
1925 Über subharmonische Funktionen und ihre Rolle in der Funktionentheorie und in der Potentialtheorie. *Acta Litt. Sci. Szeged* **2**, 87–100.

# Bibliography

F. Riesz and B. Sz.-Nagy
1955 "Functional Analysis." Ungar, New York.

A. W. Roberts
1969 The derivative as a linear transformation. *Amer. Math. Monthly* **76**, 632–638.

A. W. Roberts and D. E. Varberg
1969 Functions of bounded convexity. *Bull. Amer. Math. Soc.* **75**, 568–572.

R. T. Rockafellar
1964 A combinatorial algorithm for linear programs in the general mixed form. *SIAM J. Appl. Math.* **12**, 215–225.
1966 Level sets and continuity of conjugate convex functions. *Trans. Amer. Math. Soc.* **123**, 46–63.
1967 Convex programming and systems of elementary monotonic relations. *J. Math. Anal. Appl.* **19**, 543–564.
1970a Convex Analysis." Princeton Univ. Press, Princeton, New Jersey.
1970b On the maximal monotonicity of subdifferential mappings. *Pac. J. Math.* **33**, 209–216.

C. A. Rogers
1964 "Packing and Covering." Cambridge Univ. Press, London and New York.

H. L. Roydon
1968 "Real Analysis," 2nd ed. Macmillan, New York.

H. Rubin and O. Wesler
1958 A note on convexity in Euclidean $n$-space. *Proc. Amer. Math. Soc.* **9**, 522–523.

W. Rudin
1966 "Real and Complex Analysis." McGraw-Hill, New York.

A. M. Russell
1970 Functions of bounded second variation and Stietjes-type integrals. *J. London Math. Soc.* **2**, 193–208.

T. L. Saaty
1972 Thirteen colorful variations on Guthrie's four-color conjecture. *Amer. Math. Monthly* **79**, 2–43.

R. Salem
1949 Convexity theorems. *Bull. Amer. Math. Soc.* **55**, 851–860.

G. Sansone and J. Gerretsen
1969 "Lectures on the Theory of Functions of a Complex Variable," Vol. II. Wolters, Noordhoff, Groningen.

I. Schur
1923 Über eine Klasse von Mittelbildungen mit Anwendungen auf die Determinantentheorie. *Sitzungsber. Berlin. Math. Ges.* **22**, 9–20.

P. Sengenhorst
1952 Über konvexe Funktionen. *Math.-Phys. Semesterber.* **2**, 217–230.

O. Shisha
1965 Monotone approximation. *Pac. J. Math.* **15**, 667–671.

W. Sierpinski
1920 Sur les fonctions convexes mesurables. *Fund. Math.* **1**, 125–129.

M. Simonnard
1966 "Linear Programming." Prentice-Hall, Englewood Cliffs, New Jersey.

M. Sion
1958 On general minimax theorems. *Pac. J. Math.* **8**, 171–176.

A. Smajdor
1966 On superposition of convex functions. *Arch. Math. (Basel)* **17**, 333–335.

V. L. Smulian
1940 Sur la dérivabilité de la norm dans l'espace de Banach. *Dokl. Akad. Nauk SSSR* (N.S.) **27**, 643–649.

L. Steen and J. Seebach
1970 "Counter Examples in Topology." Holt, New York.

J. Stoer and C. Witzgall
1970 "Convexity and Optimization in Finite Dimensions," Vol. I. Springer-Verlag, Berlin and New York.

O. Stolz
1893 "Grundzüge der Differential und Integralrechnung," Vol. 1. Teubner, Leipzig.

K. Sundaresan
1967 Smooth Banach spaces. *Math. Ann.* **173**, 191–199.

G. Szekeres
1956 On a property of monotone and convex functions. *Proc. Amer. Math. Soc.* **7**, 351–353.

A. E. Taylor
1958 "Introduction to Functional Analysis." Wiley, New York.

H. P. Thielman
1941 On the convex solutions of a certain functional equation. *Bull. Amer. Math. Soc.* **47**, 118–120.

G. Thorin
1948 Convexity theorems generalizing those of M. Riesz and Hadamard with some applications. *Comm. Sem. Math. Univ. Lund.* **9**, 1–58.

E. Titchmarsh
1939 "The Theory of Functions," 2nd ed. Oxford Univ. Press, London and New York.

L. Tornheim
1950 On $n$-parameter families of functions and associated convex functions. *Trans. Amer. Math. Soc.* **69**, 457–467.

S. Vajda
1958 "Readings in Linear Programming." Wiley, New York.

F. Valentine
1964 "Convex Sets." McGraw-Hill, New York.
1967 The dual cone and Helly type theorems. *In* Proc. Colloq. Convexity (W. Fenchel, ed.) p. 492, Mat. Inst., Copenhagen.

G. Valiron
1932 Fonctions convexes et fonctions entieres. *Bull. Soc. Math. France* **60**, 278–287.

J. von Neumann
1928 Zur Theorie der Gesellschaftsspiele. *Math. Ann.* **100**, 295–320.

J. von Neumann and O. Morgenstern
1947 "Theory of Games and Economic Behavior." Princeton Univ. Press, Princeton, New Jersey.

**Bibliography**

D. V. WIDDER
1940 Functions whose even derivatives have a prescribed sign. *Proc. Nat. Acad. Sci. U.S.* **26**, 657–659.
1941 "The Laplace Transform." Princeton Univ. Press, Princeton, New Jersey.
1942 Completely convex functions and Lidstone series. *Trans. Amer. Math. Soc.* **51**, 387–398.

J. D. WILLIAMS
1954 "The Compleate Strategist." McGraw-Hill, New York.

A. WINTNER
1958 On cosine-like arches. *Amer. J. Math.* **80**, 125–130.

H. WITSENHAUSEN
1968 A minimax control problem for sampled linear systems. *IEEE Trans. Automat. Contr.* **AC-13**, 5–21.

E. M. WRIGHT
1954 An inequality for convex functions. *Amer. Math. Monthly* **61**, 620–622.

I. M. YAGLOM AND V. G. BOLTYANSKIĬ
1961 "Convex Figures." Holt, New York.

W. H. YOUNG
1912 On classes of summable functions and their Fourier series. *Proc. Roy. Soc. Ser. A* **87**, 225–229.

T. ZAMFIRESCU
1965 Sur quelques theoremes de G. Szerkeres et S. Marcus concernant les fonctions monotone et convexes, *Rev. Roumaine Math. Pures Appl.* **10**, 81–90.

V. A. ZALGALLER
1963 On the representation of a function of two variables as the difference of convex functions. *Vestn. Leningrad. Univ. Ser. Mat. Mekh.* **18**, 44–45 (Russian).

A. ZYGMUND
1968 "Trigonometric Series." Vol. I, Cambridge Univ. Press, London and New York.

# Author Index

Numbers in italics refer to the pages on which the complete references are listed.

## A

Aczel, J., 217, 269, *273*
Aggeri, J. C., 97, *273*
Alexandroff, A. D., 28, 120, *273*
Anderson, B. J., 21, *273*
Anderson, R. D., 119, *273*
Arsove, M. G., 28, *273*
Artin, E., 8, *9*, 21, 216, 222, *224*, *273*
Ascoli, G., 117, *274*
Asplund, E., 96, 118, *274*

## B

Banach, S., 62, *274*
Bary, N. K., 265, *274*
Beale, E. M. L., 147, 154, *274*
Beckenbach, E. F., *9*, 15, 20, 189, 193, 207, 208, 209, 240, 242, 244, 245, *246*, 255, *255*, 258, *259*, *274*
Bellman, R., 189, 193, 207, 208, 209, 255, *255*, 258, *259*, *274*
Bendat, J., 261, *262*, *274*

Bereanu, B., 218, *274*
Berge, C., 36, 138, 152, 176, 178, *274*
Bernstein, B., 103, *274*
Bernstein, F., 218, 220, 223, *224*, *274*
Bernstein, S., 233, *274*
Besicovitch, A. S., 22, *274*
Bing, R. H., 242, *246*
Birkoff, G., 200, *274*
Birnbaum, Z., 36, *274*
Blackwell, D., 138, *275*
Blumberg, H., 221, *225*, *275*
Boas, R. P., 234, 235, 236, *236*, *237*, 239, *240*, *275*
Boltyanskiĭ, V. G., 73, 271, *287*
Bonnesen, T., 73, *275*
Bonsall, E. F., 243, *246*, *275*
Borel, E., 138, *275*
Bourbaki, N., 219, *225*, *275*
Bracken, J., 178, *275*
Bremermann, H. J., 255, *255*, 267, 269, *275*
Brenner, J., 271, *275*
Brøndsted, A., 36, 97, 111, 187, 271, *275*, *276*

289

Brock, W. A., 104, *275*
Bruckner, A. M., 22, *276*
Brunn, H., *8*, *276*
Buck, R. C., 41, 47, 54, 66, 104, 120, 128, 185, 187, 223, *276*
Bulanov, A. P., 269, *276*
Bullen, P. S., 194, 200, 238, *240*, 248, *276*
Busemann, H., 120, 265, *276*

## C

Čebaevskaja, I. V., *253*, *276*
Charnes, A., 154, 160, *276*
Chaundy, T. W., 104, *276*
Cheney, E. W., 185, 186, 187, *276*
Ciesielski, Z., 239, *240*, *276*
Clement, P. A., 245, *246*, *276*
Cooper, L., 154, *276*
Cooper, W. W., 154, 160, *276*
Crownover, R. M., 28, *276*
Csaszar, A., 221, *225*, *276*
Cudia, D. F., 118, 270, *277*
Cummings, L., 271, *275*
Curtis, Jr., P. C., 251, *253*, *277*

## D

Dantzig, G. B., 153, *277*
Danzer, L., 80, *277*
Davies, R. O., 22, *274*
Davis, C., 260, 261, *262*, *277*
Day, M. M., 42, 53, 54, 270, *277*
Daykin, D. E., 194, *277*
Deak, E., 221, *225*, *277*
Debrunner, H., 271, *272*, *279*
de Finetti, B., 228, 229, *233*, *277*
de la Vallée Poussin, C., 28, *277*
Dieudonné, J., 70, 71, 118, 270, *277*
Dines, L. L., 112, *277*
Doetsch, G., 218, 220, 223, *224*, *274*
Dorfman, R., 154, *277*
Dresher, M., 137, 138, *277*
Duffin, R. J., 128, *277*
Dunford, N., 62, *277*

## E

Eggleston, H. G., 73, 80, *277*
Eliezer, C. J., 194, *277*, *278*
Evelyn, C. J. A., 104, *276*

## F

Fan, K., 138, *278*
Farkas, J., 152, *278*
Feller, W., 120, *276*
Fenchel, W., *9*, 36, 73, 96, 110, 111, 228, 229, *233*, 271, *272*, *275*, *278*
Ford, L. R., 154, *278*
Forsythe, G., 267, *278*
Friedman, B., 96, *278*
Fulkerson, D. R., 154, *278*

## G

Gale, D., 97, 148, 154, *278*
Gelbaum, B., 266, *278*
Ger, R., 221, *225*, 239, *240*, *278*
Gerretsen, J., 256, *256*, *285*
Ghouila-Houri, A., 36, 138, 152, 176, 178, *274*
Girshick, M. A., 138, *275*
Green, J. W., *9*, 243, 244, *246*, 257, *257*, *278*
Greenberg, H. J., 227, 231, *233*, *278*
Grünbaum, B., 73, 80, 267, 271, *272*, *277*, *278*, *279*
Guberman, I. Ja., 94, *279*

## H

Hadamard, J., *8*, *279*
Hadley, G., 154, 178, *279*
Hadwiger, H., 271, *272*, *279*
Halmos, P., 42, 69, 260, *279*
Halperin, I., 265, *279*
Hamel, G., 217, *279*
Hammer, P., 266, *279*
Hanai, S., 245, *246*, *279*
Hardy, G. H., *9*, 189, 198, 199, 224, *225*, 245, 259, 267, *279*
Hartman, P., 15, 28, 249, 251, 252, *253*, *279*
Henderson, A., 154, 160, *276*
Hewitt, E., 30, 195, 199, 266, *279*
Hirsch, W. M., 128, *279*
Hirschfield, R. A., 138, *279*
Hirschman, I. I., 224, *225*, *279*
Hölder, O., *8*, *279*
Hoffman, A. J., 128, *279*
Hopf, E., 237, *240*, *280*
Huggins, F., 28, 266, *280*

Hukuhara, M., 224, *225*, *280*
Hyers, D. H., 256, 257, *257*, *280*

## I

Ioffe, A., 36, 111, *280*

## J

Jackson, D., 185, *280*
Jensen, J. L. W. V., *8*, 211, 218, 220, *225*, *280*
John, F., 269, *280*
Jones, F. B., *225*, *280*

## K

Kaplan, 256, *256*, *280*
Karlin, S., 138, 153, 154, 176, 178, 249, 251, *253*, *280*
Kazarinoff, N. D., 189, *282*
Kelly, J. L., 85, *280*
Kemperman, J. H. B., 220, *225*, 239, 246, 248, 249, *253*, *280*
Kenyon, H., 224, *225*, *281*
Klee, V. L., 80, 85, 97, 119, 148, 158, 187, 224, *225*, 271, *272*, *273*, *277*, *278*, *279*, *281*
Köthe, G., 87, 118, 269, 270, *281*
Krasnosel'skiĭ, M. A., 9, 30, 36, *281*
Krauss, F., 261, *261*, *281*
Krekó, B., 152, *281*
Kruskal, J. B., 97, *281*
Kuczma, M. E., 218, 220, 221, *225*, 258, *281*
Kuhn, H. W., 97, 174, 176, 178, *278*, *281*
Kurepa, S., 220, *225*, *281*
Kutateladze, S., 269, *281*

## L

Landis, E. M., 28, *281*
Leeming, D., *282*
Leipnik, R., 269, *282*
Levitan, E., 269, *282*
Littlewood, J. E., 189, 198, 199, 224, *225*, 245, 259, 267, *279*
Löwner, K., 260, *261*, *282*
Lorch, E. R., 59, 62, 120, 265, *282*

Lorentz, G., 185, *282*
Lyubich, Yu. I., 128, 268, 269, *282*

## M

McCormick, G. P., 178, *275*
Maistrovskii, G. D., 128, 268, 269, *282*
Mangasarian, O. L., 153, 176, 178, *282*
Marcus, M., 189, 208, 109, 271, *282*
Marcus, S., 22, 220, 221, 224, *225*, *282*
Martin, M., 97, *281*
Mathsen, R. M., 248, 252, *253*, *282*
Mazur, S., 117, *282*
Mehdi, M. R., 221, *225*, *282*
Melzak, Z., *272*, *282*
Meschkowski, H., *272*, *282*
Miles, M. J., 14, *283*
Minc, H., 189, 208, 209, 271, *282*
Minty, G. J., 36, 111, *283*
Mitrinović, D. S., 189, *283*
Mohr, E., 221, 224, *225*, *283*
Moldovan, E., 248, *253*, *283*
Mond, B., 194, *278*
Moreau, J.-J., 36, 111, *283*
Morgenstern, O., 138, *286*
Motzkin, T. S., 251, 252, *283*

## N

Naimark, M. A., 52, 54, *283*
Namioka, I., 85, *280*
Nashed, M. Z., 117, *283*
Natanson, I. P., 7, *9*, 10, 23, 27, 116, 119, 222, 229, 265, *283*
Nevanlinna, F., 71, *283*
Nevanlinna, R., 71, *283*
Nishiura, T., 96, *283*

## O

Oberg, R., 269, *282*
Olmsted, J. M. H., 266, *278*
Opial, Z., 252, *253*, *283*
Ore, O., 271, *283*
Orlicz, W., 36, *274*
Ostrow, E., 22, *276*
Ostrowski, A., 219, 220, 224, *225*, 259, *259*, *283*
Owen, G., 138, *283*

## P

Peixoto, M. M., *9*, 243, 245, *246*, *283*, *284*
Peterson, E. L., 128, *277*
Phelphs, R. R., 118, *284*
Pierskalla, W. P., 227, 231, *233*, *278*
Polya, G., 189, 198, 199, 224, 225, 233, 235, *236*, 245, 259, 267, *275*, *279*, *284*
Polzak, B., 268, 269, *282*, *284*
Ponstein, J., 104, 231, *233*, *284*
Popoviciu, T., *9*, 28, 237, 238, *240*, 246, *252*, *284*
Protter, M. H., 235, *237*, *284*

## Q

Quintas, L., 266, *284*

## R

Rademacher, H., 119, *284*
Rado, T., 15, 254, *255*, *284*
Randolph, J. F., *9*, *284*
Reid, W., 245, *246*, *284*
Restrepo, G., 270, *284*
Rice, J. R., 185, *284*
Riesz, F., 14, 28, 254, *255*, *284*
Roberts, A. W., 28, 271, *285*
Rockafellar, R. T., 21, 36, 94, 96, 110, 111, 112, 118, 123, 127, 148, 152, 162, 170, 178, 267, *278*, *285*
Rogers, C. A., 271, *285*
Roydon, H. L., 52, 54, *285*
Rubin, H., 267, *285*
Rubinov, A., 269, *281*
Rudin, W., 187, 265, *285*
Russell, A. M., 28, 266, *285*
Rutickiĭ, Y. B., *9*, 30, *36*, *281*

## S

Saaty, T. L., 271, *285*
Salem, R., 267, *285*
Samuelson, P. A., 154, *277*
Sansone, G., 256, *256*, *285*
Schnitzer, F., 96, *283*
Schur, I., 258, 259, *259*, *285*
Schwartz, J. T., 62, *277*
Seebach, J., 266, *286*

Sengenhorst, P., *9*, *285*
Sharma, A., *282*
Sherman, S., 261, *262*, *274*
Shisha, O., 269, *285*
Sierpinski, W., 221, *225*, *285*
Simmons, C. L., 28, *276*
Simonnard, M., 152, 154, *285*
Sion, M., 138, *285*
Smajdor, A., 22, *286*
Smulian, V. L., 118, 270, *286*
Solow, R. M., 154, *277*
Steen, L., 266, *286*
Steinberg, D., 154, *276*
Stoer, J., 120, 178, *286*
Stolz, O., *8*, *286*
Stromberg, K., 30, 194, 199, 266, *279*
Studden, W. J., 249, 251, *253*, *280*
Sundaresan, K., 119, 270, *286*
Supnick, F., 266, *284*
Szekeres, G., 22, *286*
Sz.-Nagy, B., 14, 28, *284*

## T

Taylor, A. E., 52, 53, 54, 62, 107, 216, 217, 267, *286*
Thielman, H. P., 269, *286*
Thompson, R. G., 104, *275*
Thorin, G., 267, *286*
Tikhomirov, V., 36, 111, *280*
Titchmarsh, E., 267, *286*
Tornheim, L., 246, 249, 251, *252*, *286*
Toupin, R. A., 103, *274*
Tucker, A. W., 97, 174, 176, 178, *278*, *281*

## U

Ulam, S. M., 257, *257*, *280*

## V

Vajda, S., 154, *286*
Valentine, F., 29, 73, 80, 85, 187, 216, 267, 271, *272*, *286*
Valiron, G., 245, *246*, *286*
Varberg, D. E., 28, *285*
von Neumann, J., 138, *286*

# Author Index

## W

Wesler, O., 267, *285*
Widder, D. V., 224, *225*, 234, 235, *236*, 239, *240*, 265, *275*, *279*, *286*, *287*
Williams, J. D., 138, *287*
Wintner, A., 235, *237*, *287*
Witsenhausen, H., 97, *287*
Witzgall, C., 120, 178, *286*
Wright, E. M., 223, 224, *225*, *287*

## Y

Yaglom, I. M., 73, 271, *287*
Young, W. H., 36, *287*

## Z

Zalgaller, V. A., 28, *287*
Zamfirescu, T., 22, *287*
Zener, C., 128, *277*
Zygmund, A., *9*, 265, *287*

# Subject Index

## A

Absolute continuity, 4, 96
Absolutely monotonic function, 234
Additive function, 217
Affine combination, 75
Affine function (transformation)
  characterizations, 7, 55, 214
  definition of, 2, 55
Affine hull, 75
Affine set
  characterizations, 74, 75, 79
  closure properties, 75
  definition of, 73
Almost convex function, 257
Approximately convex function, 256
Approximation of convex functions, 269
Approximation problem, 181

## B

Banach space, 47
Barycentric coordinates, 77
Basis
  definition of, 39, 42
  Hamel, 217
  orthonormal, 53
  standard, 39
Bergstrom's inequality, 209
Best approximation, 181
Bilinear transformation, 68
Bolzano–Weierstrass theorem, 40, 52
Borsuk's covering problem, 271
Boundary behavior of convex functions
  continuity, 4
  derivatives, 6
  envelope function, 97
  maximum, 127
Boundary behavior of closed convex functions, 30, 95
Boundary point, 40, 47, 84
Bounded function, 3
Bounded linear transformation, 56, 61
Bounded set, 40, 47
Boundedly polyhedral, 96
Brunn–Minkowski function, 267

## C

Carathéodory's theorem, 76
Cauchy sequence, 41

295

CBS inequality, 41, 53, 191
Chain, 107
Chain rule, 66
Chebyshev set, 187
Close to convex, 256
Closed convex function, 30, 95, 111
Closed set, 40, 47, 51
Closure of a set, 40, 51
Closure properties
  convex functions on **R**, 13
  convex functions on **L**, 94
Colonel Blotto game, 137
Combination
  affine, 75
  convex, 75
  general, 217
  infinite convex, 267
  linear, 39
  rational convex, 212
Compact set, 40, 47
Complete normed linear space, 47
Completely convex function
  almost, 236
  definition of, 235
  minimal, 235
Completely monotonic, 234
Complex convexity, 255
Composition of convex functions, 16, 20
Concave function, 2
Conjugate convex function
  closed, 30
  definition of, 21, 30, 110
  properties, 28, 111
  relation to convex programming, 178
Continuous function, 54
Continuously differentiable, 68
Converge, 40, 47
Convex combination, 75
Convex function
  absolute continuity, 4, 9
  boundedness, 3
  characterizations, 8, 9, 97
  closed, 30, 95, 111
  conjugate, 21, 28, 30, 110
  continuity, 4, 7, 91, 215
  definition of, 2, 89
  derivatives, 5, 10, 11, 97
  differentiability, 7, 101, 113
  Lipschitz condition, 4, 7, 93
  maxima–minima, 14, 122

  support, 12, 104
Convex, hull, 75
Convex programming, 170
Convex sequences, 265
Convex set
  characterization, 75
  closure properties, 75, 79
  definition of, 73
  separation, 81, 83
  support, 84
  topological properties, 77
Convexity at a point, 268
Core, 80
Counterexamples, 266
Cycling, 160

## D

Derivative
  directional, 62
  distribution, 265
  Fréchet, 63, 114
  Gateau differential, 113
  generalized, 32
  generalized second, 14
  gradient, 101
  one-sided, 5
  partial, 63
  Schwarz, 265
  second Fréchet, 67, 119
Differences of convex functions, 22
Differential inequality, 245
Dimension
  of a convex set, 76, 80
  of a linear space, 42
Distance function, 95
Divided differences, 24, 237, 260, 265, 266
Doubly stochastic matrices, 86, 200, 258
Dual problem
  convex programming, 175
  linear programming, 147, 168
Duality
  conjugate convex functions, 29, 178
  convex programs, 174
  general notion, 28
  linear programs, 147
  set operations, 52
  linear spaces, 60
Duality theorem of linear programming, 148

# Subject Index

## E

Envelope, 97, 112
Epigraph, 80, 95
Euclidean $n$-space $\mathbf{R}^n$, 38
Expected payoff function, 134
Exposed point, 86
Exterior point, 40, 47
Extreme point, 84, 86, 145, 200, 208, 269
Extreme subset, 86
$\varepsilon$-neighborhood, 40, 46

## F

Farkas' lemma, 152
Feasible set, 140, 171, 175
Finite dimensional space, 42
Flat, 73
Four color problem, 271
French railroad space, 266
Functional equations, 269
Functions, *see also* Affine function, Convex function, etc.
  terminology, 54
$\mathscr{F}$-convex, 241
$\mathscr{F}$-mid-convex, 241
$\mathscr{F}_n$-convex
  definition of, 247
  weakly, 247
$\mathscr{F}\mathscr{S}$-convex, 243

## G

Games, 128
Gamma function, 21
Gateaux differential
  definition of, 113
  relation to Fréchet derivative, 114, 117
  strong, 117
Gauge function, 95, 216
Generalized convex functions, 240
Geometric mean–arithmetic mean
  condition for equality, 192
  generalized inequality, 195
  inequality, 190
Geometric programming, 128
Gradient, 101
$\mathscr{G}$-convex function, 269

## H

Hadamard matrices, 271
Hadamard's determinant theorem, 205
Hadamard's three circles theorem, 267
Hahn–Banach theorem, 105
Halfspace, 81
Harmonic function, 254
Heine–Borel theorem, 40, 52
Hessian, 103, 119, 231
Higher order convexity, 237
Hilbert space, 51
Hölder's inequality, 191, 193, 194
Hyperplane, 81

## I

I-feasible form, 158
Infinite dimensional space, 42
Inner product, 41, 50, 186
Interior point, 40, 47
Isometric isomorphism, 60
Isomorphic (isomorphism), 43

## J

Jensen function, 217
Jensen's inequality, 89, 189, 213

## K

Kantorovich's inequality, 208
Karlin's condition, 176
Krain–Millman theorem, 85
Kuhn–Tucker
  conditions, 174, 232
  duality theorem, 175

## L

Level set, 30, 228, 271
Lidstone series, 234
Limit point, 40
Limits of convex functions, 17, 20
Lindelöf property, 52
Linear combination, 39
Linear family, 249
Linear function
  characterization, 214
  continuous, 56

definition of, 2, 55
discontinuous, 61
$\epsilon$-approximately, 256
Linear functional
  continuous, 56, 62
  discontinuous, 61
  nontrivial, 57
  representation, 61
  terminology, 54, 55
Linear programming
  canonical form, 141
  constraints, 139, 154
  dual problem, 147, 168
  feasible solution, 140
  objective function, 139
  optimal solution, 140
  primal problem, 147
  problem, 140
  related problem, 142
  simplex method, 147, 154
Linear space, 41
Linearly dependent, 39
Linearly independent, 39, 42
Lipschitz condition
  definition of, 4, 92
  local, 92
  relation to convex functions, 4, 7, 93
  relation to finite second variation, 27
  upper, 96
Locally bounded, 91
Log-convex function, 18, 196
Lower bounding function, 223
$\lambda(n)$-family, 251
$\lambda(n)$-convex function, 252

## M

Majorize, 97
Matrix
  characteristic values, 69, 202, 208
  doubly stochastic, 86, 200, 258
  Hadamard, 271
  Hessian, 103
  nonnegative definite, 69, 204
  payoff, 129
  permutation, 200
  positive definite, 69, 207
  symmetric, 68, 202, 206
Matrix-convex function, 259
Matrix-monotone increasing, 260

Maximal element, 107
Maximal proper subspace, 57, 81
Maximal monotone increasing set, 32
Maximum
  of convex functions, 14, 124, 125
  of functions on $\mathbf{R}^n$, 128
  global, 123
  local, 123
Mean of order $t$, 194
Mean value theorem, 71
Measure of convexity, 264
Mid-convex function, 211
Minimax theory, 128
Minimum of convex functions, 14, 123
Minkowski's determinant theorem, 205
Minkowski's inequality, 191, 193, 194
Monotone increasing function, 99
Monotone increasing set, 32
Multiplicatively convex, 18

## N

$n$-convex function
  definition of, 238
  weakly, 239
Nearest points, 14, 187, 271
Nested set property, 52
Norm
  equivalence, 47
  Euclidean, 39
  examples of, 45, 46, 198
  general notion of, 44
  of a linear transformation, 59, 61
  $p$-norms, 198
  smooth, 118
  uniformly smooth, 118
Normed linear space, 44
  smooth, 118
  uniformly smooth, 118
Nonnegative orthant, 140
$n$-parameter family
  definition of, 246
  unrestricted, 252
$n$-simplex, 77
Nullspace, 57

## O

Objective function, 139
Open cover, 40

# Subject Index

Open set, 40, 47
Oppenheim's inequality, 208
Orthogonal vectors, 41

## P

Parallelogram law, 53
Partially monotone function, 218
Partially ordered, 107
Payoff matrix, 129
Permutation matrix, 200
Perturbation function, 178
Pivoting, 156
Point (vector), 38, 41
Polyhedral, 125, 127
Polytope, 77, 127
Pre-Hilbert space, 50
Primal problem
  convex programming, 174
  linear programming, 147
Pseudo-convex function, 104

## Q

Quadratic programming, 177
Quasi-convex function, 104, 228

## R

Real inner product space, 50
Real linear space, 41
Relative interior, 78, 80
Relatively open, 79
Riesz's convexity theorem, 266
Riesz's lemma, 52
Rotundity, 118, 270

## S

Saddlepoint, 131
Scalar multiplication, 39
Schur-convex function, 209, 258
Semi-continuous function
  lower, 112
  upper, 96
Separation of convex sets
  definition of, 81
  nice, 85
  proper, 85
  strict, 85

  strong, 81
  theorems, 81, 83, 85
Sequences of convex functions, 17, 20
Sequentially compact, 52
Simplex method, 147, 158
Slack variables, 142
Slater's condition, 176
Solution
  feasible, 140, 171
  optimal, 140
  strictly feasible, 172
  unbounded, 151
Spectrum, 259
Sphere packing, 271
Starlike function, 256
Strategies
  mixed, 134
  optimal, 134
  pure, 134
Strictly convex function
  characterizations, 9, 10, 98, 99
  definition of, 2
Strictly convex space, 184
Strictly monotone increasing, 99
Strongly convex function, 268
Subdifferential, 32, 110
Subharmonic function
  definition of, 254
  $\varepsilon$-approximately, 257
  pluri, 255
Subspace, 44, 53
Sum of order $t$, 196
Support function for a convex set, 95
Support of convex functions
  at boundary points, 14
  continuous, 108
  definition of, 12, 108
  line of support, 12
  uniqueness, 12, 110, 113, 115
Support of convex sets, 84, 112
Suprema of convex functions, 16, 30
Supremum, 40
Symmetric bilinear form, 68, 69
Symmetric function, 258
Symmetric matrix, 68, 202, 206

## T

Tableau, 155
Taylor's theorem, 70, 119

Tchebycheff system, 250
Topologically equivalent norms, 47, 199
Topologically isomorphic spaces, 50
Transformation
  affine, 55
  bilinear, 68
  bounded linear, 56, 61
  continuous linear, 56
  linear, 55
  matrix representation, 58
  nonnegative definite, 69
  positive definite, 69
  symmetric bilinear, 68, 69
Translate of a subspace, 74

## U

Uniformly convex function, 268
Uniform convex space, 186, 270
Unit ball, 45
Unit sphere, 45
  smooth, 270
Unit vector, 45

## V

Value of a game, 135
Van der Waerden's conjecture, 271
Variation
  bounded, 22, 54
  bounded $n$th, 240
  total, 23
  total $n$th, 239
Von Neumann minimax theorem, 131, 138

## W

Wright convex function, 223

## Y

Young's inequality, 29, 191

## Z

Zorn's lemma, 107